MW00846089

SUSTAINABLE ALTERNATIVES FOR AVIATION FUELS

LIBRARY OF
CONGRESS
· SURPLUS ·
DUPLICATE

SUSTAINABLE ALTERNATIVES FOR AVIATION FUELS

Edited by

ABU YOUSUF

Department of Chemical Engineering & Polymer Science, Shahjalal University of Science and Technology, Sylhet-3114, Bangladesh

CRISTINA GONZALEZ-FERNANDEZ

Biotechnological Processes Unit, IMDEA Energy, Móstoles, Spain

Department of Chemical Engineering and Environmental Technology, School of Industrial Engineering, Valladolid University, Valladolid, Spain

Institute of Sustainable Processes, Valladolid, Spain

ELSEVIER

Elsevier
Radarweg 29, PO Box 211, 1000 AE Amsterdam, Netherlands
The Boulevard, Langford Lane, Kidlington, Oxford OX5 1GB, United Kingdom
50 Hampshire Street, 5th Floor, Cambridge, MA 02139, United States

Copyright © 2022 Elsevier Inc. All rights reserved.

No part of this publication may be reproduced or transmitted in any form or by any means, electronic or mechanical, including photocopying, recording, or any information storage and retrieval system, without permission in writing from the publisher. Details on how to seek permission, further information about the Publisher's permissions policies and our arrangements with organizations such as the Copyright Clearance Center and the Copyright Licensing Agency, can be found at our website: www.elsevier.com/permissions.

This book and the individual contributions contained in it are protected under copyright by the Publisher (other than as may be noted herein).

Notices
Knowledge and best practice in this field are constantly changing. As new research and experience broaden our understanding, changes in research methods, professional practices, or medical treatment may become necessary.

Practitioners and researchers must always rely on their own experience and knowledge in evaluating and using any information, methods, compounds, or experiments described herein. In using such information or methods they should be mindful of their own safety and the safety of others, including parties for whom they have a professional responsibility.

To the fullest extent of the law, neither the Publisher nor the authors, contributors, or editors, assume any liability for any injury and/or damage to persons or property as a matter of products liability, negligence or otherwise, or from any use or operation of any methods, products, instructions, or ideas contained in the material herein.

ISBN: 978-0-323-85715-4

For information on all Elsevier publications
visit our website at https://www.elsevier.com/books-and-journals

Publisher: Charlotte Cockle
Acquisitions Editor: Peter Adamson
Editorial Project Manager: Chris Hockaday
Production Project Manager: Surya Narayanan Jayachandran
Cover Designer: Victoria Pearson Esser

Typeset by STRAIVE, India

Working together
to grow libraries in
developing countries

www.elsevier.com • www.bookaid.org

Dedication

To
My dear wife Sharmin Sultana
children Yara Shanum and Ibrah Anum
who have been deprived of my attention owing to this book

–Abu Yousuf

To
All researchers, policymakers, and stakeholders working for a more sustainable world

–Cristina Gonzalez–Fernandez

Contents

9. Life cycle assessment of biojet fuels   215

Qing Yang and Fuying Chen

10. Sustainability tensions and opportunities for aviation biofuel production in Brazil   237

Mar Palmeros Parada, Wim H. van der Putten, Luuk A.M. van der Wielen, Patricia Osseweijer, Mark van Loosdrecht, Farahnaz Pashaei Kamali, and John A. Posada

Contributors

David Bolonio
Department of Energy and Fuels, School of Mining and Energy Engineering, Universidad Politécnica de Madrid, Madrid, Spain

Laureano Canoira
Department of Energy and Fuels, School of Mining and Energy Engineering, Universidad Politécnica de Madrid, Madrid, Spain

Fuying Chen
State Key Laboratory of Coal Combustion, Huazhong University of Science and Technology, Wuhan, Hubei, PR China

Henrique dos Santos Oliveira
Department of Chemistry, Institute of Exact Sciences, Federal University of Minas Gerais, Belo Horizonte, Minas Gerais, Brazil

Marina Efthymiou
Dublin City University Business School, Dublin, Ireland

Pablo Fernández-Yáñez
School of Industrial and Aerospace Engineering of Toledo, University of Castilla-La Mancha, Ciudad Real, Spain

Reyes García-Contreras
School of Industrial and Aerospace Engineering of Toledo, University of Castilla-La Mancha, Ciudad Real, Spain

María-Jesús García-Martínez
Department of Energy and Fuels, School of Mining and Energy Engineering, Universidad Politécnica de Madrid, Madrid, Spain

Arántzazu Gómez
School of Industrial and Aerospace Engineering of Toledo, University of Castilla-La Mancha, Ciudad Real, Spain

Fernando Israel Gómez-Castro
Chemical Engineering Department, Natural and Exact Sciences Division, Campus Guanajuato, University of Guanajuato, Guanajuato, Guanajuato, Mexico

Claudia Gutiérrez-Antonio
Engineering School, Autonomous University of Querétaro, El Marqués, Querétaro, Mexico

Salvador Hernández
Chemical Engineering Department, Natural and Exact Sciences Division, Campus Guanajuato, University of Guanajuato, Guanajuato, Guanajuato, Mexico

Md. Shahadat Hossain
Department of Chemical Engineering and Polymer Science, Shahjalal University of Science and Technology, Sylhet, Bangladesh

M. Amirul Islam
Laboratory for Quantum Semiconductors and Photon-based BioNanotechnology, Department of Electrical and Computer Engineering, Faculty of Engineering, Université de Sherbrooke, Sherbrooke, QC, Canada

Ahasanul Karim
Department of Soil Sciences and Agri-food Engineering, Université Laval, Quebec, QC, Canada

Pantea Moradi
School of Chemistry, College of Science, University of Tehran, Tehran, Iran

Abdullah Nayeem
Department of Chemical Engineering, College of Engineering, Universiti Malaysia Pahang, Gambang, Pahang, Malaysia

Marcelo F. Ortega
Department of Energy and Fuels, School of Mining and Energy Engineering, Universidad Politécnica de Madrid, Madrid, Spain

Patricia Osseweijer
Department of Biotechnology, Delft University of Technology, Delft, The Netherlands

Mar Palmeros Parada
Department of Biotechnology, Delft University of Technology, Delft, The Netherlands

Vânya Marcia Duarte Pasa
Department of Chemistry, Institute of Exact Sciences, Federal University of Minas Gerais, Belo Horizonte, Minas Gerais, Brazil

Farahnaz Pashaei Kamali
Department of Biotechnology, Delft University of Technology, Delft, The Netherlands

John A. Posada
Department of Biotechnology, Delft University of Technology, Delft, The Netherlands

Md. Anisur Rahman
Department of Chemical Engineering and Polymer Science, Shahjalal University of Science and Technology, Sylhet, Bangladesh

Mohammad Jalilur Rahman
Department of Chemistry, Shahjalal University of Science and Technology, Sylhet, Bangladesh

Araceli Guadalupe Romero-Izquierdo
Engineering School, Autonomous University of Querétaro, El Marqués, Querétaro, Mexico

Tim Ryley
Griffith Aviation, School of Engineering & Built Environment, Griffith University, Brisbane, QLD, Australia

Majid Saidi
School of Chemistry, College of Science, University of Tehran, Tehran, Iran

Cristiane Almeida Scaldadaferri
College of Engineering and Physical Sciences, Energy and Bioproducts Research Institute, Aston University, Birmingham, United Kingdom

José A. Soriano
School of Industrial and Aerospace Engineering of Toledo, University of Castilla-La Mancha, Ciudad Real, Spain

Wim H. van der Putten
Department of Terrestrial Ecology, Netherlands Institute of Ecology, Wageningen, The Netherlands

Luuk A.M. van der Wielen
Department of Biotechnology, Delft University of Technology, Delft, The Netherlands; Bernal Institute, University of Limerick, Limerick, Ireland

Mark van Loosdrecht
Department of Biotechnology, Delft University of Technology, Delft, The Netherlands

Nicolas Vela-García
Department of Chemical Engineering, Institute for Development of Alternative Energies and Materials IDEMA, Universidad San Francisco de Quito, Quito, Ecuador

Qing Yang
State Key Laboratory of Coal Combustion, Huazhong University of Science and Technology, Wuhan, Hubei; China-EU Institute for Clean and Renewable Energy, Huazhong University of Science and Technology, Wuhan, PR China; John A. Paulson School of Engineering and Applied Sciences, Harvard University, Cambridge, MA, United States

Abu Yousuf
Department of Chemical Engineering and Polymer Science, Shahjalal University of Science and Technology, Sylhet, Bangladesh

Preface

Environmental pollution and global warming are great concerns caused by the combustion of fossil fuels. Biofuels are proposed as potential alternatives for fossil fuels in the transportation sector. Recently, biofuels have been introduced in the aviation transportation sector. Yet, the use of biofuels is limited due to the lack of a commercially viable and attractive production method. Biojet fuel, also known as synthetic paraffinic kerosene (SPK), is composed of renewable hydrocarbons whose properties are almost similar to that of fossil jet fuel. The combustion of SPK produces lower CO_2 emissions than fossil jet fuel. Therefore, biojet fuel has been identified by the International Air Transport Association (IATA) as the most viable alternative to substitute fossil fuels in aviation.

Biojet fuel can be synthesized from various sources of biomass through different processing routes like oleochemical, thermochemical, and biochemical. Biomass that can be used for the production of biojet fuel are triglycerides, lignocellulosic biomass, sugar, and starchy feedstock. All of the carbon sources used as feedstocks have advantages and disadvantages. Triglycerides-based feedstocks like canola oil, jatropha oil, soybean oil, and castor oil are expensive but require low-cost processing. By contrast, low-cost materials like lignocellulosic biomass, the most abundant organic materials on the earth, require long processing steps that result in increase of cost. Therefore, the selection of the best production process will depend on the availability of the raw materials, low processing cost, and industrial viability.

None of the books available in the market focus on biojet fuels from technical, economic, environmental, societal, and policy aspects under a single title. This book, *Sustainable Alternatives for Aviation Fuels*, provides an in-depth description of the main challenges for the production of alternative aviation fuels for both researchers and industries. In particular, the book focuses on the biojet fuel process pathway, feedstock availability, conceptual process design, process economics, engine performance, techno-economic analysis, environmental issues, policy design, and sustainability of alternative aviation fuel technology.

Chapter 1 explains the four predominant pathways to produce renewable jet fuels including oil-to-jet, alcohols-to-jet, sugar-to-jet, and syngas-to-jet pathways. The chapter illustrates the conceptual process and addresses the key challenges of these pathways in detail.

Chapter 2 comprehensively assesses the potential of ASTM-approved routes for biojet fuel production from agricultural wastes. It emphasizes the thermochemical and biochemical conversion routes of lignocellulosic agricultural wastes/residues from oleaginous crops and their potential use in biojet fuel production.

Chapter 3 provides an overview of the reaction network and the effect of operating parameters on the properties of lignin-derived jet fuel hydrocarbon range. To investigate the reaction network, the chapter provides a brief review of the effect of variable catalysts on product distribution and a detailed summary using the investigated primary and secondary products.

Chapter 4 presents a range of materials that can be used as feedstock for different production processes of sustainable aviation fuels (SAFs). In order to accelerate the production of SAFs, the chapter extensively discusses the biomass production chains and waste processing to convert fatty materials and forest, agriculture, industrial, and municipal wastes into biokerosene.

Chapter 5 focuses on the production of renewable aviation fuel in biorefinery schemes. It also includes the main information about the individual production processes.

Chapter 6 summarizes the different types of catalysts along with their catalytic reaction mechanism used in the catalytic conversion of different biomass feedstocks to biojet fuel. The chapter also addresses the limitations associated with some catalytic processes and suggests possible future research directions.

Chapter 7 evaluates the effect of biojet fuels on emissions in aircraft engines from the viewpoint of the composition and main physicochemical properties and the emissions associated with biofuels in jet engines and compression ignition engines.

Chapter 8 focuses on the governance and policy developments rather than the technical aspects of SAF production. The chapter starts with an introductory argument about the environmental need for SAF governance and policy. The subsequent sections outline the governance surrounding SAF and provide a more specific policy framework, followed by some ongoing policy lessons from recent developments.

Chapter 9 summarizes the biomass feedstocks available for aviation fuel and their conversion paths. Then, the technical, economic, and environmental impacts of biojet fuel are reviewed from the perspective of life cycle assessment.

Chapter 10 presents a novel context-dependent ex-ante sustainability analysis of aviation biofuel production focused on the southeast region of Brazil, which includes economic, environmental, and societal aspects. Based on inputs from local stakeholders and sustainability literature, eight aspects of sustainability are described: climate change, commercial acceptability, efficiency, energy security, investment security, profitability, social development, and soil sustainability.

Researchers and scientists with strong academic backgrounds and practical experiences have shared the thoughts and findings of their investigations in this book. We believe the book will enrich the foresight of current researchers and industrialists who are dedicatedly working on SAFs.

Abu Yousuf
Sylhet, Bangladesh

Cristina Gonzalez-Fernandez
Madrid, Spain

Acknowledgments

We express our gratitude to all the distinguished authors for their thoughtful contributions that have added to the success of this book project. Their patience and diligence in revising the first draft of the chapters after adapting the suggestions and comments are highly appreciated.

We acknowledge the solicitous contributions of all the reviewers who spent their valuable time in a constructive and professional manner to improve the quality of the book.

We are grateful to the staff at Elsevier, particularly Dr. Peter W. Adamson (Acquisitions Editor for Renewable Energy), who provided us with tremendous support throughout the project and for his great ideas and encouraging thoughts. We also acknowledge Letícia Lima (Editorial Project Manager), Chris Hockaday (Editorial Project Manager), and Srinivasan Bhaskaran (Copyrights Coordinator) for their earnest handling of the book.

CHAPTER 1

Conversion pathways for biomass-derived aviation fuels

Ahasanul Karim[a], M. Amirul Islam[b], Abdullah Nayeem[c], and Abu Yousuf[d]

[a]Department of Soil Sciences and Agri-food Engineering, Université Laval, Quebec, QC, Canada
[b]Laboratory for Quantum Semiconductors and Photon-based BioNanotechnology, Department of Electrical and Computer Engineering, Faculty of Engineering, Université de Sherbrooke, Sherbrooke, QC, Canada
[c]Department of Chemical Engineering, College of Engineering, Universiti Malaysia Pahang, Gambang, Pahang, Malaysia
[d]Department of Chemical Engineering and Polymer Science, Shahjalal University of Science and Technology, Sylhet, Bangladesh

1. Introduction

The utilization of fossil fuels such as transportation fuels has resulted in an increase of atmospheric greenhouse gases particularly carbon dioxide (CO_2) [1]. Numerous initiatives, including the expansion of biomass-based renewable energy resources have been introduced to lower the CO_2 emissions [2]. Several countries and automobile industries have a vision to partially or completely replace gasoline fueled jets by sole-electric or hybrid alternatives [3]. However, the fossil fuel resources are still being used as main source in the aviation industry. Notably, the current energy source for jet fuel is a mixture of hydrocarbon paraffins, isoparaffins, aromatics, and cycloalkanes, with defined carbon chain lengths and properties [4,5].

The demand for jet fuel is continuously increasing day by day. An annual increase of about 5% is projected by 2050 according to the prediction of International Air Transport Association (IATA) [2,6]. Furthermore, the fleet number and size will become triple in the next two decades [7,8]. In the last 20 years, the total consumption of jet fuel by airlines amplified from 260 million to 340 million per year [2]. It is projected that the fleet size will increase from 340 million to more than 500 million per year by 2026 [9]. In this context, the renewable aviation fuel could be a potential replacement to decrease CO_2 emissions and achieve the goals of decarbonization in 2050 [10,11]. The literature showed that biojet fuel (or renewable jet fuel (RJF)) produced from renewable feedstocks could drastically reduce ($\sim$80%) emission of greenhouse gases [12,13]. Therefore, the demand for RJF has stimulated for the aviation transportation [14].

Generally, the aviation-fuel is a mixture of different hydrocarbons, such as branched chain isoalkanes, n-paraffins, aromatics, and cyclic alkanes [15]. However, the carbon numbers and molecular weights of the fuel substantially depend on the refinery process and distribution of molecular size ranging from C_8 to C_{16} [16]. In general, the renewable

Copyright © 2022 Elsevier Inc.
All rights reserved.

feedstocks, especially biomass are regarded as a potential substitute of fossil for the aviation industry [17]. The common biomass feedstocks for RJF production are classified into four predominant groups, including cellulosic materials, oil crops, sugars and starch, and wastes [18]. These feedstocks can be converted into RJF using several pathways (Fig. 1), such as sugar-to-jet, gas-to-jet, alcohol-to-jet, and oil-to-jet [15,19]. These conversion techniques have been extensively used by the industries, particularly the oil-to-jet pathway, which involves the hydro-processing method. Hydro-processing technology is considered more advantageous than the other conversion pathways as it is associated with the utilization of low-quality feedstocks [20,21]. However, in biomass-based RJF production, the main problem is that the biomass inherently contains a wide variety of components ranging from low to high molecular weights in their structure. According to the European Union (EU) and American Society for Testing Materials (ASTM) standards [22], the oxygen must be separated from RJF hydrocarbons comprising a small carbon chain length, which is challenging in current conversion pathways. Thus, these constraints limit the application of biomass-based processes to produce biojet fuel.

In this context, several initiatives have been taken to enhance the production of biomass derived biojet fuel. In addition to that, biojet fuel has renewed attention due to the environmental and economic aspects. Thus, significant research works have been conducted including various processing technologies (thermochemical, bio-chemical, and chemo-catalytic conversion) and product recovery pathways. However, research focus should be given for developing efficient technology to convert the lipid feedstocks and saccharides, including complex structured lignocellulosic biomass (LCB) feedstocks into RJF [23,24]. In this chapter, the production pathways for RJF from different feedstocks are comprehensively discussed. Moreover, RJF associated products such as isoparaffin blends, aromatics, or cycloalkanes, etc. produced by the pathways are also reviewed in this chapter.

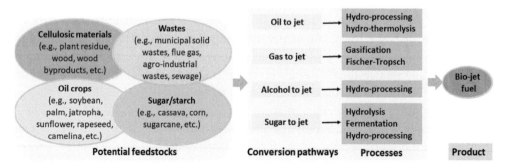

Fig. 1 The common biomass feedstocks and their conversion pathways for renewable jet fuel (RJF) production.

2. Conversion pathways of biojet fuel

2.1 Oil to jet fuel

Oil to jet (OTJ) conversion pathway can be categorized into three processes: (a) hydro-processed renewable jet (HRJ) or termed as hydro-processed esters and fatty acids (HEFA); (b) catalytic hydro-thermolysis (CHT), which is also known as hydrothermal liquefaction (HTL); and (c) hydro-treated depolymerized cellulosic jet (HDCJ) or familiarized as fast pyrolysis upgradation to jet fuel. However, only HRJ/HEFA conversion pathway products are now permitted for blending and achieved ASTM-specified certification [19]. Triglyceride-based feedstocks are used both in the HRJ and CHT pathways although the production processes for free fatty acids (FFAs) are different. The FFAs are formed via thermal hydrolysis in the CHT process while they are produced through the cleavage of propane glycerides in the HRJ process. On the other hand, the bio-oils are generated by pyrolyzing the biomass feedstocks in the HDCJ conversion process. Furthermore, the conversion pathways of jet fuel production can fluctuate based on the oil feedstock types. For instance, animal fats, cooking oil, and vegetable oils are suitable feedstocks for HEFA/HRJ process whereas algal oils or oil from plant sources are considered efficient for CHT/HTL process [13]. However, the downstream hydro-treatment methods are identical for all three processes [19].

2.1.1 Hydro-processed renewable jet

HRJ or HEFA is the process of biojet fuel production by hydro-treating the triglycerides, unsaturated or saturated fatty acids of used cooking oils, animal fats, and vegetable oils. The fuel produced by this technique could be directly used in the flight engines even without blending. Notably, the HEFA produced biojet fuels have been used in the US military flights [25]. In terms of properties, the fuel is similar to the conventional petroleum-based fuel. It contains high thermal stability and cetane number, good cold flow behaviors, low tailpipe/greenhouse gas emissions, and low sulfur and aromatic content [26,27]. A graphical representation of HEFA/HRJ process flow is presented in Fig. 2. The method can usually be split up into two phases.

(1) The first stage is to convert the unsaturated fatty acids and triglycerides into saturated fatty acids through the catalytic hydrogenation. In this step, a β-hydrogen elimination reaction of the triglycerides is occurred to produce a fatty acid [28]. Generally, renewable oils and fats that possess various degrees of unsaturation and require a hydro-generation process for saturating double bonds. To convert the glycerides and liquid-phase unsaturated fatty acids into saturated fatty acids, the catalytic hydro-generation process is used as a common technique [29]. The saturated fatty acids are then converted into alkanes containing C_{15}–C_{18} straight chain by hydro-de-oxygenation and de-carboxylation [30]. The common by-products are propane, H_2O, CO_2, and CO. The oxides and zeolites supported by noble metals

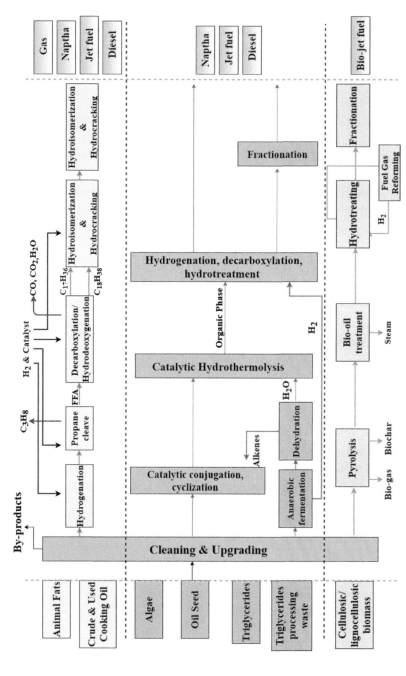

Fig. 2 Oil-to-jet (OTJ) conversion pathways including hydro-processed renewable jet (HRJ), catalytic hydro-thermolysis (CH), and hydro-treated depolymerized cellulosic jet (HDCJ) to jet fuel.

were the early-developed catalysts for this step. Later, other alternatives such as some transition metals including Co, Ni, Mo, or their supported bimetallic composites are used as well [31–33].

(2) Cracking with isomerization reactions is the second step. In this stage, the alkanes having de-oxygenated straight chains are selectively hydro-cracked and deep-isomerized to produce mixed liquid fuels containing highly branched alkanes. Various catalysts including Ni, Pt or other activated carbon supported precious alloys, i.e., Al_2O_3, zeolite molecular sieves are used to facilitate the reaction in this step [34,35]. The zeolite supported Ni catalyst showed outstanding activities, however, an over cracking and low isomers yield were seen for the strong acidic catalysts. Finally, after a fractionation process, the hydro-isomerization and hydro-cracking processes are done to isolate the mixed liquid fuels to paraffinic kerosene (HRJ-SPK) (jet fuel), naphtha, paraffinic diesel, and light gases [13]. Generally, the biojet fuels should have a superior flash point as well as excellent cold flow properties to meet the jet fuel specification [19]. Thus, it is emergent to hydro-crack and hydro-isomerize the normal paraffins that are produced from de-oxygenation to a synthetic paraffinic kerosene (SPK), a product with carbon chains ranging from C_9–C_{15} [27]. Notably, the cracking and isomerization reactions are either sequential or simultaneous [29]. The isomerization process takes the straight-chain hydrocarbons to turn them into the branched structures for reducing the freezing point to meet the jet fuel standard [36]. Primarily, the hydrocracking reactions include cracking and saturation of paraffins. However, over cracking may result in low yields of jet fuel-range alkanes and high yields of light species ranging from C_1 to C_4 and naphtha ranging from C_5 to C_8. Both are the out of jet fuel range and have lower economic value compared to diesel or jet fuel [19]. Furthermore, the high temperature (250–260°C) and pressure are usually required in this process. The glycerol by-product purification process is energy intensive that is adding cost to the overall process, however, this could be offset by glycerol selling value due to its various applications in the pharmaceutical, technical, and personal care product industries [37].

2.1.2 Catalytic hydro-thermolysis process

CHT or HTL is an alternative pathway for jet fuel production from algal or plant oil. Applied Research Associates, Inc. has developed this pathway to produce "renewable, aromatics, and drop-in" types of fuel which is well-known as "ReadiDiesel" or "ReadiJet" [38]. Triglycerides are converted into a branched, straight chain and cyclic hydrocarbon mixture during the hydrothermal process through a series of chemical reactions (Fig. 2), including cracking, hydrolysis, isomerization, decarboxylation, and cyclization [39]. The temperature range and pressure of the CHT pathway are maintained between 450°C and 475°C and around 210 bar, respectively [19]. Oxygenated species, carboxylic acids, and unsaturated molecules can be produced in this process, which are then transferred to the hydro-treatment process after completing the decarboxylation process to saturate and remove oxygen. The products (ranging from C_6 to C_{28}) including *n*-alkanes, cycloalkanes, isoalkanes, and aromatics could be obtained after treatment, which require the fractionation process to separate diesel fuel, jet fuel, and naphtha [13].

An integrated CHT processing concept was suggested by Li et al. [39] to convert triglycerides obtained from the crop oils to nonester biofuels (Table 1). It consists of three steps such as triglycerides pretreatment, CHT conversion, and postrefining processes. The pretreatments such as conjugation, cyclization, and cross-linking are applied to enhance the molecular structures. During the CHT conversion, the products go through a cracking and hydrolysis reaction in the presence of H_2O and catalysts. Thereafter, the more stable and low-molecular-weight carboxylic acids (C_7 or less) and glycerol produced from triglyceride hydrolysis undergo decarboxylation and dehydration process to form alkenes. Finally, alkanes are obtained from straight-chain, branched and cyclo-olefins through the postrefining hydrotreatment and fractionation. High level of cyclics and aromatics is one of the important characteristics of CHT biofuel. Tung-oil-based biofuels originated from the CH process contain up to 60% aromatics, which can be a desirable ingredient for fuel blends including biofuels derived from other processes [39]. ASTM and military (MIL) specifications are fulfilled for the jet fuel obtained from the CHT method due to its outstanding combustion quality, stability, and cold flow characteristics [19]. According to the previous research reports, various triglyceride feedstocks such as tung oil, soybean oil, camelina oil, and jatropha oil can be used to produce biojet fuels using the CHT process [19].

2.1.3 Hydro-processed depolymerized cellulosic jet

HDCJ is a new technology for oil upgradation by converting cellulosic biomass into diesel, gasoline, and jet fuels, which was established by Kior. It transforms bio-oils, generated from pyrolysis or hydrothermal treatment of LCB into jet fuel using hydrotreatment [40]. Biomass pyrolysis oils usually contain unattractive properties such as low energy density, inadequate thermal instability, and high corrosivity because of high oxygen content [41]. Therefore, these oils require some hydro-treating processes to be transformed into jet fuel ranged products (Fig. 2). Biojet fuel can be produced from various biomass feedstocks, such as sugarcane bagasse, forest residue, corn stover, switchgrass, algae biomass, and guinea grass, etc. can produce through this conversion pathway [42,43].

In general, a two-step hydro-processing is the key technology to upgrade bio-oil. At first, catalysts help to hydrotreat the bio-oil under mild conditions. Secondly, traditional hydrogenation set-up with catalysts is used to achieve hydrocarbon fuel under high temperature [44]. Wang et al. [45] used pyrolysis technique to covert straw stalk oils into diesel and jet fuel via a three-step reactions. Bio-oil was cracked through a catalytic pathway to obtain light olefins and low-carbon aromatics. Later, aromatic hydrocarbons

Table 1 Main processing steps to convert triglycerides into nonester biofuels.

Processing steps	Typical conditions	Reactions	Purpose
1. Catalytic conjugation	170°C, atmospheric, 4–8 h, Ni/C catalysts	– Isomerization of polyunsaturated fatty acids to shift double bonds into conjugated form	– Preparation for cyclization via Diels–Alder reaction
2. Cyclization/cross-linking	170–240°C, 15 bar, <1 h	**(a)** Diels–Alder addition of alkenes to the conjugated fatty acids **(b)** Cross-linking among different fatty acids at double bonds	– Creation of cyclics and molecular matrix with low degrees of cross-linking
3. Catalytic hydro thermolysis	240–450°C, 15–250 bar, 1–40 min, water/oil = 1:5–5:1, catalysts	**(a)** Cracking and hydrolysis of the modified oil mixture with the help of water and catalysts **(b)** Catalytic decarboxylation/dehydration (see steps 4 and 5 below)	– Formation of a spectrum of straight, branched, cyclic, and aromatic molecules, with a minimum gas products
4a. Catalytic decarboxylation/dehydration	330–450°C, 15–250 bar, 5–40 min, catalysts	– Removal of carboxylic groups of fatty acids, and removal of hydroxyl groups from the carbon backbone	– Deoxygenation to increase fuel energy content
4b. C_3 and C_4 decarboxylation and C_2, C_3, and C_4 dehydration	330–450°C, 15–250 bar, 5–40 min, catalysts	– Eradication of carboxylic groups of the volatile acids, and removal of hydroxyl groups of the alcohols	– Supply of alkenes for the cyclization step
5. Soybean waste anaerobic fermentation	37°C, atmospheric, 6–12 h, pH 5.0–5.5	– Production of microbial hydrogen	– Supply of H_2 for the hydrogenation and byproducts for alkene production
6. Postrefining hydrotreating fractionation	<400°C, <1000 psig, Co Mo catalyst, 1–2 LHSV	– Convert straight-chain, branched, and cycloolefins into corresponding alkanes	– Necessary to achieve thermal stability required for jet fuel

Reproduced with permission from L. Li, E. Coppola, J. Rine, J.L. Miller, D. Walker, Catalytic hydrothermal conversion of triglycerides to non-ester biofuels, Energy Fuels 24 (2) (2010) 1305–1315. Copyright 2010, American Chemical Society.

(ranging from C_8 to C_{15}) were synthesized through the alkylation of low-carbon aromatics with light olefins. Finally, the C_8–C_{15} cyclic alkanes were produced from C_8–C_{15} aromatic through the hydrogenation process. The characteristics of synthetic biofuels such as viscosity, combustion heat, freeze point fulfilled the technical specifications [45,46]. Nowadays, the jet fuel synthesis from hydrocarbons with derivatives (which obtain from the lignocellulose pyrolysis) has gained enormous attention to simplify the process and improve the efficiency [47,48]. Cao et al. [49] used a new strategy to convert LCB into hydrocarbons and alkylphenols through a three-step process. At first, biomass feedstock is transformed into pyrolysis oil. Then, MoC_x/C catalyzed selective hydrodeoxygenation (SHDO) reactions are performed on pyrolysis oil at 300°C and under low H_2 pressure of 5 bars. The resultant mixtures are then separated in the last step using ethyl acetate and *n*-hexane over silica. High aromatics with low oxygen content and jet fuel with few impurities could be produced by the HDCJ process. However, a high H_2 consumption and deoxygenation are required in the HDCJ. These requirements may increase the production cost. Besides, the short lifetime of catalyst and fairly small hydrocarbon yields can make it too tough to use in the aviation sector [50].

2.2 Gas-to-jet fuel

The process of converting syngas, biogas, or natural gas into biojet fuel is known as the GTJ conversion pathway. This pathway can generally be classified into two major processes: (a) Fisher-Tropsch-biomass to liquid (FT-BTL) process, shortly, Fisher-Tropsch (F-T), (b) gas fermentation process. In Fig. 3, a sketch of the process flow diagram of GTJ is shown.

2.2.1 Fisher-Tropsch biomass to liquid process

The FT-BTL or F-T process can produce liquid hydrocarbon fuels obtained from syngas. The produced fuels are usually sulfur free. They include extremely limited aromatics than diesel and gasoline, which may reduce sulfur emissions from jet engines [43]. Conversion of biomass into synthetic fuels using F-T process offers an efficient carbon neutral substitute to traditional kerosene, diesel, and gasoline [51,52]. Primarily, the biomass feedstocks are dried in the F-T process and milled to reduce the particle sizes and moisture content by pretreatment followed by the gasification of pretreated biomass [53]. It is worth mentioning that a wide range of factors such as gasifying agent, temperature, biomass types, equivalence ratio, heating rate, particle size, operating pressure, reactor configuration, and catalyst addition may impact the composition and yield of syngas [54]. Moreover, oxygen of high purity and steam are always required to be present during the F-T gasification process under an elevated temperature.

Different gasification technologies are conducted for syngas production from biomass. For example: (a) gasification process in a high-temperature, where extreme pressure convert dried biomass into raw synthesis gas under the presence of steam and highly pure

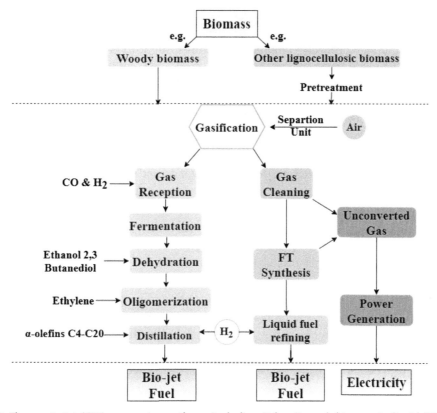

Fig. 3 The gas-to-jet (GTJ) conversion pathway including Fisher-Tropsch-biomass-to-liquid (FT-BTL) process and gas fermentation process.

O_2 at around 1300°C [53]; (b) an indirect gasification and tar reforming, the endothermic gasification is occurred by an indirect heating process through hot olivine circulation. While the material is in the gasifier, it becomes fluid using steam during gasification under the atmospheric condition at 880°C [55]. In comparison with the high temperature gasification, the latter is considered more beneficial in terms of energy and capital cost [56]. Generally, the F-T tail gas includes H_2, H_2O, CH_4, CO, CO_2, N_2, Ar, and other heavier hydrocarbons which are reprocessed back to the production system of syngas. The pressure swing absorber purifies H_2 present in the tail gas. Then it becomes usable during the isomerization/hydrocracking process [19]. The syngas penetrates into the acid gas elimination system after the gasification process for removing the acid gas, such as sulfide, CO_2, and H_2S. Later the clean gas enters the gas conditioning step for adjusting the proportion of CO to H_2 by the water-gas-reaction. The ratio between CO and H_2 possesses an imperative role in the F-T technique. Thereafter, the gas mixture containing CO and H_2 enters the F-T reactor for initiating F-T synthesis. During F-T, the key reactions are

CO and H_2 reacted to generate $C_nH_{2n}O_2$, $C_nH_{2n}O$, $C_nH_{2n+2}O$, C_nH_{2n+2}, and C_nH_{2n}. Some F-T gas including unconverted syngas might be recycled to the F-T reactor following an upgrading process. The liquid outcomes ought to be purified so that a wide variety of fuel could be obtained whereas the surplus gas can be employed to produce electricity [13].

2.2.2 Gas fermentation process

Syngas can be fermented into liquid biofuels rather than catalytical upgradation of biojet fuel from F-T syngas (Fig. 3). First, the syngas is usually produced via gasification using LB as the feedstock. Thereafter, the fermentation of cooled syngas is carried out to obtain ethanol or butanol using acetogenic bacteria [57]. The most common bacteria is *Clostridium*, which produces 2,3-butanediol and ethanol by consuming H_2 and CO [57]. However, acetone, acetate, isopropanol, butanol, and other organic products could be generated through other biosynthetic routes by several microbial strains [58]. Ethanol or 2,3-butanediol and other mixed alcohols are transformed into jet fuel through the alcohol-to-jet technology (ATJ) which is illustrated in the subsequent section. This process consists of a number of steps including dehydration, oligomerization, distillation, and hydrogenation [19]. The gas fermentation process possesses some potential benefits, such as producing more products than the conventional thermochemical or biochemical pathways [57]. It holds a total energy efficiency of 57%, which is only 45% in the F-T process [58]. The process also demands cost-effective enzymes, lower temperature, and pressure [57]. It not only utilizes the agricultural wastes and energy crops, but also industrial and municipal organic wastes as feedstock [59].

2.3 Alcohol to jet fuel

Alcohol feedstocks, such as butanol, ethanol, methanol, and other long-chain fatty alcohols can be utilized for biofuel production following a sequence of reactions including dehydration, oligomerization and hydro-processing, and distillation [60]. The process of producing fuel from conversion of alcohols, is called ATJ, also termed as alcohol oligomerization. Currently, a maximum of 10%–15% ethanol could be utilized in the gasoline-powered motor vehicles. So, it becomes challenging to get market access to ethanol as a gasoline blend stock [19]. Therefore, an upgradation of ethanol to jet fuel blend stock could be a promising pathway to develop fungible fuels for the aviation industry. In the industrial production process, ethanol, butanol, and isobutanol are always used as the intermediate while biomass is converted into the jet fuel. Generally, the hydrocarbon (in the jet fuel range) production process from the alcohols follows an upgrading process of four steps (Fig. 4). First, the alcohol dehydration, where the alcohol is dehydrated to produce olefins. Thereafter, the olefins are oligomerized in the presence of catalysts for producing a middle distillate. The next steps are to hydrogenate the middle distillates for creating the jet fuel-ranged hydrocarbons and the distillation [61,62].

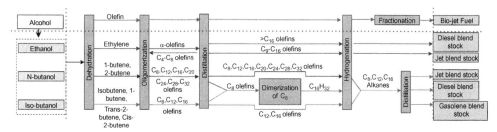

Fig. 4 The conversion pathway of alcohol-to-jet (ATJ) fuel including the process flow diagram of ethanol, butanol, isobutanol to produce biojet fuels [19].

2.3.1 Ethanol to jet

In the ethanol-to-jet conversion process, anhydrous ethanol (purity level ≈ 99.5%–99.9%) is typically needed for blending with gasoline to avert separation [63]. Though, it is still uncertain to use highly pure ethanol for upgrading the jet fuel products [19]. The complete process diagram for ethanol to jet fuel is shown in Fig. 4. In the ethanol dehydration process, Al_2O_3, transition metal oxides, heteropoly acid, and zeolite are used as common catalysts [64]. The H-FER and faujasite exhibit the highest ethylene yield (about 99.9%) at 300°C [65]. Andrei et al. [66] formulated a Ni-AlSBA-15 catalyst, which was heterogeneous by nature. It was also selective, stable, and highly active for ethylene oligomerization. Branched alkanes can be obtained from these oligomers after a hydrotreating, and isomerization process and jet fuel might be achieved via distillation. Generally, a wide range of carbon number distributions like 5% C_4; 50% C_6–C_{10}; 30% C_{12} and C_{14}; 12% C_{16} and C_{18}; and 3% C_{20} and C_{20+} are produced at 200°C and 250 bar in the industrial oligomerization processes [67]. The resultant olefins are distilled to light olefins, diesel- and jet-range fuels [68]. Distillation technique is used to separate light olefins (C_4–C_8) and then recycled back to the oligomerization process. Jet fuel range products (C_9–C_{16}) are subjected to hydrogenation at 370°C temperature, WHSV of 3 h^{-1} with feeding H_2 over 5% by weight of Pd or Pt on activated carbon catalyst [69]. The alkanes (C_9–C_{16}) obtained from the hydrogenation step are suitable products for RJF. The advantage of this process is that all these process steps are commercially applicable. However, it is still challenging to develop an integrated process on biomass-derived intermediates [19].

2.3.2 Butanol to jet

Butanol can also be transformed into a jet fuel following the same protocol by applying catalysts. The mostly used catalysts to dehydrate butanol contain zirconia, zeolite, HPW ($H_3PW_{12}O_{40}$), solid acid catalysts, and mesoporous silica group. Jeong et al. [70] investigated a novel mesoporous material as catalyst which was generated from ferrierite to synthesize butene via butanol dehydration. They observed an excellent performance of the catalyst along with the high stability, and selectivity [70]. In comparison with the direct conversion of 1-butanol to 1-butene, John et al. [71] found the lower energy barriers in the consecutive reaction system of the dehydration of 1-butanol to ether followed by ether decomposition. Butene then undergoes the oligomerization process after the dehydration. Wright et al. [72] applied Cp_2ZrCl_2 as an oligomerization catalyst which was dissolved in methylaluminoxane solution. It could bring $\geq$95% carbon use. The products were then transformed into jet fuel through hydrogenation and distillation. Wright [73] showed that N-butanol could be dehydrated to 1-butene at 380°C and 2.1 bar over a γ-alumina catalyst. The maximum yield of biobutenes was 98%, with 95% selectivity of 1-butene [73]. The remaining product was 2-butene that was isomerized from 1-butene. Thereafter, the 1-butene undergoes the oligomerization step to generate olefins (C_8–C_{32}) with an attractive conversion rate (97%) [74]. Considering the unreacted olefins, the 2-butene consisting of *cis-* and *trans-*2-butene are separated via distillation at a controlled temperature [19]. Finally, the jet fuel is blended with the paraffins (C_{12}–C_{16}) and the remaining alkanes (C_{20}–C_{32}) are separated to sell as lubricants [74].

2.3.3 Iso-butanol to jet

A mixture of olefins such as 1-butene, *cis-*2-butene, *trans-*2-butene, and isobutene could be produced by the dehydration of isobutanol [75]. Moreover, oligomers, trimmers, and tetramers are obtained from the isobutene at 100°C in presence of Amberlyst-35 as catalyst. The products contain 20% C_8, 70% C_{12}, and 10% C_{16} olefins [68,69]. The C_8 olefins may either be dimerized into $C_{16}H_{32}$ or reacted with butenes to produce C_{12} olefins, which lead to an increase of C_{12} and C_{16} for the jet-range chemicals [68]. Iso–butanol is mostly dehydrated over a mildly acidic γ-Al_2O_3 catalyst through several catalysts including metal oxides, inorganic acids, acidic resins, zeolites have been widely studied for this process in the past few years. Taylor et al. [75] investigated the isobutanol dehydration using a γ-Al_2O_3 catalyst under 60 psi and 325°C, the isobutylene selectivity and feedstock conversion were obtained as 95.0% and 98.8%, respectively. Then, the isobutene undergoes an oligomerization process to jet range alkenes after dehydration. The phosphoric acid (H_3PO_4) impregnation on solid supports is an initial industrial catalyst for the oligomerization of light olefins. Among the different catalysts such as sulfated titania, sulfated zirconia, nickel supported on sulfated zirconia, nickel-doped zeolites, sulfonic acid resins, and NiO-W_2O_3/Al_2O_3, zeolites are considered more perpetual acid catalysts as it contains both Lewis and Bronsted acid sites [13].

2.4 Sugar to jet fuel

The sugar feedstocks can be converted into jet fuel, which is known as sugar-to-jet (STJ) fuel. The process mainly includes (a) catalytic upgrading of sugars to hydrocarbons; (b) fermentation of sugars to hydrocarbons; and (c) sugars-to-sugar intermediates and upgrading to fuel [19]. The whole process can be categorized into two pathways according to the reports described by the Energy efficiency & Renewable energy bioenergy technologies office of the United States Department of Energy [76]: biochemical and thermochemical pathways. In the biochemical route, sugars can directly be transformed to alkane type fuels through fermentation rather than initially converting to ethanol intermediate. This process is known as direct sugars-to-hydrocarbons (DSH) or direct fermentation of sugars-to-jet (DFSTJ), which is termed as synthetic isoparaffin from fermented hydro processed sugar (SIP) by the ASTM. Besides, sugars could also be transformed to the jet fuel in the biochemical pathway via a thermochemical route, such as aqueous-phase-reforming (APR) [13].

1. Fermentation of sugars-to-hydrocarbons: the sugars obtained from biomass can be fermented directly to hydrocarbon fuels or hydrocarbon intermediates, recovered, purified, and further upgraded to drop-in hydrocarbon fuels. This process has been operated by two companies such as LS9 and Amyris. The RJF produced from a sugarcane feedstock has been used in an Embraer E195 jet operated by Azul Brazilian Airline [77].

2. Catalytic upgradation of sugars/sugar intermediates-to-hydrocarbons: the sugars are collected from biomass feedstocks (e.g., milled corn stover) through a series of chemical and biochemical processes and upgraded into hydrocarbon fuels through APR. The process has been operated by few companies such as Virent and Virdia. The jet fuel produced from the process has been tested in U.S. Air Force Research Laboratory (AFRL) and passed the relevant specifications of Jet A1 fuel.

2.4.1 Direct sugar-to-hydrocarbons

The direct production of alkane type fuels from sugars through the fermentation is known as DSH process. Like the feedstocks used in ethanol production, the common feedstocks for DSH process include beets and maize, sugarcane, etc. LCB is also considered as a potential feedstock after several pretreatments [78]. However, this production pathway is different from the ATJ (where an alcohol intermediate is required). Basically, this technology is associated with genetic engineering and screening technology, which allows altering the conversion pathway of microbes to metabolize sugars [79,80]. Typically, cell walls of biomass are broken down to release the hemicellulose sugars. Then, the water is eliminated from the liquid sugars to concentrate it for producing hydrocarbon intermediates through the aerobic fermentation (continuous or fed-batch mode). Finally, the hydrocarbon fuels are produced through the phase separation step [81]. The products of fermentation can be different based on the fermentation pathways as regulated by the species of microbes and types of feedstocks, [80,82]. In a recent study conducted by Davis et al. [82], a complete conversion technique for DSH was proposed involving six main steps, such as (i) pretreatment and conditioning, (ii) enzymatic hydrolysis, (iii) hydrolysate clarification, (iv) biological conversion, (v) product purification, (vi) hydro-processing. It is worth noting that the DSH process requires low energy. However, the fuel blend is very limited (~10%), thus the produced fuels are not meeting several performance standards [83]. A process flow diagram for DSH technique is presented in Fig. 5.

2.4.2 Catalytic upgrading of sugar to hydrocarbons

In the catalytic aqueous phase reforming (APR) process, the soluble plant sugars are first converted into several intermediate chemicals such as aldehydes, alcohols, ketones, furans, acids, and other oxygenated hydrocarbons. These intermediates are subsequently transformed into the jet fuel range hydrocarbons [84,85]. Traditionally, the method begins with enzymatic hydrolysis after a suitable pretreatment of LCB to obtain the small chain carbon (C_5–C_6) sugars (Fig. 5). Thereafter, the slurry of hydrolysis (hydrolysate) is purified and concentrated to meet the requirements of upgrading processes. Subsequently, APR transformation of sugars is performed before supplying the products to a fractionation step to obtain jet fuel as a final product [85]. It is worth noting that the lighter alkanes, lignin, and unconverted solids from APR can be used to supply heat for the process [84].

Generally, the solvent sugars are transformed into the polyhydric alcohols via hydrogenation or short chain oxygenated compounds through hydrogenolysis process before the APR process. Based on the literature, three major pathways are followed to convert oxygenate intermediates to hydrocarbons (jet fuel range) during the APR [86].

1. Acid condensation, which converts the oxygenates including ketones, alcohols, aldehydes, and acids into alkanes, isoalkanes, and aromatics using acid catalysts, such as solid acids and zeolites/ZSM-5 catalyst. The reactions include dehydration of oxygenates to alkenes; oligomerization of the alkenes to larger alkenes; cracking, cyclization and dehydrogenation of larger alkenes to aromatics isomerization; and hydrogen-transfer to form alkanes [87,88]. The heavier species of the products can be distilled and blended into jet fuel.

2. Base condensation, which transforms ketones, aldehydes, and alcohols to the alkanes via direct catalytic condensation using multifunctional catalysts [89,90]. The condensation reactions can be occurred in several ways: (i) oxygen can be converted into β-hydroxy aldehyde or β-hydroxy ketone through the aldol condensation; (ii) the conjugated enone is formed from β-hydroxy aldehyde or β-hydroxy ketone via the dehydration process; (iii) the enone is transformed into aldehydes or ketones or alcohol by the hydrogenation process; and (iv) finally, alkanes are formed by removing hydroxyls via dehydration/hydrogenation or hydrogenolysis, i.e., the alcohols are converted to jet fuel [86,89]. Most of the products produced through this pathway could be used as jet fuel.

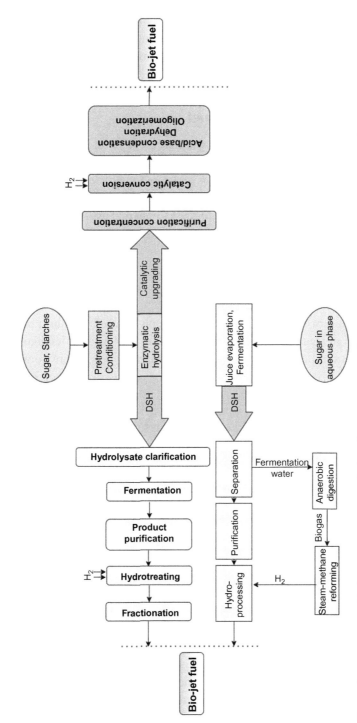

Fig. 5 The sugars-to-jet (STJ) conversion route including direct sugars-to-hydrocarbons (DSH) and catalytic upgrading of sugars-to-hydrocarbons production process.

3. Dehydration/oligomerization, which is to transform the oxygenates derived from APR into alkenes and alkanes by forming reactions of dehydration and dehydrogenation-dehydration. The kerosene is usually produced by oligomerizing alkenes using solid phosphoric catalysts or zeolite [91,92].

3. Lignocellulosic biomass to jet fuel

LCBs are the most abundantly available biomass source in nature [93,94]. LCB is mainly composed of hemicellulose, cellulose, and lignin. Several techniques have been used to convert LCB into fuel. Usually, the bio-oils generated via hydrothermal or pyrolysis of the lignocellulose are converted to jet fuel using HDCJ pathway. Moreover, DHS or APR are also the potential processes for producing biojet fuel using LCB. These processes are already discussed in the previous sections. Briefly, after the pretreatment and fractionation steps, LCB is converted and separated into cellulose, hemicellulose, and lignin. The fractionated cellulose and hemicellulose are usually converted into sugars (C_5 or C_6) using acid or enzymatic hydrolysis. The carbohydrates are turned into polyhydric alcohols via hydrogenation or short-chain oxygenates via hydrogenolysis [19]. Lignin, in this process, is sent to the combustor to provide process heat simply through combustion, which is a waste of resources and harmful for the system economy [84,95]. Besides, lignin has abundant sources and it can be easily converted into aromatic hydrocarbons compared to the other biomass. Aromatic hydro-carbons are indispensable component of commercial grade jet fuel as they are associated with safety and quality of fuel [96]. Therefore, it is crucial to design the pathway of lignin-to-jet. In this section, the processes of lignin conversion for jet fuel production are discussed, which predominantly include four steps such as lignin depolymerization; lignin extraction, and purification; improving lignin derived bio-oils into jet hydrocarbons; and distillation to biojet fuel [96]. The overview of the process is presented in Fig. 6.

Generally, lignins are obtained from the final residuals of hemicellulose and cellulose hydrolysis in the traditional process of jet fuel production. However, it is hard to attain pure lignin. The ionic liquid extraction and Organosolv processes are well-known methods for the extraction and purification of lignin. The depolymerization technique is used to gather lignin-derived bio-oils. Subsequently, such bio-oils go through the hydrodeoxygenation (HDO) and alkylation process. Finally, the jet fuel is obtained after distillation process [96]. In general, several methods including fast pyrolysis, hydrogenolysis, and hydrolysis are used for lignin depolymerization. The lignin polymers are converted into phenolic dimers and monomers due to very high heat in the fast pyrolysis method. Different factors such as sources of lignin, reaction conditions, and lignin pretreatment methods, etc. severely influence the product distribution and bio-oil yield from lignin [96,97]. Hydrolysis is another method for lignin depolymerization, which

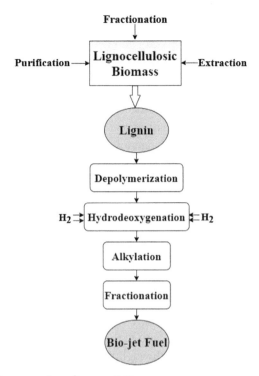

Fig. 6 Lignin to jet fuel process flow diagram [13].

includes organic solvolysis, ionic liquid solvolysis, and hydrothermal liquefaction. Among them, ionic liquid solvolysis has gained attention due to several benefits, including controllable physical properties, low partial pressure requirement, and solvent or catalyst serving ability during the process [96]. Hydrogenolysis is also considered as an attractive process to produce monomeric phenols with lower amount of chars [98]. Based on hydrogen sources, hydrogenolysis lignin depolymerization can be categorized into three groups such as in situ hydrogen produced by a solvent of hydrogen donor, pressurized gaseous hydrogen, and hydrogen produced from the solvent using selective catalyst [96].

The bio-oils generated from lignin are predominantly composed of dimeric and monomeric alkoxy phenols. Several factors such as acidity, high oxygen content, complexity of compounds, and high viscosity instability limit them to be directly used as jet fuels [99]. For upgradation of lignin-derived bio-oils, the focus has been given to the cycloalkanes and aromatic hydrocarbons due to the difficulty of cycloalkane and aromatic ring-opening reaction [96]. The HDO is the mostly used upgrading technique of lignin-derived bio-oils. In this process, the bio-oil quality is upgraded via saturation and deoxidization process under moderate temperature and high hydrogen pressure. Several reactions including hydrogenation, trans-alkylation, decarboxylation, deoxygenation and hydrogenolysis are involved in this method [96]. Two basic reactions such as hydrogenation and direct HDO occur simultaneously during the HDO process. In the direct HDO process, the aromatic properties of the molecules can be maintained by modifying C—O bonds in the side chains through the hydrogenolysis and aromatic hydrocarbons are the primary products. In hydrogenation reaction process, the benzene ring is broken down along with the C—O bonds hydrogenolysis in the side chains. Thus, few products of cycloalkanes, such as cyclohexane and methyl-cyclohexane can be generated through the process [97,100,101]. Numerous catalysts such as metal oxides, sulfides, nitrides, phosphides, and carbides can be used to stimulate the HDO process [102,103]. After the upgradation, the final products obtained as aromatic hydrocarbons and alkylated cycloalkanes, and they could generate iso- and n-alkanes through the alkylation process.

4. Conclusion and future perspectives

In this chapter, we have provided an outline of four predominant conversion techniques of biojet fuels (Table 2). The major portion of the renewable biojet fuel is generated from the complex structure liquid biomass through the hydro-processing technology. However, several challenges still need to be addressed for the development of this technology. It has been observed that the F–T synthesis and HEFA techniques are the most promising methods to produce biojet fuels. However, future works can be considered in few aspects, such as exploring the suitable feedstocks to make the technology practically applicable. The uses of compatible feedstocks (biomass-based feedstocks) could be advantageous to achieve large scale biojet fuel production and to meet emission reduction targets. Furthermore, the problems associated with high production cost and low yield of carbon must be addressed by enhancing conversion efficiency and exploring cost effective feedstocks [114]. Nevertheless, the hybrid pathway by integrating thermo-catalytic steps and biochemicals could be recommended to address some challenges [114]. For instance, target intermediates (e.g., organic acids, alcohols) could be selectively produced from biomass through the biological processes utilizing the selective nature of biochemical techniques using the hybrid pathway [19]. Subsequently, the intermediates can be transformed into jet-range hydrocarbons using thermal catalytic process. Since these intermediates contain a less complicated structure than the initial biomass, the conversion efficiency and product selectivity of the thermo-catalytic step can easily be controlled to produce targeted products [115]. We argue that the pathways reviewed in this chapter would be helpful to design new hybrid process for enhancing low-cost bio-based jet fuel production.

Table 2 Summary of jet fuel conversion pathways.

Category	Pathway	Key conversion step	Catalyst	Companies	Feedstocks
Oil to jet (OTJ)	HEFA	Catalytic hydrogenation Cracking and isomerization	Noble and transitional metals Ni, Pt, or other noble metals	UOP; SG Biofuels, UOP, AltAir Fuels, SG, Agrisoma Biosciences, PetroChina, Neste Oil; Sapphire Energy (Syntroleum/Tyson Food), ASA, PEMEX	Camelina oil [104], Soybean oil [105], Jatropha [106], Waste oils and animal fats [105,106], Microalgae [107]
	CH	Catalytic hydrothermolysis Decarboxylation/ Hydrotreating	Zinc acetate Nickel	Aemetis/Chevron Lummus Global, Applied Research Assoc.	Camelina Oil [108], Lignocellulose [83]
	HDCJ	Hydro-deoxygenation	Pd-Mo, MoC$_x$/C	Envergent, Kior/Hunt Refining/Petrotech, Dynamotive, GTI	Lignocellulose [83]
Gas to jet (GTJ)	FT	F–T synthesis	Ni, Co, Ru, and Fe	SynFuels, Syntroleum, Shell, Rentech, Solena	Lignocellulose [107,109]
Alcohol to jet (ATJ)	Ethanol to jet	Ethanol dehydration	Al$_2$O$_3$, zeolite catalysts, metal (transitional) oxides, and heteropoly acid catalysts	Coskata, Lanza Tech/ Swedish Biofuels, Terrabon/MixAlco	Sugar cane (Biochemistry) [108–110], Corn grain (Biochemistry) [108,110], Lignocellulose (Biochemistry) [108–111]
	Butanol to jet	Butanol dehydration	Zeolite, zirconia, solid acid catalyst, HPW (H$_3$PW$_{12}$O$_{40}$) and mesoporous silica group	Byogy; Albemarle/ Colbalt, Gevo, Solazyme	
	Isobutanol to jet	Iso-butanol dehydration	Acidic resins, zeolites, metal oxides, inorganic acids	—	

Continued

Table 2 Summary of jet fuel conversion pathways—cont'd

Category	Pathway	Key conversion step	Catalyst	Companies	Feedstocks
Sugar to jet (STJ)	DSH	Enzymatic hydrolysis Fermentation	K_3PO_4, palladium catalyst, and several inorganic catalysts [112]	Solazyme, LS9, Amyris/Total	Sugar cane [77,107], Lignocellulose [83]
	APR	Acid condensation Hydrodeoxygenation	Acid catalysis Ru/C	Virdia, Virent/Shell	Lignocellulose [113]
Lignin to jet		Hydrodeoxygenation	Metal oxides, metal phosphides, metal sulfides, Transition metals, metal nitrides, carbides	—	Lignocellulose [83,96,113]
		Hydrogenation	Ru/C		

References

[1] S. Chu, Y. Cui, N. Liu, The path towards sustainable energy, Nat. Mater. 16 (1) (2017) 16–22.

[2] M. Wang, R. Dewil, K. Maniatis, J. Wheeldon, T. Tan, J. Baeyens, et al., Biomass-derived aviation fuels: challenges and perspective, Prog. Energy Combust. Sci. 74 (2019) 31–49.

[3] D. Connolly, H. Lund, B.V. Mathiesen, Smart Energy Europe: the technical and economic impact of one potential 100% renewable energy scenario for the European Union, Renew. Sustain. Energy Rev. 60 (2016) 1634–1653.

[4] M.D. Boot, M. Tian, E.J. Hensen, S.M. Sarathy, Impact of fuel molecular structure on auto-ignition behavior—design rules for future high performance gasolines, Prog. Energy Combust. Sci. 60 (2017) 1–25.

[5] J.G. Speight, Handbook of Petroleum Product Analysis, John Wiley & Sons, 2015.

[6] I. Iata, Sustainable Aviation Fuel Roadmap, International Air Transport Association, Montreal and Geneva, 2015.

[7] N. Yilmaz, A. Atmanli, Sustainable alternative fuels in aviation, Energy 140 (2017) 1378–1386.

[8] H. Hao, Y. Geng, J. Sarkis, Carbon footprint of global passenger cars: scenarios through 2050, Energy 101 (2016) 121–131.

[9] A.W. Schäfer, A.D. Evans, T.G. Reynolds, L. Dray, Costs of mitigating CO2 emissions from passenger aircraft, Nat. Clim. Chang. 6 (4) (2016) 412–417.

[10] R. Mawhood, E. Gazis, S. de Jong, R. Hoefnagels, R. Slade, Production pathways for renewable jet fuel: a review of commercialization status and future prospects, Biofuels Bioprod. Biorefin. 10 (4) (2016) 462–484.

[11] M. Hassan, H. Pfaender, D.N. Mavris, Feasibility analysis of aviation CO2 emission goals under uncertainty, in: 17th AIAA Aviation Technology, Integration, and Operations Conference, 2017, p. 3267.

[12] W. Wong, T. Cheung, A. Zhang, Y. Wang, Is spatial dispersal the dominant trend in air transport development? A global analysis for 2006–2015, J. Air Transp. Manag. 74 (2019) 1–12.

[13] H. Wei, W. Liu, X. Chen, Q. Yang, J. Li, H. Chen, Renewable bio-jet fuel production for aviation: a review, Fuel 254 (2019) 115599.

[14] B.H.H. Goh, H.C. Ong, M.Y. Cheah, W.-H. Chen, K.L. Yu, T.M.I. Mahlia, Sustainability of direct biodiesel synthesis from microalgae biomass: a critical review, Renew. Sustain. Energy Rev. 107 (2019) 59–74.

[15] J. Yang, Z. Xin, K. Corscadden, H. Niu, An overview on performance characteristics of bio-jet fuels, Fuel 237 (2019) 916–936.

[16] D. Kim, A. Violi, Hydrocarbons for the next generation of jet fuel surrogates, Fuel 228 (2018) 438–444.

[17] A. AlNouss, G. McKay, T. Al-Ansari, A techno-economic-environmental study evaluating the potential of oxygen-steam biomass gasification for the generation of value-added products, Energy Convers. Manag. 196 (2019) 664–676.

[18] H. Cantarella, A.M. Nassar, L.A.B. Cortez, R.B. Junior, Potential feedstock for renewable aviation fuel in Brazil, Environ. Dev. 15 (2015) 52–63.

[19] W.-C. Wang, L. Tao, Bio-jet fuel conversion technologies, Renew. Sustain. Energy Rev. 53 (2016) 801–822.

[20] K.W. Waldron, Advances in Biorefineries: Biomass and Waste Supply Chain Exploitation, Elsevier, 2014.

[21] M. Molefe, D. Nkazi, H.E. Mukaya, Method selection for biojet and biogasoline fuel production from castor oil: a review, Energy Fuels 33 (7) (2019) 5918–5932.

[22] D. Chiaramonti, Sustainable aviation Fuels: the challenge of decarbonization, Energy Procedia 158 (2019) 1202–1207.

[23] S.S. Hassan, G.A. Williams, A.K. Jaiswal, Moving towards the second generation of lignocellulosic biorefineries in the EU: drivers, challenges, and opportunities, Renew. Sustain. Energy Rev. 101 (2019) 590–599.

[24] T. Wang, S. Qiu, Y. Weng, L. Chen, Q. Liu, J. Long, et al., Liquid fuel production by aqueous phase catalytic transformation of biomass for aviation, Appl. Energy 160 (2015) 329–335.

[25] L. Rye, S. Blakey, C.W. Wilson, Sustainability of supply or the planet: a review of potential drop-in alternative aviation fuels, Energy Environ. Sci. 3 (1) (2010) 17–27.

[26] T.K. Hari, Z. Yaakob, N.N. Binitha, Aviation biofuel from renewable resources: routes, opportunities and challenges, Renew. Sustain. Energy Rev. 42 (2015) 1234–1244.

[27] M.N. Pearlson, A Techno-Economic and Environmental Assessment of Hydroprocessed Renewable Distillate Fuels, Massachusetts Institute of Technology, 2011.

[28] T. Morgan, E. Santillan-Jimenez, A.E. Harman-Ware, Y. Ji, D. Grubb, M. Crocker, Catalytic deoxygenation of triglycerides to hydrocarbons over supported nickel catalysts, Chem. Eng. J. 189 (2012) 346–355.

[29] T.N. Kalnes, M.M. McCall, D.R. Shonnard, Renewable diesel and jet-fuel production from fats and oils, in: Thermochemical Conversion of Biomass to Liquid Fuels and Chemicals, RSC Energy and Environment Series, RSC Publishing, 2010, pp. 468–495 (1).

[30] Y. Yang, Q. Wang, X. Zhang, L. Wang, G. Li, Hydrotreating of C18 fatty acids to hydrocarbons on sulphided NiW/SiO2–Al2O3, Fuel Process. Technol. 116 (2013) 165–174.

[31] H. Wang, S. Yan, S.O. Salley, K.S. Ng, Support effects on hydrotreating of soybean oil over NiMo carbide catalyst, Fuel 111 (2013) 81–87.

[32] T.K. Hari, Z. Yaakob, Production of diesel fuel by the hydrotreatment of jatropha oil derived fatty acid methyl esters over γ-Al2O3 and SiO2 supported NiCo bimetallic catalysts, React. Kinet. Mech. Catal. 116 (1) (2015) 131–145.

[33] A. Galadima, O. Muraza, Catalytic upgrading of vegetable oils into jet fuels range hydrocarbons using heterogeneous catalysts: a review, J. Ind. Eng. Chem. 29 (2015) 12–23.

[34] N. Mo, P.E. Savage, Hydrothermal catalytic cracking of fatty acids with HZSM-5, ACS Sustain. Chem. Eng. 2 (1) (2014) 88–94.

[35] M. Lu, X. Liu, Y. Li, Y. Nie, X. Lu, D. Deng, et al., Hydrocracking of bio-alkanes over Pt/Al-MCM-41 mesoporous molecular sieves for bio-jet fuel production, J. Renew. Sustain. Energy 8 (5) (2016) 053103.

[36] J.H. Gary, G.E. Handwerk, M.J. Kaiser, Petroleum Refining: Technology and Economics, CRC Press, 2007.

[37] F. Yang, M.A. Hanna, R. Sun, Value-added uses for crude glycerol—a byproduct of biodiesel production, Biotechnol. Biofuels 5 (1) (2012) 1–10.

[38] G. Sivakumar, J. Xu, R.W. Thompson, Y. Yang, P. Randol-Smith, P.J. Weathers, Integrated green algal technology for bioremediation and biofuel, Bioresour. Technol. 107 (2012) 1–9.

[39] L. Li, E. Coppola, J. Rine, J.L. Miller, D. Walker, Catalytic hydrothermal conversion of triglycerides to non-ester biofuels, Energy Fuels 24 (2) (2010) 1305–1315.

[40] A. Elgowainy, J. Han, M. Wang, J. Hileman, N. Carter, Development of life-cycle analysis module of aviation fuel/aircraft systems in GREET, in: GREET Training Workshop, 2012.

[41] M. Ajam, C. Woolard, C. Wiljoen, Biomass pyrolysis oil as a renewable feedstock for bio-jet fuel. 2013, in: Proceedings of the 13th International Conference on Stability, Handling and Use of Liquid Fuels (IASH2013), Rhodes, Greece, 2013, pp. 6–10.

[42] D. Elliott, M. Olarte, T. Hart, Pilot-Scale Biorefinery: Sustainable Transport Fuels From Biomass and Algal Residues via Integrated Pyrolysis, Catalytic Hydroconversion and Co-Processing With Vacuum Gas Oil, UOP LLC, Des Plaines, IL, 2016.

[43] M. Olarte, A. Zacher, D. Elliott, D. Santosa, G. Neuenschwander, T. Hart, et al., Bio-oil upgrading and stabilization at PNNL, in: Proceedings of the 2011 Harvesting Clean Energy Conference, 2011.

[44] X. Xu, C. Zhang, Y. Liu, Y. Zhai, R. Zhang, Two-step catalytic hydrodeoxygenation of fast pyrolysis oil to hydrocarbon liquid fuels, Chemosphere 93 (4) (2013) 652–660.

[45] J. Wang, P. Bi, Y. Zhang, H. Xue, P. Jiang, X. Wu, et al., Preparation of jet fuel range hydrocarbons by catalytic transformation of bio-oil derived from fast pyrolysis of straw stalk, Energy 86 (2015) 488–499.

[46] D. Radlein, A. Quignard, A short historical review of fast pyrolysis of biomass, Oil Gas Sci. Technol. Rev. IFP Energ. nouv. 68 (4) (2013) 765–783.

[47] G. Li, N. Li, X. Wang, X. Sheng, S. Li, A. Wang, et al., Synthesis of diesel or jet fuel range cycloalkanes with 2-methylfuran and cyclopentanone from lignocellulose, Energy Fuels 28 (8) (2014) 5112–5118.

[48] V. Chubukov, A. Mukhopadhyay, C.J. Petzold, J.D. Keasling, H.G. Martín, Synthetic and systems biology for microbial production of commodity chemicals, NPJ Syst. Biol. Appl. 2 (1) (2016) 1–11.

[49] Z. Cao, J. Engelhardt, M. Dierks, M.T. Clough, G.H. Wang, E. Heracleous, et al., Catalysis meets nonthermal separation for the production of (alkyl) phenols and hydrocarbons from pyrolysis oil, Angew. Chem. Int. Ed. 56 (9) (2017) 2334–2339.

[50] R. Mawhood, A.R. Cobas, R. Slade, Establishing a European renewable jet fuel supply chain: the technoeconomic potential of biomass conversion technologies, in: Biojet Fuel Supply Chain Development and Flight Operations (Renjet), 2014.

[51] J. Hu, F. Yu, Y. Lu, Application of Fischer–Tropsch synthesis in biomass to liquid conversion, Catalysts 2 (2) (2012) 303–326.

[52] M.E. Dry, Practical and theoretical aspects of the catalytic Fischer-Tropsch process, Appl. Catal. A Gen. 138 (2) (1996) 319–344.

[53] F. You, B. Wang, Life cycle optimization of biomass-to-liquid supply chains with distributed–centralized processing networks, Ind. Eng. Chem. Res. 50 (17) (2011) 10102–10127.

[54] G.A. Richards, K.H. Casleton, T. Lieuwen, V. Yang, R. Yetter, Gasification Technology to Produce Synthesis Gas, CRC Press, Boca Raton, FL, 2009.

[55] A. Dutta, M. Talmadge, J. Hensley, M. Worley, D. Dudgeon, D. Barton, et al., Process Design and Economics for Conversion of Lignocellulosic Biomass to Ethanol: Thermochemical Pathway by Indirect Gasification and Mixed Alcohol Synthesis, National Renewable Energy Lab. (NREL), Golden, CO, 2011.

[56] S. Phillips, Technoeconomic analysis of a lignocellulosic biomass indirect gasification process to make ethanol via mixed alcohols synthesis, Ind. Eng. Chem. Res. 46 (26) (2007) 8887–8897.

[57] J. Daniell, M. Köpke, S.D. Simpson, Commercial biomass syngas fermentation, Energies 5 (12) (2012) 5372–5417.

[58] D.W. Griffin, M.A. Schultz, Fuel and chemical products from biomass syngas: a comparison of gas fermentation to thermochemical conversion routes, Environ. Prog. Sustain. Energy 31 (2) (2012) 219–224.

[59] B. Acharya, P. Roy, A. Dutta, Review of syngas fermentation processes for bioethanol, Biofuels 5 (5) (2014) 551–564.

[60] D. Chiaramonti, M. Prussi, M. Buffi, D. Tacconi, Sustainable bio kerosene: process routes and industrial demonstration activities in aviation biofuels, Appl. Energy 136 (2014) 767–774.

[61] J.C. Serrano-Ruiz, E.V. Ramos-Fernández, A. Sepúlveda-Escribano, From biodiesel and bioethanol to liquid hydrocarbon fuels: new hydrotreating and advanced microbial technologies, Energy Environ. Sci. 5 (2) (2012) 5638–5652.

[62] V. Tret'yakov, Y.I. Makarfi, K. Tret'yakov, N. Frantsuzova, R. Talyshinskii, The catalytic conversion of bioethanol to hydrocarbon fuel: a review and study, Catal. Ind. 2 (4) (2010) 402–420.

[63] D. Rutz, R. Janssen, Biofuel Technology Handbook, WIP Renewable Energies, 2007, p. 95.

[64] M. Zhang, Y. Yu, Dehydration of ethanol to ethylene, Ind. Eng. Chem. Res. 52 (28) (2013) 9505–9514.

[65] T.K. Phung, L.P. Hernández, A. Lagazzo, G. Busca, Dehydration of ethanol over zeolites, silica alumina and alumina: Lewis acidity, Brønsted acidity and confinement effects, Appl. Catal. A Gen. 493 (2015) 77–89.

[66] R.D. Andrei, M.I. Popa, F. Fajula, V. Hulea, Heterogeneous oligomerization of ethylene over highly active and stable Ni-AlSBA-15 mesoporous catalysts, J. Catal. 323 (2015) 76–84.

[67] K. Weissermel, H.-J. Arpe, Industrial Organic Chemistry, John Wiley & Sons, 2008.

[68] M. Peters, J. Taylor, Renewable Jet Fuel Blendstock From Isobutanol, Gevo, Inc., 2011.

[69] P.R. Gruber, M.W. Peters, J.M. Griffith, Y. Al Obaidi, L.E. Manzer, J.D. Taylor, et al., Renewable compositions, Google Patents, 2012.

[70] S. Jeong, H. Kim, J.-h. Bae, D.H. Kim, C.H. Peden, Y.-K. Park, et al., Synthesis of butenes through 2-butanol dehydration over mesoporous materials produced from ferrierite, Catal. Today 185 (1) (2012) 191–197.

[71] M. John, K. Alexopoulos, M.-F. Reyniers, G.B. Marin, Reaction path analysis for 1-butanol dehydration in H-ZSM-5 zeolite: ab initio and microkinetic modeling, J. Catal. 330 (2015) 28–45.

[72] M.E. Wright, B.G. Harvey, R.L. Quintana, Diesel and jet fuels based on the oligomerization of butene. Google Patents, 2013.

[73] M.E. Wright. Process for the dehydration of aqueous bio-derived terminal alcohols to terminal alkenes, Google Patents, 2014.

[74] M. Wright, Biomass to Alcohol to Jet/Diesel, Interne Präsentation, Australia, 2012.

[75] J.D. Taylor, M.M. Jenni, M.W. Peters, Dehydration of fermented isobutanol for the production of renewable chemicals and fuels, Top. Catal. 53 (15–18) (2010) 1224–1230.

[76] R.D. Cortright, Catalytic conversion of sugars to conventional liquid fuels, in: Abstracts of Papers of the American Chemical Society. 238, American Chemical Society, Washington, DC, 2009.

[77] L.A. Cortez, F.E. Nigro, L.A. Nogueira, A.M. Nassar, H. Cantarella, M.A. Moraes, et al., Perspectives for sustainable aviation biofuels in Brazil, Int. J. Aerosp. Eng. 2015 (2015).

[78] S. Al-Zuhair, K. Ahmed, A. Abdulrazak, M.H. El-Naas, Synergistic effect of pretreatment and hydrolysis enzymes on the production of fermentable sugars from date palm lignocellulosic waste, J. Ind. Eng. Chem. 19 (2) (2013) 413–415.

[79] A. Milbrandt, C. Kinchin, R. McCormick, Feasibility of Producing and Using Biomass-Based Diesel and Jet Fuel in the United States, National Renewable Energy Lab.(NREL), Golden, CO, 2013.

[80] F. Zhu, X. Zhong, M. Hu, L. Lu, Z. Deng, T. Liu, In vitro reconstitution of mevalonate pathway and targeted engineering of farnesene overproduction in *Escherichia coli*, Biotechnol. Bioeng. 111 (7) (2014) 1396–1405.

[81] U. Doe, National Algal Biofuels Technology Roadmap, US Department of Energy, Office of Energy Efficiency and Renewable Energy. Biomass Program, 2010, p. 140.

[82] R. Davis, M.J. Biddy, E. Tan, L. Tao, S.B. Jones, Biological Conversion of Sugars to Hydrocarbons Technology Pathway, Pacific Northwest National Lab. (PNNL), Richland, WA, 2013.

[83] S. De Jong, R. Hoefnagels, A. Faaij, R. Slade, R. Mawhood, M. Junginger, The feasibility of short-term production strategies for renewable jet fuels–a comprehensive techno-economic comparison, Biofuels Bioprod. Bioref. 9 (6) (2015) 778–800.

[84] R. Davis, L. Tao, E. Tan, M. Biddy, G. Beckham, C. Scarlata, et al., Process Design and Economics for the Conversion of Lignocellulosic Biomass to Hydrocarbons: Dilute-Acid and Enzymatic Deconstruction of Biomass to Sugars and Biological Conversion of Sugars to Hydrocarbons, National Renewable Energy Lab. (NREL), Golden, CO, 2013.

[85] M. Biddy, S. Jones, Catalytic Upgrading of Sugars to Hydrocarbons Technology Pathway, National Renewable Energy Lab. (NREL), Golden, CO, 2013.

[86] P. Blommel, R. Cortright, Production of Conventional Liquid Fuels From Sugars, vol. 25, Virent Energy Systems, Inc., Madison, WI, 2008, pp. 1–14.

[87] P.W. Goguen, T. Xu, D.H. Barich, T.W. Skloss, W. Song, Z. Wang, et al., Pulse-quench catalytic reactor studies reveal a carbon-pool mechanism in methanol-to-gasoline chemistry on zeolite HZSM-5, J. Am. Chem. Soc. 120 (11) (1998) 2650–2651.

[88] A. de Klerk, R.J. Nel, R. Schwarzer, Oxygenate conversion over solid phosphoric acid, Ind. Eng. Chem. Res. 46 (8) (2007) 2377–2382.

[89] J.N. Chheda, J.A. Dumesic, An overview of dehydration, aldol-condensation and hydrogenation processes for production of liquid alkanes from biomass-derived carbohydrates, Catal. Today 123 (1–4) (2007) 59–70.

[90] F. King, G. Kelly, Combined solid base/hydrogenation catalysts for industrial condensation reactions, Catal. Today 73 (1–2) (2002) 75–81.

[91] E.I. Gürbüz, J.A. Dumesic, Catalytic Strategies and Chemistries Involved in the Conversion of Sugars to Liquid Transportation Fuels, Max Planck Institute for the History of Science, 2013.

[92] E.I. Guerbuez, E.L. Kunkes, J.A. Dumesic, Dual-bed catalyst system for C–C coupling of biomass-derived oxygenated hydrocarbons to fuel-grade compounds, Green Chem. 12 (2) (2010) 223–227.

[93] F.H. Isikgor, C.R. Becer, Lignocellulosic biomass: a sustainable platform for the production of bio-based chemicals and polymers, Polym. Chem. 6 (25) (2015) 4497–4559.

[94] A. Yousuf, S. Sultana, M.U. Monir, A. Karim, S.R.B. Rahmaddulla, Social business models for empowering the biogas technology, Energy Sources Part B Econ. Plan. Policy 12 (2) (2017) 99–109.

[95] J.Q. Bond, A.A. Upadhye, H. Olcay, G.A. Tompsett, J. Jae, R. Xing, et al., Production of renewable jet fuel range alkanes and commodity chemicals from integrated catalytic processing of biomass, Energy Environ. Sci. 7 (4) (2014) 1500–1523.

[96] F. Cheng, C.E. Brewer, Producing jet fuel from biomass lignin: potential pathways to alkyl-benzenes and cycloalkanes, Renew. Sustain. Energy Rev. 72 (2017) 673–722.

[97] J.-Y. Kim, J.H. Lee, J. Park, J.K. Kim, D. An, I.K. Song, et al., Catalytic pyrolysis of lignin over HZSM-5 catalysts: effect of various parameters on the production of aromatic hydrocarbon, J. Anal. Appl. Pyrolysis 114 (2015) 273–280.

[98] M.P. Pandey, C.S. Kim, Lignin depolymerization and conversion: a review of thermochemical methods, Chem. Eng. Technol. 34 (1) (2011) 29–41.

[99] P.M. Mortensen, J.-D. Grunwaldt, P.A. Jensen, K. Knudsen, A.D. Jensen, A review of catalytic upgrading of bio-oil to engine fuels, Appl. Catal. A Gen. 407 (1–2) (2011) 1–19.

[100] D. Garcia-Pintos, J. Voss, A.D. Jensen, F. Studt, Hydrodeoxygenation of phenol to benzene and cyclohexane on Rh (111) and Rh (211) surfaces: insights from density functional theory, J. Phys. Chem. C 120 (33) (2016) 18529–18537.

[101] Y. Shao, Q. Xia, L. Dong, X. Liu, X. Han, S.F. Parker, et al., Selective production of arenes via direct lignin upgrading over a niobium-based catalyst, Nat. Commun. 8 (1) (2017) 1–9.

[102] E. Budsberg, J.T. Crawford, H. Morgan, W.S. Chin, R. Bura, R. Gustafson, Hydrocarbon bio-jet fuel from bioconversion of poplar biomass: life cycle assessment, Biotechnol. Biofuels 9 (1) (2016) 1–13.

[103] C. Li, X. Zhao, A. Wang, G.W. Huber, T. Zhang, Catalytic transformation of lignin for the production of chemicals and fuels, Chem. Rev. 115 (21) (2015) 11559–11624.

[104] E. Mupondwa, X. Li, L. Tabil, K. Falk, R. Gugel, Technoeconomic analysis of camelina oil extraction as feedstock for biojet fuel in the Canadian Prairies, Biomass Bioenergy 95 (2016) 221–234.

[105] M. Pearlson, C. Wollersheim, J. Hileman, A techno-economic review of hydroprocessed renewable esters and fatty acids for jet fuel production, Biofuels Bioprod. Biorefin. 7 (1) (2013) 89–96.

[106] W.-C. Wang, Techno-economic analysis of a bio-refinery process for producing hydro-processed renewable jet fuel from Jatropha, Renew. Energy 95 (2016) 63–73.

[107] D. Klein-Marcuschamer, C. Turner, M. Allen, P. Gray, R. Dietzgen, P. Gresshoff, et al., Technoeconomic analysis of renewable aviation fuel from microalgae, *Pongamia pinnata*, and sugarcane, Biofuels Bioprod. Biorefin. 7 (2013) 416–428.

[108] R.H. Natelson, W.-C. Wang, W.L. Roberts, K.D. Zering, Technoeconomic analysis of jet fuel production from hydrolysis, decarboxylation, and reforming of camelina oil, Biomass Bioenergy 75 (2015) 23–34.

[109] C. Gutiérrez-Antonio, F. Gómez-Castro, J. de Lira-Flores, S. Hernández, A review on the production processes of renewable jet fuel, Renew. Sustain. Energy Rev. 79 (2017) 709–729.

[110] K. Atsonios, M.-A. Kougioumtzis, K.D. Panopoulos, E. Kakaras, Alternative thermochemical routes for aviation biofuels via alcohols synthesis: process modeling, techno-economic assessment and comparison, Appl. Energy 138 (2015) 346–366.

[111] G.W. Diederichs, M.A. Mandegari, S. Farzad, J.F. Görgens, Techno-economic comparison of biojet fuel production from lignocellulose, vegetable oil and sugar cane juice, Bioresour. Technol. 216 (2016) 331–339.

[112] P. Anbarasan, Z.C. Baer, S. Sreekumar, E. Gross, J.B. Binder, H.W. Blanch, et al., Integration of chemical catalysis with extractive fermentation to produce fuels, Nature 491 (7423) (2012) 235–239.

[113] Y. Li, L. Chen, X. Zhang, Q. Zhang, T. Wang, S. Qiu, et al., Process and techno-economic analysis of bio-jet fuel-range hydrocarbon production from lignocellulosic biomass via aqueous phase deconstruction and catalytic conversion, Energy Procedia 105 (2017) 675–680.

[114] S.P. Adhikari, J. Zhang, Q. Guo, K.A. Unocic, L. Tao, Z. Li, A hybrid pathway to biojet fuel via 2, 3-butanediol, Sustain. Energy Fuels 4 (8) (2020) 3904–3914.

[115] A.A. Fuels, Overview of Challenges, Opportunities, and Next Steps, US Department of Energy, 2017.

CHAPTER 2

Thermochemical conversion of agricultural waste to biojet fuel

Nicolas Vela-García[a], David Bolonio[b], María-Jesús García-Martínez[b], Marcelo F. Ortega[b], and Laureano Canoira[b]

[a]Department of Chemical Engineering, Institute for Development of Alternative Energies and Materials IDEMA, Universidad San Francisco de Quito, Quito, Ecuador
[b]Department of Energy and Fuels, School of Mining and Energy Engineering, Universidad Politécnica de Madrid, Madrid, Spain

1. Introduction

As far as energy and fuel are concerned, the past century was predominantly denoted by technological improvements regarding the production of petroleum-derived fuels and chemicals. Because of the impending effects of global warming and many countries' dependence on fossil fuels, industry and science are increasingly focused on producing biomass-derived fuels as fossil-based fuels surrogates [1]. One of the main challenges of the 21st century is energy security and the clean, affordable, reliable, sustainable, and renewable (CARSR) liquid fuels production capacity to meet the global demand [2]. The question arises: what kind of biofuel can be produced to satisfy the CARSR liquid fuels demand in the medium term?

In the context of a future circular economy based on liquid biofuels, significant changes are required to achieve a low-carbon transition based on the logical development of alternative fuels while gradually replacing fossil fuels. Energy security is an essential pivotal in the development of a country. Biomass plays a potential role in the energy matrix, encompassing economic, political, technical, food security, and environmental considerations [3]. The biomass-to-liquid (BTL) pathway is considered one of the leading green alternatives for producing biobased energy, fuels, and chemicals through biochemical and thermochemical technologies [4]. Alternative jet fuel (AJF) has received much attention among such products because air transport is an essential component of today's globalization, moving nearly 55 Mt goods and over 4 billion travelers in 2019 before the severe acute respiratory syndrome coronavirus 2 (SARS-CoV-2) [5].

The AJF or biojet fuel positively impacts sustainability since it is produced from renewable feedstocks and contains no sulfur or nitrogen. In contrast, traditional aviation fuel affects the atmosphere through pollutant emissions such as CO_2, NO_x, sulfate, aerosols, soot, and, increasing the cloudiness, contributing to ozone layer depletion and climate change. More than 99% of jet fuel generated corresponds to traditional

Sustainable Alternatives for Aviation Fuels
https://doi.org/10.1016/B978-0-323-85715-4.00002-1

Copyright © 2022 Elsevier Inc.
All rights reserved.

fuels, i.e., petroleum-derived Jet A1 and Jet-A. Jet fuel accounts for up to 40% of the air transport industry's operational costs [4]. In 2035, the number of flights is expected to double from 2015, resulting in increased fuel consumption and greenhouse gas emissions. According to the International Civil Aviation Organization (ICAO), passenger air transportation accounts for 2% of global CO_2 pollution. Furthermore, the aviation sector (passenger and cargo) anticipates 10%–32% of overall CO_2 emissions by 2050 [6].

In this regard, the aviation industry has committed to the third step of the Carbon Offsetting and Reduction Scheme for International Aviation (CORSIA) to cut net emissions in half by 2050 relative to 2005 levels. Nevertheless, the improved efficiency of modern engines and effective air traffic control are insufficient to compensate for expected emissions and foster a carbon-neutral expansion [7]. Even though the primary biofuels used in gasoline blending for ground transportation, such as isobutanol and ethanol, have been widely implemented, their usage with aviation fuels is not technically feasible because of their increased volatility and lower energy density [8]. Therefore, technological development in biojet fuel production from renewable sources as a fossil-based aviation fuel surrogate is needed.

Biojet fuel should replace traditional jet fuel without affecting the aircraft engines design or fuel delivery systems; thus, it consists of a "drop-in" blending component. Since biojet fuel characteristics differ from traditional jet fuel, the American Society for Testing and Materials (ASTM) certified routes for producing aromatic-free biojet fuel ensure homogeneity, combustion behavior, and pumping requirements [9]. Approved technologies involve the upgrading of lignocellulose-derived alcohol into jet fuel (ATJ); triglycerides and vegetable oil catalytic hydrothermolysis jet (CHJ); hydroprocessed esters and fatty acids (HEFA); power-to-liquid conversion of syngas through a Ficher-Tropsch (FT) reactor; synthesized kerosene with increased aromatics content (FT-SPK/A); and hydroprocessed synthesized isoparaffins via hydroprocessed fermented sugars (SIP-HFS), as detailed in Fig. 1.

Fossil jet fuel comprises approximately 20% paraffin, 40% isoparaffins, 20% naphthenes, and 20% aromatics, determining its physicochemical properties (see Table 1). On the other hand, biojet fuel composition comprehends renewable hydrocarbons within the boiling range of fossil-based jet fuel. At the same time, the aromatics content can vary depending on the selected ASTM route. If aromatic compounds are not present, particles released by burning biojet fuel are lower than those emitted by fossil jet fuel [10]. However, the absence of aromatics can cause leakage, wear, and damages in specific engine types. As a result, the biojet fuel must be blended up to 50 vol% with the traditional jet fuel to achieve the established specifications [11].

For drop-in biojet fuel production, thermochemical processing of biomass from BTL pathways is a promising technology [12]. The biomass transformation can be done either by gasification, yielding syngas as an intermediate stream, or fast pyrolysis, which outputs biochar, bio-oil and noncondensable gas. Syngas is an important industrial

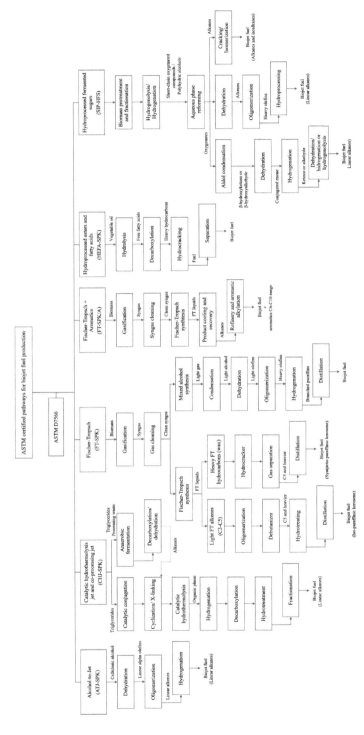

Fig. 1 Certified ASTM pathways for biojet fuel processing.

Table 1 Physicochemical properties of conventional aviation fuel.

Property	Unit	Limit	Kerosene type Jet-A	Jet A1
Aromatics content	vol%	Max	25	
Boiling range	°C		170–300	
Freezing temperature	°C		−40	−47
Flash temperature	°C	Min	38	
Density at 15°C	kg/m^3		775–840	
Viscosity at −20°C	mm^2/s	Max	8.0	
Lower heating value	MJ/kg	Min	42.8	
Smoke point	mm	Min	25	

feedstock; its global production in 2004 was 6 EJ, accounting for about 2% of the world energy consumption. While the forecasted output for 2040 is 50 EJ, corresponding to 10% of the world energy consumption, with BTL technologies providing near 88% of the global syngas market [13]. Syngas is converted into a wide range of bio-based fuels and chemicals through FT synthesis, a well-known and established catalytic chemical process, to meet rising and diverse market demand, including biojet fuel [14]. Biochemical conversion of biomass, on the contrary, has defined several limitations, such as byproduct inhibition to fermentation, crystallinity, size reduction, ethanol tolerance by yeast, cofermentation of various sugars, and deactivation of cellulase [15]. Nevertheless, the main disadvantage is the elevated cost compared with thermochemical routes. Intermediates from the biochemical process are converted into biojet fuel via the ATJ route [16].

Moving from the current linear "take, make, dispose of" paradigm to the systemic, circular alternative "reduce, reuse, recycle, regenerate" is needed for a sustainable agricultural sector. Public and private technological initiatives in circular economy development have made significant efforts to improve biorefineries' performance and long-term viability [1,17]. Agricultural waste biomass, in this context, should be considered a renewable resource rather than waste. Thus, this study aims to provide a vision to explore critical issues in ensuring sustainable use of agri-waste through a transdisciplinary perspective, taking into account the three foundations of sustainability: technical, economic, and environmental. Besides, contributing to innovative holistic methods in support of eco-efficient conversion routes and efficient management strategies [18].

The present work is a concise overview of lignocellulosic agricultural wastes/residues from oleaginous crops and their potential use in biojet fuel production through thermochemical and biochemical conversions. Their subsequent upgrade to biojet fuel integrates the ASTM D7566-certified technologies for aviation turbine fuel containing synthesized hydrocarbons.

2. Agricultural waste generation and composition

Agricultural waste and byproducts are generally described as residues of plants or animals that are not further transformed into food or feed, commonly representing environmental and economic burdens in farming's primary processing sectors. The agriculture and agri-food industry generates large amounts of organic waste, with the EU28 countries producing up to 90 million tonnes a year in 2017. Nevertheless, the 2050 forecasted world population (9 billion people) will undoubtedly increase food production demand, yielding a proportional increase in agricultural waste [19]. Environmental and economic concerns associated with primary agri-waste are correlated with local expertise regarding animal feed or production (e.g., facilities, waste handling and processing and energy supply technologies, etc.). In regions dedicated to animal husbandry, immense quantities of manure residue are generated, resulting in intensive odor and pollution of bacteria, elevated greenhouse gases (GHG) emissions, and high loads of organic matter and nutrients (e.g., nitrogen). Instead, there is a depletion of nutrients and organic matter in regions predominantly dedicated to producing vegetable crops, resulting in a global disparity [20].

On the other hand, lignocellulosic waste resources represented around 50% of the fresh weight of harvested and gathered crops, accounting for a potential of 90 million-tonne oil equivalent (MTOE), considerably more than any other residual streams such as wood chip production (57 MTOE), municipal and various other wastes (42 MTOE), and tertiary woodland residues (32 MTOE) [21]. Furthermore, food wastes and debris from the agri-food industry are estimated to account for about 30% of the global food processing, making proper waste management critical for global sustainability [22]. The organic waste noncontrolled decomposition can contaminate surface and groundwater sources, soil, and the atmosphere on a large scale. The decay of one metric ton of organic solid waste can release 90–140 m^3 of methane and 50–110 m^3 of CO_2 into the atmosphere [23].

In addition, it is convenient to mention the vegetable oil obtained from the lignocellulosic biomass as another valuable residue from agriculture. Vegetable oil is considered among the primary food and feedstock components to develop alternative fuels. The variety of oilseeds is categorized into edible and nonedible. The nonedible segment involves residual oils as possible sources for processing oleochemicals byproducts such as biojet fuel. Biojet fuel produced from lignocellulosic biomass and vegetable oil is exceptional among the renewable sources due to its compatibility with conventional Jet A1 [9,24].

2.1 Lignocellulosic biomass

Among renewable energy feedstocks, lignocellulosic biomass is the most significant, with a 13% share in the global international energy mix. In contrast, hydropower at 3% and all other sustainable sources (geothermal, solar, wind, and tidal) add to just 2%. Nonfood lignocellulosic biomass from agriculture waste is a valuable source for energy purposes due to its carbon-free footprint, abundance, and versatility to produce bio-based chemicals [25]. Throughout palm oil fresh fruit bunches (FFB) harvesting, the crop's planted area generates lignocellulosic biomass byproducts such as palm fronds (PF) and palm trunks (PT). In comparison, the extraction process of vegetable oil by pressing FFB produces, to a greater extent, empty fruit bunch (EFB) (24 wt%), palm mesocarp fiber (PMF) (20 wt%), palm kernel (10 wt%), and shell (9 wt%). Besides, the milling stage during extraction requires a significantly large quantity of water, producing palm oil mill effluent (POME) (0.87 m^3/t_{FFB}). Lignocellulosic biomass as a direct ignition power source is not technically or economically viable due to its high moisture content. Given the pressure on the environment's ecological integrity caused by the drastic increment in air transport operations, renewable resources are pivotal in producing synthetic fuels with high calorific content.

The chemical composition of the oil palm biomass corresponds to cellulose (40–50 wt %), hemicellulose (25–35 wt%), lignin (15–25 wt%), and to a lesser extent, extractives and inorganics (ash) (3–7 wt%) [26]. Bioethanol is the most common intermediate and one of the most relevant liquid biofuels produced from various lignocellulosic biomass. According to an estimation, considering the feedstock used, it will reduce greenhouse gas emissions by around 30%–85% relative to gasoline [24]. Unlike land transport, air transport requires more elaborated and sophisticated pathways to produce a suitable Jet A1-blending component. Thus, biochemical and thermochemical pathways have been developed to obtain a homogeneous biojet fuel from cellulosic alcohol.

2.2 Vegetable oil from residues

Vegetable oil has recently gained growing recognition as an environmentally sustainable and promising energy source to replace petroleum-based fuels. Vegetable oil is considered a pivotal alternative to mitigate global warming and avoid atmospheric pollution by reducing carbon dioxide emissions. New jet engines have been shown to operate satisfactorily on fuels derived from vegetable oils rather than petroleum-based fuels. Several vegetable oil types are used to produce biojet fuel, including virgin and residual vegetable oils. Furthermore, the rent and employment rise in agricultural areas and the effect on related sectors benefit the production of vegetable oil crops for biojet fuel. In Europe, it is worth emphasizing that it is most economical for farmers to obtain subsidies specified in the EU Agricultural Policy by producing energy crops on set-aside land. It is necessary to provide a high-quality, low-cost fuel to compete with traditional kerosene (Jet A1) to expand biojet fuel use [27]. The top vegetable oil category is palm oil in production volume and consumption; hence, this research focuses on its waste/residue. Global palm oil production yielded approximately 65.5 Mt in 2017/2018, and about 89 Mt of palm oil crop wastes are projected for 2021. Asia accounted for about 89% of the global palm oil supply in 2013. That year, the top manufacturers of palm oil were Indonesia and Malaysia, followed by Thailand, Nigeria, and Colombia [28].

The vegetable oil from lignocellulosic biomass waste suits better as biojet fuel feedstock than virgin oils due to its economic feasibility, sustainability, greater net energy ratio, and waste management enhancement [29]. Waste vegetable oil is obtained from lignocellulosic biomass with high oil content after harvesting and extracting processes. On the other hand, palm fatty acid distillates (PFAD) are generated during alkali refining of vegetable oil at the separation and deodorization last stages. The PFAD is a complex blend composed of free fatty acids (FFA), sterol esters, tocopherols, sterols, aldehydes, hydrocarbons, ketones, breakdown products of fatty acids, and acyl glycerol [30,31]. Depending on the refining process's raw material and operating conditions, FFA content comprises 25%–75% of the extract. FFA separation from the PFAD blend is achieved through molecular distillation due to the differences between vapor pressures and molecular weights of FFA (lighter) and the significant compounds [32]. The PFAD feasibility lies in its ease of integration as a drop-in feedstock during hydroprocessing pathways in the decarboxylation stage.

The waste oil is primarily composed of triglycerides, diglycerides, monoglycerides, free fatty acids, and aldehydes. Among the nonedible vegetable oils, jatropha and palm kernel oil-derivates are potential candidates for biojet fuel production because of their composition dominated by lauric fatty acid (C12:0) and linoleic fatty acid (C18:2), respectively [29]. Besides, waste oil has a more straightforward structure than oleaginous crop lignocellulosic biomass; thus, it has a higher effective hydrogen-carbon ratio ideal for high-calorific fuel upgrades.

However, considering its fatty acid composition, vegetable oil suitability as feedstock is assessed. The most representative sorts of vegetable oil are described in Table 2.

3. Conversion of agri-waste

Agricultural wastes/residues predominantly refer to lignocellulosic biomass, characterized by its considerable production rate from different sources, and biodegradability can be transformed through thermochemical and biochemical conversions to generate potentially renewable energy and fertilizers materials and molecules. The transformation of agricultural residues is essential for decoupling economic development and human wellness from primary resources, avoiding pressure on land, and producing detrimental impacts on biodiversity and endangering global food security.

Table 2 Composition of vegetable oil.

Type	Annual production in 2019 (Mt)	Saturated fatty acids (wt%)	Monounsaturated fatty acids Oleic acid (ω-9)	Polyunsaturated fatty acids		Ref.
				α-Linoleic acid	Linoleic acid (ω-6)	
Coconut	2.8	83	6	–	–	[33]
Corn	4.1	13	27	1	58	[34]
Cottonseed	4.9	26	19	1	54	[35]
Grapeseed	0.4	11	14	–	75	[36,37]
Palm	75.5	49	40	<1	9	[38]
Palm kernel oil	8.8	83	15	<1	2	[33]
Sesame	0.9	14	39	<1	41	[39]
Soybean	57.9	16	23	7	51	[40]
Sunflower (<60% linoleic)	20.0	10	45	<1	40	[41]

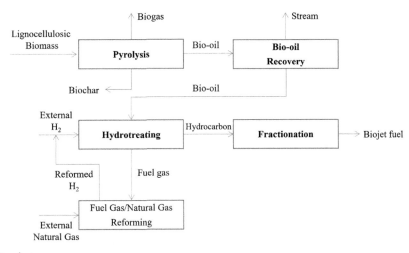

Fig. 2 Pyrolysis process.

3.1 Thermochemical

3.1.1 Pyrolysis

Pyrolysis thermally decomposes the lignocellulosic biomass into the liquid (bio-oil), solid (biochar), and gas (noncondensable gas) phases, as shown in Fig. 2. The reaction is categorized in torrefaction or mild pyrolysis, slow pyrolysis, intermediate pyrolysis and gasification, fast pyrolysis, and flash pyrolysis, depending on the heating temperature residence time [42]. Besides, hydropyrolysis is an additional classification, which transforms biomass in the presence of hydrogen, yielding bio-oil enhanced in hydrocarbons [43]. The bio-oil is a potential feedstock for renewable energy. It is impregnated with numerous oxygenated aromatic and carbonyl compounds from lignocellulosic biomass depolymerization and can be separated into light and heavy fractions [44]. The light fraction stands for an aqueous condensate from the lignocellulosic biomass's moisture content, released volatile organic compounds, and evaporated water during pyrolysis. On the other hand, the heavy fraction comprises fragments stemming from lignin (pyrolytic lignin) [45]. Nevertheless, bio-oil production through fast pyrolysis requires a relatively high temperature (500–560°C) for high organic yields, making it economically and environmentally less attractive [46].

3.1.2 Gasification

Since the moisture content in lignocellulosic biomass (65–75 wt%), its suitability as a direct energy source is constrained. Instead, EFB or mesocarp fibers can be used as feedstock on direct-fired power sources such as anaerobic codigestion to produce heat and steam for electricity generation as well as the thermochemical conversion of solids to the gas phase (lignocellulosic biomass gasification) to produce syngas as a potential natural

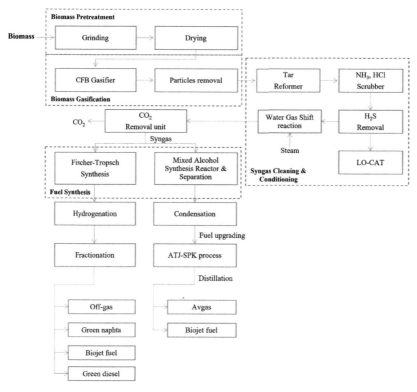

Fig. 3 Conceptual process overview of branched paraffin production through lignocellulosic biomass gasification.

gas surrogate [47]. Gasification is defined as the thermal conversion process of lignocellulosic biomass into raw synthetic gas (syngas), primarily composed of CO, H_2, CO_2, O_2, CH_4, and various other compounds, including tar residue and particulate matter (see Fig. 3) [4].

After cleaning and conditioning stages, the effluent stream is composed in a greater extent of H_2 and CO in an H_2/CO molar ratio specified for each downstream enforcement, such as liquid fuels Fischer-Tropsch synthesis and chemicals [48]. Several variables associated with lignocellulosic biomass physicochemical characteristics directly influence the gasification reaction yield, such as porosity, shape, density, and composition (cellulose, hemicellulose, lignin, extractives, and ash). Apart from using lignocellulosic biomass for gasification, the slurries from fast pyrolysis are commonly employed; their composition varies depending on the mixing of biomass and glycerol, biomass and water, bio-oil and biochar [49,50].

Syngas cleaning stage is demanded to eliminate contaminants and impurities to reach the quality requirements for the gas synthesis of CARSR biofuels and biobased chemicals. Gas impurities are classified into particulate matter (biochar and ash), inorganic impurities (alkali metal halides, nitrogen, and sulfur), and organic contaminants (aromatic, organic sulfur compounds, and tar) [51]. The gas conditioning adjusts the specific H_2/CO molar ratio through a Methane Steam Reforming, Water Gas Shift, or Methane CO_2 reforming reaction. Finally, the upgrade into biojet fuel es performed using the Ficher-Tropsch pathway [50].

3.2 Biochemical

Biochemical conversion, also known as anaerobic digestion, fermentation, or composting, breaks down biomass using enzymes generated by bacteria or other microorganisms. Lignocellulose is a structural material of the plant cell wall, containing a heterogeneous mixture of cellulose, hemicellulose and lignin. Cellulose is composed of β-1,4-linked homopolymers of glucose, while hemicellulose is a heteropolymer of pentoses (mainly arabinose and xylose) and hexoses (glucose). Lignin is a heterogeneous polymer of phenylpropanoid units. Bio-based alcohols production requires hydrolyzing cellulose and hemicellulose from lignocellulosic biomass substrate before mixed sugars fermentation [52]. Hemicellulose is chemically degraded during lignocellulosic biomass pretreatment at ambient pressure by dilute acid or high temperature and high pressure by chemicals such as sulfuric acid. Pretreatment technologies for solubilizing hemicellulose and lignin content, increasing polymeric sugar release and maximizing fermentation yield are classified as physical, chemical, physicochemical, and biological [53]. In this context, lignin solubility will vary depending on the selected pretreatment methods, considering the biorefinery's productive configuration given the high added value of lignin as a potential byproduct. For example, when a diluted acid-catalyzed steam explosion (DASE) mechanism is used, hemicellulose degradation achieves 90%, whereas lignin's solubility does not occur to any appreciable extent. Lignin remains an essential compound in the valorization of lignocellulosic biomass due to its suitability as an in situ energy source to produce high-pressure steam and satisfy the biorefinery utility demand [54].

Although cellulose fiber is hydrolyzed to glucose enzymatically, the DASE pretreatment increases the contact surface between cellulose and hydrolysis enzymes (cellulases and cellobiases) by separating lignocellulose from the matrix as the rigid structure of lignin and carbohydrates tight linkage hinder the enzymatical attack. Lignocellulosic palm oil biomass's chemical composition contains a significantly higher amount of cellulose (40–50 wt%) and hemicellulose (25–35 wt%) than other sorts of lignocellulosic biomass. Hence, maximum productivity and product yield is achieved when mixed sugars derived from hemicellulose are utilized. The integrated configuration of simultaneous hydrolysis (saccharification) and fermentation (SSF) uses cellulosic fiber as a sole carbon source, removing glucose by fermentation, and reducing feedback inhibition of cellulases and cellobiases. Nevertheless, to prevent glucose from remaining in the medium and achieving maximum efficiency, the microorganism utilization rate must be higher than the glucose production rate through enzymatic hydrolysis.

In contrast, glucose maximizes cellulolytic activity by removing feedback inhibition in simultaneous saccharification and cofermentation (SSCF) microbial utilization. During SSCF, pentose utilization and SSF of cellulose occur in a single reactor independently, without interferences from each process. Simultaneously, pentose metabolism guarantees microorganism optimal cellular activity. Cofermentation employs specific host strains engineered explicitly for simultaneous glucose and xylose consumption, enabling fermentation efficiency and greater productivity. *Saccharomyces cerevisiae* and *Zymomonas mobilis* have a long history as host strains for mixed sugar utilization to produce cellulosic ethanol from lignocellulosic biomass. On the other hand, hydrothermal conversion involved gasifying the high moisture content lignocellulosic biomass to produce syngas, obtaining low carbon liquid biofuels as an intermediate compound.

4. Thermochemical production pathways of biojet fuel

The certified ASTM pathways for the production of synthesized paraffinic kerosene (SPK) as a Jet A1 blending component embraces linear, branched, and cyclic hydrocarbons, ensuring combustion efficiency and fulfilment with the standard specifications [55,56]. Waste conversion pathways into biojet fuel are shown in Fig. 1.

4.1 Fischer-Tropsch (FT-SPK)

The Fischer-Tropsch (FT) biojet fuel is a sulfur-free mixture with relatively low aromatics content relative to gasoline and diesel, resulting in lower emissions when used in jet engines. Biomass feedstocks are first dried to reduce particle sizes throughout pretreatment [57]. There are several gasification technologies to transform biomass into syngas. The dried biomass is pressurized in a high-temperature gasification process and converted into raw synthetic gas at a temperature ranging from 880°C to 1300°C, in the presence of high purity oxygen and steam [58]. After syngas is further conditioned in a fluid catalytic cracker, it is polished with activated carbon and zinc oxide. Finally, the syngas is processed by FT synthesis to obtain liquid hydrocarbons.

FT synthesis involves a set of catalytic reactions for transforming syngas into liquid fuels. Two well-known FT operating modes are available; low temperature and high temperature [59]. The low-temperature process runs between 200°C and 240°C in the presence of iron or cobalt catalysts, producing linear high-molecular-mass waxes. The high temperature, on the other hand, operates with iron-based catalysts at 300–350°C. In this process are generated gasoline and linear olefins due to their lower molecular mass primarily. Linear waxes produced at low-temperature processes have a higher molecular mass than those generated at the high-temperature process—the FT process product ranges from methane to long-chain hydrocarbons. However, oxygenated compounds such as alcohols, aldehydes, and carboxylic acids are likewise formed in addition to alkanes and alkenes. In the high-temperature process, aromatics and ketones are also generated. The FT process is very exothermic; thus, to avoid catalyst overheating and subsequent deactivation and undesired methane production, the reaction heat must be removed rapidly. The FT synthesis product is upgraded to high-quality, low-aromatic, and almost zero-sulfur-content fuels through conventional refinery processes, such as hydrocracking, fractionation, isomerization, and hydrogenation [60].

Instead of catalytic upgrading, the FT route utilizes syngas from the gasification stage to conduct a mixed alcohol synthesis (MAS) based on a power-to-liquid conversion. The clean syngas is separated from CO_2 through the acid gas removal method Rectisol, removing 95% of the CO_2 with high partial pressure. The MAS reactor and the separation unit produce, to a large extent, higher alcohols with a minor amount of unconverted syngas, light hydrocarbons and CO_2. The MAS is catalytically performed at a pressure above 4 MPa and temperature ranging from 250°C to 320°C, favoring the alcohol synthesis. The mechanism is described in Reaction (R1).

$$n\text{CO} + 2n\text{H}_2 \rightarrow \text{C}_n\text{H}_{2n+1}\text{OH} + (n-1)\text{H}_2\text{O} \tag{R1}$$

Ethanol is the MAS reactor's primary alcohol product in the presence of a modified FT catalyst, even considering that the formed methanol is reinjected in the reactor for higher alcohols formation. Ethanol condensation for butanol synthesis is the first phase of fuel upgrading.

$$2\text{C}_2\text{H}_5\text{OH} \rightarrow \text{C}_4\text{H}_9\text{OH} + \text{H}_2\text{O} \tag{R2}$$

The reinjected methanol successively reacts with the ethanol and propanol, resulting in isobutanol.

$$\text{CH}_3\text{OH} + \text{C}_2\text{H}_5\text{OH} \rightarrow \text{C}_3\text{H}_7\text{OH} + \text{H}_2\text{O} \tag{R3}$$

$$\text{CH}_3\text{OH} + \text{C}_3\text{H}_7\text{OH} \rightarrow \text{C}_4\text{H}_9\text{OH} + \text{H}_2\text{O} \tag{R4}$$

Finally, the condensed product is upgraded via the Alcohol-to-Jet process to obtain a branched paraffinic mixture. The main disadvantage is the limited gas-to-liquid mass transfer rate requiring specific reactor designs.

4.2 Hydroprocessed esters and fatty acids (HEFA-SPK)

HEFA conversion technology is at a relatively advanced maturity level; its commercial production is feasible and used in commercial and military flights. In contrast to petroleum-derived Jet A1, the HEFA fuel has advantages in its physicochemical and combustion properties attributed to higher cetane numbers, lower aromatic content, lower sulfur content, and theoretically lower GHG emissions. The hydrogenation process fully saturates the double bonds of the renewable fats and oils depending on their degree of unsaturation. Catalytic hydrogenation transforms unsaturated fatty acids or glycerides into saturated ones in the liquid phase, followed by propane cleaving to produce three FFA moles. By incorporating hydrogen, the glycerol part of the triglyceride molecule is converted into propane. On the other hand, thermal hydrolysis is the alternative route for converting glycerides to FFA. By treating the oils and fats with 3 mol of water, its primary triglyceride content is transformed into 3 mol of FFA and one glycerol mole.

$$\text{Triglyceride}+ 3H_2O \leftrightarrow \text{Glycerol}+ 3FFA \qquad \text{(R5)}$$

On the glycerol backbone, the hydrogen atom from water is attached, forming 1 mol of glycerol. Simultaneously, the hydroxyl group from water is added to the ester group, and 3 mol of FFAs are formed. For water to dissolve in the oil process, a high temperature (250–260°C) is needed.

Besides a high flash point, the HEFA biojet fuel must have good cold flow characteristics to meet the ASTM standard specification for alternative jet fuel. Hence, straight-chain alkanes yielded during deoxygenation must be either parallel or sequentially hydro-isomerized and hydrocracked to produce isoparaffins with carbon chains primary varying from C_9 to C_{15}. The branched-chain mixture from isomerization has a reduced freezing point compared with normal alkane hydrocarbons. Subsequent hydrocracking reactions are exothermic, resulting in lighter liquids and gas products being formed. These reactions are relatively slow, and much of the hydrocracking occurs in the last portion of the reactor, including paraffin saturation. Low yields of jet-fuel-range alkanes and high light species products ranging from C_1 to C_4 and naphtha from C_5 to C_8 would result from over cracking. A fractionation distillation complemented the hydro-isomerization and hydrocracking stages to separate the mixtures of biojet fuel, paraffinic diesel, naphtha, and light gases.

4.3 Synthesized isoparaffin via hydroprocessed fermented sugars (SIP-HFS)

Synthesized isoparaffin from hydroprocessed biomass fermentation product (farnesene) was approved in 2015 as a Jet A1 blending component to less than 10 vol%. Farnesene is a branched alkene with a chemical composition of $C_{15}H_{24}$ obtained by fermentation of lignocellulosic and sugarcane biomass in the presence of microorganisms or biocatalyst. Analogous to the HEFA pathway, the farnesene obtained must be further hydrogenated.

The standard requirements for SIP are significantly different from those for FT SPK, HEFA, and FT SPK/A attributed to its peculiar chemical and carbon chain structure (a minimum of 97 wt% farnesane containing the hydrocarbon chain of C_{15}). Due to high viscosity and low combustion efficiency caused by the relative long carbon chains length, only up to 10 vol% of SIP fuel can be mixed with fossil jet fuels. However, efficient engineering of *S. cerevisiae* has been accomplished to produce farnesene. Compared with the FT–SPK technique, the extraction of synthetic paraffins from sugar represents a comparatively low-cost pathway for biojet fuel production that usually involves high investment costs.

4.4 Fischer-Tropsch with increased aromatic content (FT-SPK/A)

The process involves transforming the lignocellulosic biomass into biojet fuel and diesel fuel range hydrocarbons, including the directional production of C_8–C_{15} aromatics through catalytic depolymerization and deoxygenation of lignin to low carbon monomers (C_6–C_8) with the alkylation of aromatics and directional production of cycloparaffins (C_8–C_{15}) by hydrogenation of aromatics [61]. High C_8–C_{15} aromatics selectivity (94%) is accomplished through the lignin-derived monomers alkylation, utilizing ionic liquid catalyst, and the gaseous mixture of light olefins (C_2–C_4) as the alkylating agent under mild reaction conditions.

The standard specifications for FT-SPK/A resemble FT-SPK and HEFA-SPK, excepting the aromatic content and slightly different density requirements given the FT-SPK/A composition containing less than 20 wt% of aromatics. The relevance of including an appropriate quantity of aromatics compounds right into *n*-paraffins was determined through a comprehensive kerosene kinetic model to estimate its shooting tendency and molecular growth accurately [62].

4.5 Alcohol-to-jet synthetic paraffinic kerosene (ATJ-SPK)

Lignocellulosic-based alcohol suitability as drop-in jet fuel blending components is limited due to low energy density, high water absorption, corrosivity, elastomer decomposition propensity, high volatility, and reduced flash point. Oxygenated compounds, such as ethanol, methanol, butanol (C_4), and long-chain fatty alcohols (C_{4+}) are upgraded to a higher heating value biojet fuel by the ATJ pathway, so-called alcohol oligomerization [63]. The ATJ process involves three steps, including alcohol dehydration, oligomerization, and hydroprocessing. Oxygen is removed in the form of H_2O during alcohol dehydration, forming a double bond between the carbons associated with the removed *OH* and *H* under high pressure at a temperature above 250°C over Amberlyst acidic resins and ZSM-5 zeolites, as Reaction (R6) shows.

$$C_nH_mOH \rightarrow C_nH_{m-1} + H_2O \qquad (R6)$$

After dehydration, α-olefins oligomerization increases the carbon atoms chain length.

$$iC_nH_m \rightarrow C_{ni}H_{mi}; i = 2, 3, 4 \qquad (R7)$$

The reaction occurs in the liquid phase within stirred reactors to combine unsaturated molecules (alkenes) to form their corresponding oligomers (dimmers, trimers, and even, tetramers) in the presence of specific catalysts enhances oligomers formation compared with cracking, dehydrogenation, and polymerization reactions. The transformation of alkenes into paraffins by adding hydrogen is needed to create an organic mixture treated as jet fuel blending. In the presence of *Pd* or *Pt* catalysts at relatively high pressure (>2 MPa), at 200–350°C, multiple studies have documented oligomers hydrogenation achievement [16].

$$C_nH_{2n} + H_2 \rightarrow C_{2n}H_{2n+2}, n = 8, 12, 16 \qquad (R8)$$

4.6 Catalytic hydrothermolysis jet and coprocessing jet (CHJ-SPK)

Catalytic hydrothermolysis (CH), also called hydrothermal liquefaction, is a unique method developed and patented by Applied Research Associates, Inc. to produce aromatic and drop-in biojet fuel (ReadiJet), as well as hydrocarbons in the diesel range (ReadiDiesel) from biomass-derived waste oil. The hydrothermal stage includes a series of reactions: cracking, hydrolysis, decarboxylation, isomerization, and cyclization, transforming triglycerides into a cyclic chain, branched, and straight hydrocarbons blend [64]. The CH reaction is catalytically performed (or without a catalyst) in the presence of water at a temperature ranging from 450°C to 475°C and 21 MPa pressure. The product, including unsaturated molecules and carboxylic acids, is subsequently sent to decarboxylation and hydro-treating processes for oxygen removal and saturation, respectively. The treated hydrocarbons mixture (C_6–C_{28}) contains a significant amount of *n*-alkanes, isoalkanes, cycloalkanes, and a lesser extent of aromatics (military purposes require an increased aromatics content) that must be fractionated and separated to diesel, jet fuel and naphtha. The produced biojet fuel from the CH pathway satisfies the ASTM and military requirements due to its exceptional cold fluidity characteristics, stability, and combustion quality [65].

5. Future challenges

5.1 Feedstock challenge

The feedstock source for biofuel processing is a primary challenge, even when biomass is an inexhaustible resource. Nonfood feedstock for second-generation biofuel processing competes with crops for land. In developing countries, where most land is used for cultivation, the land issue would worsen. Biofuel processing on an industrial scale would replace fiber, forage, and food, increasing local prices. Nonetheless, the noncropland enhancement of energy crops is a viable way to prevent this crisis, taking into account the local economy and local policies. Because of biomass highly scattered nature, logistics is also a problem for biofuel production. Biomass collection, packaging, and transportation from fields to central processing plants may be highly energy-intensive. Transportation costs and emissions will increase due to the unequal feedstock distribution through districts and countries. Besides, storing a significant amount of biomass feedstock will affect the economy, increasing the cost of bioenergy production. On an energy-equivalent basis, second-generation biofuels are expected to double or triple fossil fuels prices. A complete techno-economic and logistics analysis of the conversion techniques must determine the cost-effective biofuel processing methods. Also, a solution such as dispersed biomass production infield will reduce transportation and feedstock storage costs.

5.2 The technical challenge in thermochemical conversion of biomass

The characteristics of biomass composition, such as high moisture content, low energy content, and high oxygen content, cause the most technical challenges in biofuel processing. Since most thermochemical conversion processes require low moisture content in the biomass, drying is typically needed before biofuel output. The hydrothermal system seems to be ideal for coping with elevated moisture levels in biomass without drying it. The process's high pressure and water vaporization, on the other hand, are much more energy-intensive. Both bio-oil and biogas need additional catalytic upgrades to meet traditional fuel requirements in terms of fuels. Although catalysts contribute to a higher selectivity when directly contacting biomass, they are rapidly deactivated during the reactions.

Furthermore, the most effective catalysts designed for reforming are noble metals-based, which are expensive and are seldom used in large-scale biorefineries. Several studies are currently being done to effectively improve the primary biofuel while maintaining a long catalyst life. On an industrial scale, a significant concern is the large number of mineral acids produced throughout the biomass thermochemical conversion, particularly at higher temperate (gasification, pyrolysis, and combustion), resulting in severe corrosion and erosion of pipelines and reactors within the system. As a result, heat transfer inside the reactor is considerably hampered. In practice, after many cycles, the reaction mechanism for thermochemical biomass conversion must be washed. Before large-scale biofuel processing, the above issues must be appropriately resolved.

5.3 Future perspectives of thermochemical conversion for biofuel production

From the preceding discussion, it can be inferred that biojet fuel can fulfil air transportation needs while also promoting petroleum-derived jet fuel independence and greener development. Biomass selection and properties are essential for optimal biojet fuel processing. Biomass selection is determined by its availability, required technology for pretreatment, the intermediate desired for biojet fuel upgrading, and economics. Exploring different biomass sources and enhancing the biomass collection and handling systems will contribute to increasing biomass supply. Techniques that can enrich the biomass structure, whether chemical, physical, or microbial, are highly desirable. Reactor design, operating conditions, and other engineering aspects are all helping to optimize the biofuel development process. Another significant field of focus is the adaptation and improvement of engine systems for biofuel use and developing newer ones. Biotechnological methods such as genetic engineering and recombinant DNA technology may be used to solve the bottlenecks of biochemical conversion and microbial metabolic routes.

6. Summary and conclusions

This chapter comprehensively assessed the approved ASTM routes potential for biojet fuel production from agricultural waste as feedstock. Due to its renewability and availability, lignocellulosic biomass is the most promising feedstock. Approved technologies involve the upgrading of lignocellulose-derived alcohol into jet fuel (ATJ); triglycerides and vegetable oil catalytic hydrothermolysis jet (CHJ); hydroprocessed esters and fatty acids (HEFA); power-to-liquid conversion of syngas through a Ficher-Tropsch (FT) reactor; synthesized kerosene with increased aromatics content (FT–SPK/A); and hydroprocessed synthesized isoparaffins via hydroprocessed fermented sugars (SIP-HFS). Depending on the processing pathway and feedstock sort employed, the produced biojet fuel contains linear, branched, and cyclic hydrocarbons, ensuring the physicochemical homogeneity and high-quality combustion required for a "drop-in" Jet A1 blending component.

Jet fuel costs, energy security, energy supply, and aviation emissions have driven fuel and aircraft technologies innovation throughout the past decade of the air industry's significant development. Lignocellulosic biomass-derived jet fuel from agricultural wastes offers a sustainable opportunity to solve these problems. Several conversion processes have been developed to convert lignocellulosic biomass into an economically competitive synthetic jet fuel. Thermochemical conversion efficiently converts biomass into a wide range of biofuels. Nevertheless, to achieve large-scale industrial development of biofuels by thermochemical conversion of biomass, a better comprehension of the conversion mechanism of a single compound and the techno-economic assessment of the biofuel industry is needed.

This section described and reviewed the ASTM certified pathways considering agricultural residues from oleaginous crops and their respective upgrading procedures. Not only does renewable biojet fuel represent environmental benefits for aviation but it also aims to enhance a continually growing industry. Before alternative jet fuels are commercially feasible, there is still a long way to move forward. Through the partnership of experts in the aviation sector, biofuel firms, agricultural organizations, governments and the educational systems, development is being made toward an efficient process employing sustainable feedstock sources, is amenable with existing facilities and provides renewable jet fuel. Also, encouraging a circular economy approach is of utmost importance to further increase resource quality and improve direct production waste management.

References

[1] B.C. Klein, et al., Techno-economic and environmental assessment of renewable jet fuel production in integrated Brazilian sugarcane biorefineries, Appl. Energy 209 (2018) 290–305, https://doi.org/10.1016/j.apenergy.2017.10.079.

[2] P. Bains, P. Psarras, J. Wilcox, CO2 capture from the industry sector, Prog. Energy Combust. Sci. 63 (2017) 146–172, https://doi.org/10.1016/j.pecs.2017.07.001.

[3] S. Heidenreich, P.U. Foscolo, New concepts in biomass gasification, Prog. Energy Combust. Sci. 46 (2015) 72–95, https://doi.org/10.1016/j.pecs.2014.06.002.

[4] R.C. Neves, et al., A vision on biomass-to-liquids (BTL) thermochemical routes in integrated sugarcane biorefineries for biojet fuel production, Renew. Sustain. Energy Rev. 119 (2020) 109607, https://doi.org/10.1016/j.rser.2019.109607.

[5] International Air Transport Association, Annual Review, 76th Annual General Meeting, 2020.

[6] C. Gutiérrez-Antonio, A.G. Romero-Izquierdo, F.I. Gómez-Castro, S. Hernández, A. Briones-Ramírez, Simultaneous energy integration and intensification of the hydrotreating process to produce biojet fuel from *Jatropha curcas*, Chem. Eng. Process. Process Intensif. 110 (2016) 134–145, https://doi.org/10.1016/j.cep.2016.10.007.

[7] R. Mawhood, E. Gazis, S. de Jong, R. Hoefnagels, R. Slade, Production pathways for renewable jet fuel: a review of commercialization status and future prospects, Biofuels Bioprod. Biorefin. 10 (4) (2016) 462–484, https://doi.org/10.1002/bbb.1644.

[8] R. Pujan, S. Hauschild, A. Gröngröft, Process simulation of a fluidized-bed catalytic cracking process for the conversion of algae oil to biokerosene, Fuel Process. Technol. 167 (July) (2017) 582–607, https://doi.org/10.1016/j.fuproc.2017.07.029.

[9] ASTM D7566, Standard Specification for Aviation Turbine Fuel Containing Synthesized Hydrocarbons, 2020, https://doi.org/10.1520/D7566-19.

[10] M. Wise, M. Muratori, P. Kyle, Biojet fuels and emissions mitigation in aviation: an integrated assessment modeling analysis, Transp. Res. Part D Transp. Environ. 52 (2017) 244–253, https://doi.org/10.1016/j.trd.2017.03.006.

[11] G. Liu, B. Yan, G. Chen, Technical review on jet fuel production, Renew. Sustain. Energy Rev. 25 (2013) 59–70, https://doi.org/10.1016/j.rser.2013.03.025.

[12] C. Gutiérrez-Antonio, F.I. Gómez-Castro, J.A. de Lira-Flores, S. Hernández, A review on the production processes of renewable jet fuel, Renew. Sustain. Energy Rev. 79 (January) (2017) 709–729, https://doi.org/10.1016/j.rser.2017.05.108.

[13] L.G. Pereira, H.L. MacLean, B.A. Saville, Financial analyses of potential biojet fuel production technologies, Biofuels Bioprod. Biorefin. 11 (4) (2017) 665–681, https://doi.org/10.1002/bbb.1775.

[14] H. Wei, W. Liu, X. Chen, Q. Yang, J. Li, H. Chen, Renewable bio-jet fuel production for aviation: a review, Fuel 254 (2019) 115599, https://doi.org/10.1016/j.fuel.2019.06.007.

[15] S. Mahalaxmi, C. Williford, Biochemical conversion of biomass to fuels, in: Handbook of Climate Change Mitigation and Adaptation, Springer International Publishing, Cham, 2017, pp. 1777–1811, https://doi.org/10.1007/978-3-319-14409-2_26.

[16] N. Vela-García, D. Bolonio, A.M. Mosquera, M.F. Ortega, M.-J. García-Martínez, L. Canoira, Techno-economic and life cycle assessment of triisobutane production and its suitability as biojet fuel, Appl. Energy 268 (2020) 114897, https://doi.org/10.1016/j.apenergy.2020.114897.

[17] L. de Souza Noel Simas Barbosa, E. Hytönen, P. Vainikka, Carbon mass balance in sugarcane biorefineries in Brazil for evaluating carbon capture and utilization opportunities, Biomass Bioenergy 105 (2017) 351–363, https://doi.org/10.1016/j.biombioe.2017.07.015.

[18] M.D.B. Watanabe, et al., Sustainability assessment methodologies, in: Virtual Biorefinery, 2016, pp. 155–188, https://doi.org/10.1007/978-3-319-26045-7_6.

[19] N. Gontard, et al., A research challenge vision regarding management of agricultural waste in a circular bio-based economy, Crit. Rev. Environ. Sci. Technol. 48 (6) (2018) 614–654, https://doi.org/10.1080/10643389.2018.1471957.

[20] Y. Dai, et al., Utilizations of agricultural waste as adsorbent for the removal of contaminants: a review, Chemosphere 211 (2018) 235–253, https://doi.org/10.1016/j.chemosphere.2018.06.179.

[21] R.E.H. Sims, "Energy-Smart" Food for People and Climate – Issue Paper, Food and Agriculture Organization of the United Nations (FAO), 2011.

[22] F.G. Fermoso, A. Serrano, B. Alonso-Fariñas, J. Fernández-Bolaños, R. Borja, G. Rodríguez-Gutiérrez, Valuable compound extraction, anaerobic digestion, and composting: a leading biorefinery approach for agricultural wastes, J. Agric. Food Chem. 66 (32) (2018) 8451–8468, https://doi.org/10.1021/acs.jafc.8b02667.

[23] M. Macias-Corral, et al., Anaerobic digestion of municipal solid waste and agricultural waste and the effect of co-digestion with dairy cow manure, Bioresour. Technol. 99 (17) (2008) 8288–8293, https://doi.org/10.1016/j.biortech.2008.03.057.

[24] J.K. Saini, R. Saini, L. Tewari, Lignocellulosic agriculture wastes as biomass feedstocks for second-generation bioethanol production: concepts and recent developments, 3 Biotech 5 (4) (2015) 337–353, https://doi.org/10.1007/s13205-014-0246-5.

[25] K. Li, H. Bian, C. Liu, D. Zhang, Y. Yang, Comparison of geothermal with solar and wind power generation systems, Renew. Sustain. Energy Rev. 42 (2015) 1464–1474, https://doi.org/10.1016/j.rser.2014.10.049.

[26] S.-H. Kong, S.-K. Loh, R.T. Bachmann, S.A. Rahim, J. Salimon, Biochar from oil palm biomass: a review of its potential and challenges, Renew. Sustain. Energy Rev. 39 (2014) 729–739, https://doi.org/10.1016/j.rser.2014.07.107.

[27] M.P. Dorado, F. Cruz, J.M. Palomar, F.J. López, An approach to the economics of two vegetable oil-based biofuels in Spain, Renew. Energy 31 (8) (2006) 1231–1237, https://doi.org/10.1016/j.renene.2005.06.010.

[28] M. El-Hamidi, F.A. Zaher, Production of vegetable oils in the world and in Egypt: an overview, Bull. Natl. Res. Cent. 42 (1) (2018) 19, https://doi.org/10.1186/s42269-018-0019-0.

[29] S. Baroutian, M.K. Aroua, A.A.A. Raman, A. Shafie, R.A. Ismail, H. Hamdan, Blended aviation bio-fuel from esterified *Jatropha curcas* and waste vegetable oils, J. Taiwan Inst. Chem. Eng. 44 (6) (2013) 911–916, https://doi.org/10.1016/j.jtice.2013.02.007.

[30] S. Ramamurthi, A.R. McCurdy, Enzymatic pretreatment of deodorizer distillate for concentration of sterols and tocopherols, J. Am. Oil Chem. Soc. 70 (3) (1993) 287–295, https://doi.org/10.1007/BF02545310.

[31] R. Piloto-Rodríguez, et al., Conversion of fatty acid distillates into biodiesel: engine performance and environmental effects, Energy Sources Part A Recover. Util. Environ. Eff. 42 (4) (2020) 387–398, https://doi.org/10.1080/15567036.2019.1587085.

[32] P.F. Martins, V.M. Ito, C.B. Batistella, M.R.W. Maciel, Free fatty acid separation from vegetable oil deodorizer distillate using molecular distillation process, Sep. Purif. Technol. 48 (1) (2006) 78–84, https://doi.org/10.1016/j.seppur.2005.07.028.

[33] A. Llamas, M.J. García-Martínez, A.M. Al-Lal, L. Canoira, M. Lapuerta, Biokerosene from coconut and palm kernel oils: production and properties of their blends with fossil kerosene, Fuel 102 (2012) 483–490, https://doi.org/10.1016/j.fuel.2012.06.108.

[34] N. Rasimoglu, H. Temur, Cold flow properties of biodiesel obtained from corn oil, Energy 68 (2014) 57–60, https://doi.org/10.1016/j.energy.2014.02.048.

[35] X. Shang, C. Cheng, J. Ding, W. Guo, Identification of candidate genes from the SAD gene family in cotton for determination of cottonseed oil composition, Mol. Gen. Genomics 292 (1) (2017) 173–186, https://doi.org/10.1007/s00438-016-1265-1.

[36] D. Bolonio, M.-J. García-Martínez, M.F. Ortega, M. Lapuerta, J. Rodríguez-Fernández, L. Canoira, Fatty acid ethyl esters (FAEEs) obtained from grapeseed oil: a fully renewable biofuel, Renew. Energy 132 (2019) 278–283, https://doi.org/10.1016/j.renene.2018.08.010.

[37] D. Donoso, D. Bolonio, M. Lapuerta, L. Canoira, Oxidation stability: the bottleneck for the development of a fully renewable biofuel from wine industry waste, ACS Omega 5 (27) (2020) 16645–16653, https://doi.org/10.1021/acsomega.0c01496.

[38] S. Braipson-Danthine, V. Gibon, Comparative analysis of triacylglycerol composition, melting properties and polymorphic behavior of palm oil and fractions, Eur. J. Lipid Sci. Technol. 109 (4) (2007) 359–372, https://doi.org/10.1002/ejlt.200600289.

[39] J. Ji, Y. Liu, L. Shi, N. Wang, X. Wang, Effect of roasting treatment on the chemical composition of sesame oil, LWT 101 (2019) 191–200, https://doi.org/10.1016/j.lwt.2018.11.008.

[40] J. Corach, E.F. Galván, P.A. Sorichetti, S.D. Romano, Estimation of the composition of soybean bio-diesel/soybean oil blends from permittivity measurements, Fuel 235 (2019) 1309–1315, https://doi.org/10.1016/j.fuel.2018.08.114.

[41] D.B. Konuskan, M. Arslan, A. Oksuz, Physicochemical properties of cold pressed sunflower, peanut, rapeseed, mustard and olive oils grown in the Eastern Mediterranean region, Saudi J. Biol. Sci. 26 (2) (2019) 340–344, https://doi.org/10.1016/j.sjbs.2018.04.005.

[42] A.V. Bridgwater, Review of fast pyrolysis of biomass and product upgrading, Biomass Bioenergy 38 (2012) 68–94, https://doi.org/10.1016/j.biombioe.2011.01.048.

[43] V. Dhyani, T. Bhaskar, A comprehensive review on the pyrolysis of lignocellulosic biomass, Renew. Energy 129 (2018) 695–716, https://doi.org/10.1016/j.renene.2017.04.035.

[44] G. Kabir, A.T. Mohd Din, B.H. Hameed, Pyrolysis of oil palm mesocarp fiber and palm frond in a slow-heating fixed-bed reactor: a comparative study, Bioresour. Technol. 241 (2017) 563–572, https://doi.org/10.1016/j.biortech.2017.05.180.

[45] W.T. Tsai, M.K. Lee, Y.M. Chang, Fast pyrolysis of rice straw, sugarcane bagasse and coconut shell in an induction-heating reactor, J. Anal. Appl. Pyrolysis 76 (1–2) (2006) 230–237, https://doi.org/10.1016/j.jaap.2005.11.007.

[46] K. Papadikis, S. Gu, A.V. Bridgwater, CFD modelling of the fast pyrolysis of biomass in fluidised bed reactors: modelling the impact of biomass shrinkage, Chem. Eng. J. 149 (1–3) (2009) 417–427, https://doi.org/10.1016/j.cej.2009.01.036.

[47] M.F. Awalludin, O. Sulaiman, R. Hashim, W.N.A.W. Nadhari, An overview of the oil palm industry in Malaysia and its waste utilization through thermochemical conversion, specifically via liquefaction, Renew. Sustain. Energy Rev. 50 (2015) 1469–1484, https://doi.org/10.1016/j.rser.2015.05.085.

[48] F. Trippe, M. Fröhling, F. Schultmann, R. Stahl, E. Henrich, Techno-economic assessment of gasification as a process step within biomass-to-liquid (BtL) fuel and chemicals production, Fuel Process. Technol. 92 (11) (2011) 2169–2184, https://doi.org/10.1016/j.fuproc.2011.06.026.

[49] N. Dahmen, et al., The bioliq process for producing synthetic transportation fuels, Wiley Interdiscip. Rev. Energy Environ. 6 (3) (2017) e236, https://doi.org/10.1002/wene.236.

[50] M.L. de Souza-Santos, Proposals for power generation based on processes consuming biomass-glycerol slurries, Energy 120 (2017) 959–974, https://doi.org/10.1016/j.energy.2016.12.005.

[51] N. Abdoulmoumine, S. Adhikari, A. Kulkarni, S. Chattanathan, A review on biomass gasification syngas cleanup, Appl. Energy 155 (2015) 294–307, https://doi.org/10.1016/j.apenergy.2015.05.095.

[52] Y. Sánchez-Borroto, M. Lapuerta, E.A. Melo-Espinosa, D. Bolonio, I. Tobío-Perez, R. Piloto-Rodríguez, Green-filamentous macroalgae Chaetomorpha cf. gracilis from Cuban wetlands as a feedstock to produce alternative fuel: a physicochemical characterization, Energy Sources Part A Recover. Util. Environ. Eff. 40 (10) (2018) 1279–1289, https://doi.org/10.1080/15567036.2018.1476931.

[53] B. Kumar, N. Bhardwaj, K. Agrawal, V. Chaturvedi, P. Verma, Current perspective on pretreatment technologies using lignocellulosic biomass: an emerging biorefinery concept, Fuel Process. Technol. 199 (2020) 106244, https://doi.org/10.1016/j.fuproc.2019.106244.

[54] M.D. Staples, et al., Lifecycle greenhouse gas footprint and minimum selling price of renewable diesel and jet fuel from fermentation and advanced fermentation production technologies, Energy Environ. Sci. 7 (5) (2014) 1545–1554, https://doi.org/10.1039/C3EE43655A.

[55] M. Lapuerta, L. Canoira, The suitability of fatty acid methyl esters (FAME) as blending agents in jet A-1, in: Biofuels for Aviation, Elsevier, 2016, pp. 47–84, https://doi.org/10.1016/B978-0-12-804568-8.00004-4.

[56] K.K. Gupta, A. Rehman, R.M. Sarviya, Bio-fuels for the gas turbine : a review, Renew. Sustain. Energy Rev. 14 (9) (2010) 2946–2955, https://doi.org/10.1016/j.rser.2010.07.025.

[57] F. You, B. Wang, Life cycle optimization of biomass-to-liquid supply chains with distributed–centralized processing networks, Ind. Eng. Chem. Res. 50 (17) (2011) 10102–10127, https://doi.org/10.1021/ie200850t.

[58] G.W. Diederichs, M. Ali Mandegari, S. Farzad, J.F. Görgens, Techno-economic comparison of biojet fuel production from lignocellulose, vegetable oil and sugar cane juice, Bioresour. Technol. 216 (2016) 331–339, https://doi.org/10.1016/j.biortech.2016.05.090.

[59] K. Atsonios, M. Kougioumtzis, K.D. Panopoulos, E. Kakaras, Alternative thermochemical routes for aviation biofuels via alcohols synthesis: process modeling, techno-economic assessment and comparison, Appl. Energy 138 (2015) 346–366, https://doi.org/10.1016/j.apenergy.2014.10.056.

[60] S.T. Sie, R. Krishna, Fundamentals and selection of advanced Fischer–Tropsch reactors, Appl. Catal. A Gen. 186 (1–2) (1999) 55–70, https://doi.org/10.1016/S0926-860X(99)00164-7.

[61] P. Bi, et al., From lignin to cycloparaffins and aromatics: directional synthesis of jet and diesel fuel range biofuels using biomass, Bioresour. Technol. 183 (2015) 10–17, https://doi.org/10.1016/j.biortech.2015.02.023.

[62] C. Vovelle, J.-L. Delfau, M. Reuillon, Formation of Aromatic Hydrocarbons in Decane and Kerosene Flames at Reduced Pressure, 1994, pp. 50–65, https://doi.org/10.1007/978-3-642-85167-4_4.

[63] N. Vela-García, D. Bolonio, M.-J. García-Martínez, M.F. Ortega, D. Almeida Streitwieser, L. Canoira, Biojet fuel production from oleaginous crop residues: thermoeconomic, life cycle and flight performance analysis, Energy Convers. Manag. 244 (2021) 114534, https://doi.org/10.1016/j.enconman.2021.114534.

[64] W.-C. Wang, L. Tao, Bio-jet fuel conversion technologies, Renew. Sustain. Energy Rev. 53 (2016) 801–822, https://doi.org/10.1016/j.rser.2015.09.016.

[65] L. Li, E. Coppola, J. Rine, J.L. Miller, D. Walker, Catalytic hydrothermal conversion of triglycerides to non-ester biofuels, Energy Fuels 24 (2) (2010) 1305–1315, https://doi.org/10.1021/ef901163a.

CHAPTER 3

Conversion of lignin-derived bio-oil to bio-jet fuel

Majid Saidi and Pantea Moradi
School of Chemistry, College of Science, University of Tehran, Tehran, Iran

1. Introduction

1.1 Fossil and renewable fuels

As reported by Zhao et al. [1], conventional aviation fuel emitted 705 million metric tons of CO_2 in 2013, which emphasizes the requirement of a clean and renewable bio-based fuel. Moreover, depletion of fossil fuels accompanied with climate changes, due to the increasing Green House Gasses (GHG) emissions, has prompted researchers to find a clean substitute for conventional fossil fuels [2–5]. As illustrated in Fig. 1, bio-oils as a renewable and clean fuel has attracted many researchers' attention [6]. Miscibility with fossil fuels, lowering harmful GHG emissions, providing high cetane number along with high heating value (HHV), are some of the advantages of bio-oils [7]. Feedstocks used for bio-oil production are generally categorized into edible and residual biomasses. Variable edible oils such as corn oil have been used for bio-oil production in the past decades [8]. Residual biomasses are known as an appropriate substitute for edible oils due to high carbon content in their structure [9]. As biomass is a CO_2 sequestration feedstock, application of residual and nonedible biomasses makes bio-oil production as a green process that is compliant with the aid of Life Cycle Assessment (LCA) [10,11]. Biomass is generally composed of polymeric components including cellulose (35%–50%), hemicellulose (25%–30%), and lignin (15%–30%) [12,13]. As shown in Mars-van Krevelan diagram (Fig. 2), cellulosic biomass possesses the highest O/C ratio in comparison with other biomasses used for fuel production. H/C ratio of cellulosic biomass is approximately equal to lignin. Hence, due to the high C/O and H/C ratios and high energy density consequently, lignin-derived biomass is mostly utilized for biofuel production among other polymeric structures. Auersvald et al. [14] have claimed that aromatics meet combustion properties according to ASTM D1655 of the aviation fuel standards, which are presented in Table 1. Different biological and thermochemical processes have been utilized for converting biomass to valuable chemicals such as bio fuels. Gasification, pyrolysis, and torrefaction are the main thermochemical processes for bio-oil production. Biochar and syngas are produced along with bio-oil production in thermochemical processes.

Sustainable Alternatives for Aviation Fuels
https://doi.org/10.1016/B978-0-323-85715-4.00001-X
Copyright © 2022 Elsevier Inc.
All rights reserved.

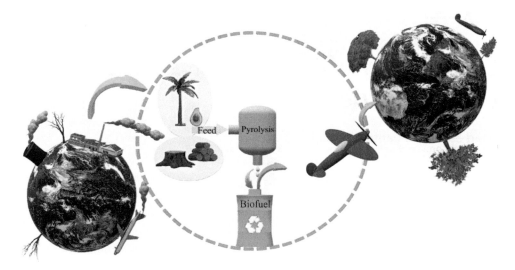

Fig. 1 Conversion of lignin-derived pyrolysis bio-oil to jet fuel hydrocarbon range.

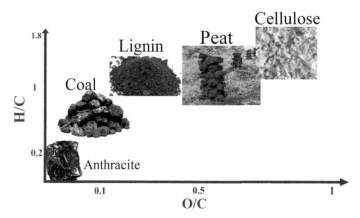

Fig. 2 Mars-van Krevelan diagram for biomass-derived polymers.

Distribution of syngas, bio-oil, and biochar, is mainly dependent to the operating temperature and time. Bio-oil is produced as the main product during fast pyrolysis. Based on the operating conditions, wide range of valuable oxygenated monomers are produced. Generally, at temperatures in the range of 200–400°C, lignin is fractured to syringol and guaiacols. In the mentioned range of temperature, direct homolysis of C—C and C—O doesn't occur and methoxyl groups of aromatics are quite stable. By elevating the temperature in the range of 400–450°C, catechols, o-cresol, and phenols are produced due to the cleavage of C—C and CH$_3$—O bonds. Phenol and O-cresol are relatively stable products at higher pyrolysis temperatures [15,16]. The produced

Table 1 Characterization of pyrolytic and aviation fuel properties.

Property	Unit	Value	
		Crude biooil	Aviation fuel
Acidity	Total KOH mg/g of sample	pH = 2.4–4.3	0.1
Heat of Combustion	MJ/kg	17.4–32.4	<42.8
Flash Point	°C	–[a]	<38
Density	kg/m^3	980–1190	775–840

Elemental composition	Unit	Value	
C	wt%	41.7–69.5	Not Reported
H	wt%	5.5–9.4	Not Reported
S	wt%	<0.5	0.003
O	wt%	19.4–50.30	Not Reported
Water	wt%	20–30	–

[a]Flash point cannot be correctly measured due to the presence of water.

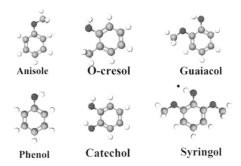

Anisole   O-cresol   Guaiacol

Phenol   Catechol   Syringol

Fig. 3 Lignin-derived bio-oil components.

components during depolymerization are shown at Fig. 3. As illustrated in Fig. 4, the produced monomers have the capability of being used in different pharmaceutical, polymer, and transportation industries such as aviation fuel. Many researchers have investigated the high yield and selective bio-oil production from lignin over heterogeneous catalysts. As an example, Milovanović et al. [17] investigated the selectivity of product stream over NiO on different zeolite types as the heterogeneous support. They succeeded to produce phenolic compounds with the selectivity of 75% over NiO/HY catalyst.

1.2 Concerns about bio-oil

Although wide range of chemicals are produced in depolymerization of lignin, the obtained bio-oil is highly composed of functionalized aromatics such as methoxylated and hydroxylated compounds. According to ASTM standards, jet fuels are hydrocarbons without oxygen [10]. As presented in Table 1, lignin-derived bio-oils have high oxygen

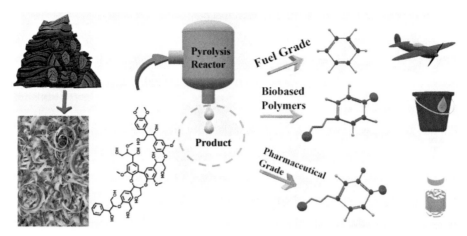

Fig. 4 The application of the produced monomers during lignin depolymerization.

content in their structure which extremely lowers HHV. Eliminated miscibility with fossil fuels is another concern caused by the polar structure of crude bio-oil. Moreover, polymerization of oxygenated compounds as a side reaction, readily occurs at combustion temperature. Phenolic and acidic compounds cause corrosion in the engine at high temperatures [18]. Different chemical and physical treatments such as hydrotreating, hydrocracking, supercritical treatment, steam reforming, emulsion, and solvent addition are used for upgrading bio-oils [1,19]. Among the mentioned pathways, HydroDeOxygenation (HDO) is one of the most applicable hydrotreating approaches for the production of C_8–C_{15} aromatics and cycloalkanes in the range of aviation fuel.

1.3 HDO as an appropriate process for upgrading bio-oils

As shown in Fig. 5, to valorize biogas, bio-oil and biochar-derived pyrolysis process may undergo variable upgrading methods. HDO is a hydrotreating process that takes place in presence of H_2 pressure and elevated temperatures in the range of 500–700 K over homogenous or heterogeneous catalysts [20]. Different reaction pathways including decarboxylation, decarbonylation, hydrogenolysis, and direct deoxygenation (DDO) result in simultaneous enhancement of C/O and H/O ratio. H_2O, CO, CO_2, and methanol are produced in the reaction media, as a consequence of hydrogenolysis, dehydration, decarboxylation, and demothoxylation reactions. Also, esterification, transesterification, and saturation are other reaction pathways as reported in some researches [21]. These reactions are shown in Table 2. Transalkylation reactions results in alkylphenols or alkyl benzene production as long-chain chemicals, which can be used as aviation fuels [19]. Phenolic compounds are reactive chemicals for transalkylation and alkylation reactions [22]. He et al. [23] demonstrated phenolic and oxygenated compounds as representative of lignin-derived bio-oil. As these compounds are primary

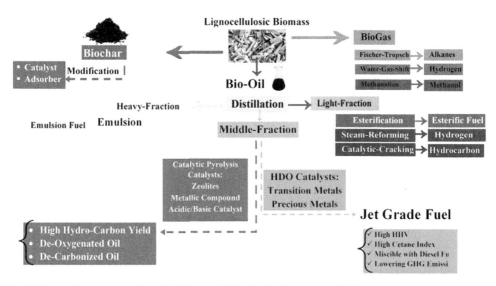

Fig. 5 Upgrading process for valorizing pyrolysis-derived components.

products of lignin degradation, therefore, many researchers pursue HDO process through these samples. Yu et al. [24] investigated the effect of zeolites on bio-oil distribution. Based on their reports, aromatic production is promoted over H-ZSM-5 followed by H-Beta and H-Mordenite, while H-Beta and H-Y catalyze the deoxygenation reactions. Saidi et al. [25] presented a research on anisole upgrading through HDO over γ-Al$_2$O$_3$. They perceived raise in the selectivity for phenol production by enhancing the reaction temperature. In another study, Saidi and Safaripour [26] claimed that, benzene production is enhanced at high operating temperatures and low pressures of H$_2$. Hence, it can be concluded that for optimizing individual chemical formation via parameter optimization studies, reaction pathways should be clarified [27]. Khromova et al. [28] studied anisole upgrading process. Based on the reported data, different HDO routes over NiCu/SiO$_2$ catalyst, are first-order reactions. However, high-order reactions for upgrading anisole are reported. Thus, kinetic data and reaction network are required. As a purpose number of researchers have focused on revealing reaction network.

As a purpose of investigating reaction network, a brief review on the effect of variable catalysts on products distribution will be discussed. Afterward, a detailed summary with the aid of investigating primary and secondary products will be surveyed.

2. Catalyst utilization in HDO process for production of aviation fuels

HDO is generally composed of two main reaction pathways including: direct deoxygenation and hydrogenolysis of C$_{aromatic}$—O bond. These reaction routes are improved in presence of

Table 2 Reaction pathways through HDO.

Reaction	Reactants		Products	
			Main product	Byproduct
Saturation		H_2		—
Hydrolysis		H_2		CH_4
Direct hydrodeoxygenation		H_2		CH_4OH
Decarbonylation		H_2		
Decarboxylation		H_2		

hydrogen activator active phase and catalyst supports that enhance $C_{aromatic}$—O bond cleavage [29]. Lup et al. [30] expressed the requirement of acidic supports for to increasing deoxygenation reactions of oxygenated compounds. Mäki-Arvela and Murzin [15] claimed that the formation of catechol resulted from hydrogenolysis reaction is improved in presence of acidic catalyst. Therefore, the application of an oxophilic supports for enhancing the cleavage of C=O and C—O, such as TiO_2, carbon and ZrO_2 is promoted. Hence the catalyst used in HDO process needs to have properties as oxophilicity, moderate metal-oxygen bond strength, and high H_2 activation [31].

Noble metals, transition metals, metal sulfide, and metal phosphide catalysts are some of the commonly used catalyst in HDO [29]. Among noble metals, Pt, Pd, Ru, and Rh are highly active for improving HDO process, due to the activation of H_2 and hence increasing H_2 and bio-oil interactions [32]. Also, in presence of noble metals, saturation reactions are enhanced resulting in cycloalkanes formation. Elumalai et al. [32] stated that, dispersion of noble metals on Al_2O_3 or SiO_2, enhances deoxygenation of guaiacol toward cyclohexane formation. Mu et al. [33] examined the HDO process of guaiacol over Rh/C catalyst. According to the reported data, at, 40 bar and 250°C, phenol was produced with the highest selectivity (35%) via direct deoxygenation. And the calculated selectivity for cyclohexanone and cyclohexanol formation was 13% and 25%, respectively. Ouedraogo and Bhoi [34] discerned Ru as a better active phase for deoxygenation reactions in comparison with Pt and PD as a result of binding to oxygen. Some of the HDO researches catalyzed by noble metals are presented in Table 3. The constraining fact about the application of noble metals is their high cost and low selectivity.

Therefore, the application of transition metals as Ni, Co, Cu, and Mo are being investigated to validate their activity and selectivity in HDO process [42]. Lower price and deactivation at higher operating temperatures are some of the advantages of transition metals over noble metals [34]. Liu et al. introduced transition metals with Lewis acid sites such as $M^{\delta+}$ as promoters for deoxygenation reactions [43]. Nakagawa et al. [44] surveyed the upgrading processes over Ni as a transition metal and over precious metals (Pd and Ir). According to their results, Ni has high efficiency for replacing precious metals. A brief review on HDO researches catalyzed by transition metals is presented in Table 4.

Different methods such as investigating kinetic data and thermodynamic analysis have been explored for revealing the reaction mechanism and network during the HDO process. A review on these methods is given in the following section with the aid of disclosing main reaction pathways.

2.1 Proposed reaction networks for bio aviation fuels production

Ranga et al. [56] provided a reaction network for anisole upgrading at 340°C, at hydrogen pressure of 0.5 MPa over $CoMo/Al_2O_3$ catalyst. According to their presented

Table 3 A brief review on noble metals catalyzed HDO.

Feed	Catalyst	Operating condition	Product	Reference
Pyrolysis oil	Pd-W/ZrO$_2$	240°C, Carrier gas = H$_2$, 2.75 MPa	Saturated cyclic rings (C$_{14-15}$), aromatics (C$_{18-20}$)	[35]
Pyrolytic lignin	Pd/C	400°C, Carrier gas = H$_2$, 10 MPa	Aromatics and alkyl phenolics 19.1 wt%	[36]
Anisole	Pt/ZrP	300–400°C, flowrate of bio-oil: 5–8 mL/min, flow rate of H$_2$: 185–300 g H$_2$/100 g bio-oil feed ratio	Monoaromatics (C$_{7-24}$), naphthalenes, and naphthenes	[37]
Guaiacol	Pt/SiO$_2$ Pt/H–MFI-90	180°C, Carrier gas = H$_2$, 5 MPa	86% and 100% conversion to Cyclohexane over Pt/SiO$_2$ and Pt/H–MFI-90, 90% selectivity toward cyclohexane	[38]
Phenol	Pt/NiO–Al$_2$O$_3$@Fe$_2$O$_3$	300°C, Carrier gas = H$_2$, 200 kPa	Cyclohexane and Benzene from guaiacol, Methyl Cyclohexane and Toluene from 4-methyl phenol	[39]
Guaiacol	Rh/C	250°C, Carrier gas = H$_2$, 4 MPa	The selectivity for Phenol production = 35%, the selectivity for Cyclohexanol and Cyclohexanone production = 25%	[33]
m-Cresol	Pt/γAl$_2$O$_3$	300°C, Carrier gas = H$_2$, 100 kPa	Toluene with 86% selectivity for production	[40]
Anisole	Pt/SiO$_2$	400°C, Carrier gas = H$_2$, 100 kPa	Benzene with 69.2% selectivity for production	[41]

Table 4 A brief review on transition metals catalyzed HDO.

Feed	Catalyst	Operating condition	Product	Reference
Anisole	Fe/Ni/H-Beta	220°C, Carrier gas = H_2, 100 kPa	Benzene, Toluene, 26.85% of the biofuel is consisted of oxygenated compounds including phenol and xylene	[45]
Phenol	Ni_3P	150–350°C, Carrier gas = H_2, Gas flow = 100 mL/min, 4 MPa	The selectivity for Cycloalkane production is 97%, Cycloalkenes, Cycloketone, and Cycloalcohols	[46]
Phenolics	Ni/SiO_2	300°C, Carrier gas = H_2, 5 MPa, reactor = autoclave	Ni/SiO_2 has produced 90% of cyclohexane and small traces of Methylpentane, Benzene, Cyclohexene, and Cyclohexanol were produced	[47]
Anisole	Ni/Al-SBA15	290°C, Carrier gas = H_2, 300 MPa	Cyclohexene with selectivity of 60%	[48]
Anisole	Ni/TiO_2	290°C, Carrier gas = H_2, 300 MPa	Aromatics with selectivity of 75%	[48]
Phenol	MoO_2C/TiO_2	400°C, Carrier gas = H_2, 2.5 MPa	Benzene with selectivity of 90%	[49]
Anisole	$NiMoO-SiO_2$	300°C, Carrier gas = H_2, 6 MPa	Cyclohexane with selectivity of 90%	[50]
Guaiacol	Co/SiO_2	300°C, Carrier gas = H_2, 1 MPa	Benzene with selectivity of 53.1%	[51]
Guaiacol	Co-MCM-41	400°C, Carrier gas = H_2, 100 kPa	Benzene with selectivity of 20%	[52]
Guaiacol	ReO_x/SiO_2	300°C, Carrier gas = H_2, 5 MPa	Cyclohexene with selectivity of 54.3%	[53]
Guaiacol	Fe/SiO_2	400°C, Carrier gas = H_2, 100 kPa	Benzene, Toluene, and Xylene with selectivity of 51.4%	[54]
Phenol Anisole Guaiacol	MoO_3/ZrO_2	350°C, Carrier gas = H_2, 6 MPa	46% Benzene 40% Benzene 23% phenolic compounds	[55]

reaction network, shown in Fig. 5, they expressed demethylation (hydrogenolysis) as the primary reaction pathway as methane was detected in the reaction. They approved that the activation energy for the hydrogenolysis of C_{methyl}—O bond is reduced over their acidic catalyst. Moreover, hydrogenation occurred rarely with the selectivity of <1% for cyclohexene production. Byproduction of 2-methyl phenol, 2,4-dimethyl phenol, and 2-methyl anisole indicated intramolecular methane shift of phenol and anisole, respectively. They presented higher selectivity values for transalkylation and demethylation reactions rather than deoxygenation reaction due to the acidic properties of the catalyst. He et al. [57] studied HDO of guaiacol at temperatures in the range of 150–350°C, 7 MPa and over Rh/ZrO_2 catalyst at batch reactor. Their interpreted reaction network is illustrated in Fig. 6. Based on the rection network, cyclohexane is produced with the selectivity of 87% as the main product at 300°C. Based on the reaction mechanism, hydrogenation and two steps of hydrogenolysis in a row leads in cyclohexane production. The produced cyclohexane may take part in some transesterification and hydrocracking reactions. The presented HDO reactions at the four operating temperatures are pseudo-first-order reactions. Resende et al. [58] presented detailed reaction network for HDO of anisole at 200–400°C and atmospheric pressure over Pd/ZrO_2 through the application of Lagrange multiplayer calculation in thermodynamic studies. According to the revealed reaction network at 400°C, elucidated in Fig. 7, dehydration of 1,3-cyclohexadien1-ol for benzene production is thermodynamically a favored reaction route rather hydrogenolysis of phenol. Liu et al. [59] proposed a reaction network by

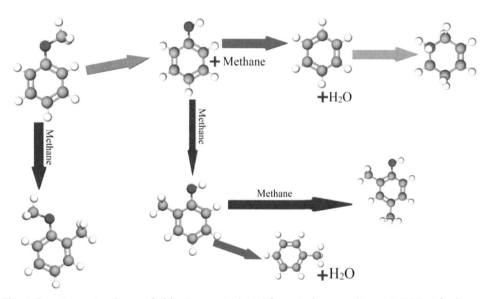

Fig. 6 Reaction network provided by Ranga et al. [56] for anisole upgrading at 340°C, at hydrogen pressure of 0.5 MPa over $CoMo/Al_2O_3$ catalyst.

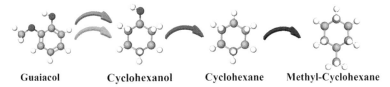

Guaiacol   Cyclohexanol   Cyclohexane   Methyl-Cyclohexane

Fig. 7 Reaction network provided by He et al. [57] for guaiacol at 150, 250, 300, and 350°C, 7 MPa, and over Rh/ZrO$_2$ catalyst at batch reactor.

microkinetic modeling for *m*-cresol upgrading process over NiFe (1 1 1) catalyst as represented in Fig. 8. They suggested the length of C—OH bond, as an important parameter affecting the deoxygenation reactions. As, deoxygenation reactions are promoted when the length of C—OH bond is higher, although the *m*-cresol is adsorbed on the oxophilic sites of Fe. Moreover, tautomerization is a crucial step, which may result in ketone or alcohols formation in the HDO process. As noticed in Fig. 8, two steps of tautomerization and a hydrogenation step, respectively, ends in 3-methyl-cyclohexanone production. Yang et al. [60] studied HDO of *m*-cresol at 300°C, 1 atm, 0.5 h, and over Ni-Re/SiO$_2$. According to the recommended reaction network, observed at Fig. 9, *m*-cresol is initially hydrogenated to methyl cyclohexanone and methylcyclohexanol as the reaction intermediates. Due to the partially hydrogenation of these compounds on the catalyst surface, these compounds undergo hydrodeoxygenation readily for toluene production. Toluene is also produced through the DDO reaction. Toluene undergoes different reactions. Methylcyclohexane is produced with the selectivity of 7.4% from toluene via hydrogenation route. Phenol is produced through the hydrogenolysis of C—O bond or demethylation. Benzene is produced in the reaction media from deoxygenation of phenol and demethylation of toluene along with H$_2$O and methane production, respectively. Venkatesan et al. [61] provided a quantitative reaction network for syringol upgrading at 350–500°C, over H-ZSM-5 catalyst. Total syringol conversion increased from 20% to 65% by raising the operating temperature in the range of 350–500°C. According to the illustrated reaction network, in Fig. 10, hydrogenolysis

Anisole   Benzene

1,3-Cyclohexadien1-ol

Fig. 8 Reaction network provided by Resende et al. [58] for HDO of anisole at 200–300°C and atmospheric pressure over Pd/ZrO$_2$.

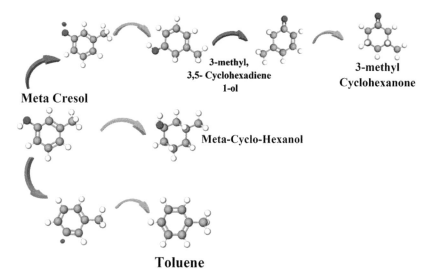

Fig. 9 Reaction network provided by Liu et al. [59] by microkinetic modeling for *m*-cresol upgrading process over NiFe (1 1 1) catalyst.

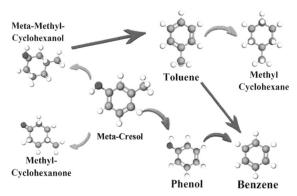

Fig. 10 Reaction network provided by Yang et al. [60], for *m*-cresol upgrading at 300°C, 0.1 MPa, 0.5 h and over Ni-Re/SiO$_2$.

of syringol to guaiacol and guaiacol to phenol are the main reaction pathways as they possess the lowest activation energy. While hydrogenolysis and DDO for anisole production from syringol, hydrogenation and methylation of anisole for cresol production and DDO of cresol to toluene are the slowest routes due to their high activation energy. The activation energy required for upgrading syringol into aviation fuels is presented at Table 5. Based on the calculated activation energies it can be concluded that syringol and guaiacol are the least stable components in HDO process. Saidi et al. [18] provided

Table 5 Presented activation energies from syringol [61], anisole [18], 4-methyl anisole [18] in HDO process, and cyclohexanone [27].

Activation energy for chemicals production from syringol [61]		Activation energy for chemicals production from anisole [18]		Activation energy for chemicals production from 4-methyl anisole [9]		Activation energy for chemicals production from cyclohexanone [27]	
Product	Ea (kJ/mol)	Product	Ea (kJ/mol)	Product	Ea (kJ/mol)	Product	Ea (kJ/mol)
Guaiacol	58.7	Phenol	25.3	Toluene	9.7	Benzene	35.6
Anisole	151	2-Methyl Phenol	55.4	Cyclohexanone	36.5	Cyclohexene	26.2
Phenol	86	2,6-Dimethyl Phenol	43.4	Phenol	55.1	2-Cyclohexan-1-one	42.4
Anisole (from guaiacol)	104	Hexamethyl Benzene	70.1	4-Methyl Phenol	31.9	Phenol	43.3
Phenol (from guaiacol)	75.5	2,4,6-Trimethyl Phenol	40.2	2,4-Di methyl Phenol	32.9	2-Methyl Phenol	91.3
Cresol	100	2,3,5,6-Tetra methyl Phenol	93.6	2,4,6-Trimethyl Phenol	35.8	2-Cyclohexyl cyclohexan-1-one	39.5
Cresol (from anisole)	118.7			2,4,5,6-Tetra methyl Phenol	28.2	Biphenyl	54.5
Toluene	124			2-Tert-butyl-4-methyl Phenol	25.5	2-Phenyl Phenol	97.8
Benzene	928.4					Cyclohexylbenzene	105.4
						2-Cyclohexyl Phenol	76.6

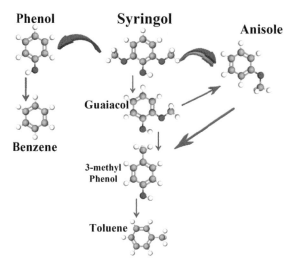

Fig. 11 Reaction network provided by Venkatesan et al. [61], for syringol upgrading at 350–500°C, over H-ZSM-5 catalyst.

proper kinetics data and reaction network for anisole upgrading at 8–14 bar, 300–400°C and over Pt/γAl$_2$O$_3$. Based on the revealed data, phenol, benzene, and 2–methyl phenol are produced as primary chemicals. As conveyed at Fig. 11, other phenolic compounds are formed through transalkylation and alkylation of phenol and 2-methyl phenol. And hexamethyl benzene is formed via alkylation of benzene as secondary products. The calculated activation energies are summarized in Table 5. According to the calculated data, benzene formation is not a second–order reaction. While other reaction pathways obey first-order reaction kinetics. Phenol, 2-methyl phenol and benzene are also produced as primary products in anisole upgrading over CuCrO$_4$·CuO as demonstrated by Deutsch and Shanks [62]. Sankaranarayanan et al. [63] studied anisole upgrading process over Ni–Co catalyst. They introduced hydrogenolysis to phenol, DDO to benzene, hydrogenation to methyl cyclohexyl ether and isomerization to methylphenol, and di-methyl phenol as main reaction routes from anisole. Saidi et al. [9] surveyed upgrading of 4-methyl anisole at 800–2000 kPa and 300–400°C, over CoMo/γAl$_2$O$_3$ catalyst. As observed in Fig. 12, hydrogenolysis to 4-methyl anisole, hydrodeoxygenation to toluene, hydrogenation to cyclohexanone and methyl cyclohexanone, and alkylation and trans-alkylation to methylated phenols as primary reactions. The calculated activation energies are reported in Table 5. Due to the low required activation energy, toluene production is favored due to the low required activation energy. Shetty et al. [64] studied the HDO of anisole over MoO$_3$/ZrO catalyst at 300–340 K and 1 bar. As illustrated in Fig. 13, phenol, cresol, and methyl anisole are primary products with relative activation energies of 93 ± 16, 100 ± 25, and 89 ± 21 kJ/mol, respectively. They reported alkylation reactions as nonselective reaction pathways at elevated operating temperatures. Moreover,

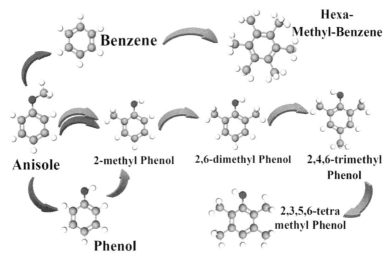

Fig. 12 Reaction network provided by Saidi et al. [18], for anisole upgrading at 0.8–1.4 MPa, 300–400° C and over Pt/γAl₂O₃.

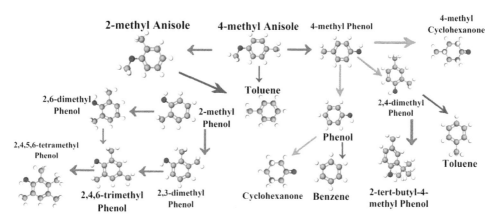

Fig. 13 Reaction network provided by Saidi et al. [9] for upgrading of 4-methyl anisole at 800–2 MPa and 573–673 K over CoMo/γAl₂O₃ catalyst.

based on their studies, benzene is produced from anisole and phenol through deoxygenation with activation energy of 116 ± 2 kJ/mol. (See Fig. 14.)

2.2 Future prospective for bio aviation fuel production

Lignin is left as a large waste biomass worldwide along with development of lignocellulosic biomass to ethanol production. Moreover, lignin biomass has high potential for bio aviation fuel production as it is mainly composed of aromatics. Therefore, variable

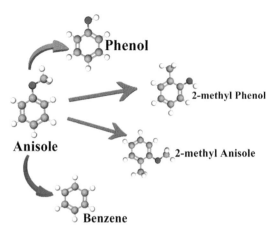

Fig. 14 Reaction network provided by Shetty et al. [64], for upgrading of anisole over MoO$_3$/ZrO catalyst at 300–340°C and 0.1 MPa.

processes are used for converting lignin to fuel grade products. The mentioned products are mostly oxygenated compound, which require deoxygenation pretreatment before usage.

As a purpose, effect of different active phases and supports on reaction pathways through the prevalent HDO process of lignin-derived-pyrolytic oil is reviewed in this study. HDO process is generally proceeded through deoxygenation, hydrogenation, and alkylation reactions. Kinetic data as an important information are still remained not detected. By acknowledging the kinetic data, parameters affecting reaction pathways can be controlled for selective hydrocarbon formation in the range of aviation fuels or fuel blending components. Therefore, investigating the kinetic data may improve the upgrading process.

Extracted saponifiable oils including free fatty acid and triacyl-glyceride from different crops and glycerol as the byproduct of biodiesel production can also be used as feedstock for producing renewable fuel through HDO process. During the HDO of the mentioned feedstocks, branched and linear alkane are produced over variable catalysts.

For decreasing the cost of HDO process, different procedures can be studied as expressed: (a) catalyst development should be investigated for selective production of the desired products and hence, reaching an economically feasible process in industrial scale, (b) utilizing procedures that enable concurrent H$_2$ production and HDO process, (c) using hydrogen-donors such as alcohols and acids as alternatives for the external H$_2$ that mollifies the operating conditions of HDO, and (d) application of renewable and waste materials such as chitosan and biochar lead to greener process for bio aviation fuel production in addition to the lower cost of their production.

During the HDO process, catalyst deactivation is usually proceeded by three main routes: (a) formation of carbonaceous deposit (coke) as a consequence of depolymerization and thermal cracking of the phenolic compounds in the pyrolytic oil, (b) decomposition of active phase via the reaction with the internal water of the pyrolytic oil or the produced water, and (c) pore construction and blockage. Catalyst deactivation is not studied in the current study, but is a crucial concern which, results in yield drop. Therefore, further studies are required for investigating the deactivation phenomena for designing metastable and reusable catalyst to improves the economic feasibility of the HDO process.

3. Conclusions

Jet range hydrocarbons are successfully produced from lignin-derived bio-oil via HDO approach over variable catalysts including noble and transition metals. The selection of the proper catalyst varies based on the processing goals, so as to utilize the produced fuel as promoters or blends with the conventional aviation fuels. Chemical structure of bio-oil is another crucial parameter in selection of the catalyst. As electron donating groups such as alkyl and alcohols promote proton transfer for deoxygenation reactions. However polarized bonds are resistant to bond scission (deoxygenation). Reactivity of oxygenated compounds in deoxygenation reactions lowers in the order of: alcohols > ketones > di alkyl ethers > phenols > di aryl ethers. In metallic catalysts, moderate metal—O bond strength is required for ensuring the reactant and product adsorption and desorption. Selection of proper support is another important step in HDO process. As they lead to selective deoxygenation or multiple side reactions, development of one-pot multisteps reactor may improve the economic feasibility of bio aviation fuel production.

References

[1] X. Zhao, X. Sun, X. Cui, D. Liu, Production of Biojet Fuels From Biomass, Sustainable Bioenergy, Elsevier Inc., 2019, pp. 127–165.

[2] F. Siavashi, M. Saidi, M.R. Rahimpour, Purge gas recovery of ammonia synthesis plant by integrated configuration of catalytic hydrogen-permselective membrane reactor and solid oxide fuel cell as a novel technology, J. Power Sources 267 (2014) 104–116.

[3] M. Saidi, M.A.R. Fallah, N. Nemati, M.R. Rahimpour, Model-based design of experiments for kinetic study of anisole upgrading process over Pt/γAl2O3: model development and optimization by application of response surface methodology and artificial neural network, Chem. Prod. Process. Model. 3 (2017).

[4] M. Faraji, M. Saidi, Hydrogen-rich syngas production via integrated configuration of pyrolysis and air gasification processes of various algal biomass: process simulation and evaluation using Aspen Plus software, Int. J. Hydrog. Energy 46 (36) (2021) 18844–18856.

[5] P. Moradi, M. Saidi, A.T. Najafabadi, Biodiesel production via esterification of oleic acid as a representative of free fatty acid using electrolysis technique as a novel approach: non-catalytic and catalytic conversion, Process Saf. Environ. Prot. 147 (2021) 684–692.

[6] A. Korshunov, B. Kichatov, V. Sudakov, A. Kolobov, V. Gubernov, A. Golubkov, P.A. Libet, A. Kireynov, S.O. Yurchenko, Hygroscopic property of biofuel obtained by torrefaction of wood in a quiescent layer of bentonite, Fuel 282 (2020) 118766.

[7] M.M. Ambursa, J.C. Juan, Y. Yahaya, Y.H. Taufiq-Yap, Y.-C. Lin, H.V. Lee, A review on catalytic hydrodeoxygenation of lignin to transportation fuels by using nickel-based catalysts, Renew. Sustain. Energy Rev. 138 (2021) 110667.

[8] F. Cheng, H. Bayat, U. Jena, C.E. Brewer, Impact of feedstock composition on pyrolysis of low-cost, protein- and lignin-rich biomass: a review, J. Anal. Appl. Pyrolysis 147 (2020) 104780.

[9] M. Saidi, H.R. Rahimpour, B. Rahzani, P. Rostami, B.C. Gates, M.R. Rahimpour, Hydroprocessing of 4-methylanisole as a representative of lignin-derived bio-oils catalyzed by sulphided CoMo/γ-Al2O3: a semi-quantitative reaction network, Can. J. Chem. Eng. 94 (8) (2016) 1524–1532.

[10] M. Wang, R. Dewil, K. Maniatis, J. Wheeldon, T. Tan, J. Baeyens, Y. Fang, Biomass-derived aviation fuels: challenges and perspective, Prog. Energy Combust. Sci. 74 (2019) 31–49.

[11] M. Saidi, M.H. Gohari, A.T. Ramezani, Hydrogen production from waste gasification followed by membrane filtration: a review, Environ. Chem. Lett. 18 (5) (2020) 1529–1556.

[12] E. Paone, T. Tabanelli, F. Mauriello, The rise of lignin biorefinery, Curr. Opin. Green Sustain. Chem. 24 (2020) 1–6.

[13] M. Saidi, M.R. Rahimpour, S. Raeissi, Upgrading process of 4-methylanisole as a lignin-derived bio-oil catalyzed by Pt/γ-Al2O3: kinetic investigation and reaction network development, Energy Fuels 29 (5) (2015) 3335–3344.

[14] M. Auersvald, B. Shumeiko, M. Staš, D. Kubička, J. Chudoba, P. Šimáček, Quantitative study of straw bio-oil hydrodeoxygenation over a sulfided NiMo catalyst, ACS Sustain. Chem. Eng. 7 (7) (2019) 7080–7093.

[15] A. Agarwal, M. Rana, J.-H. Park, Advancement in technologies for the depolymerization of lignin, Fuel Process. Technol. 181 (2018) 115–132.

[16] B. Rahzani, M. Saidi, H.R. Rahimpour, B.C. Gates, M.R. Rahimpour, Experimental investigation of upgrading of lignin-derived bio-oil component anisole catalyzed by carbon nanotube-supported molybdenum, RSC Adv. 7 (17) (2017) 10545–10556.

[17] J. Milovanović, R. Luque, R. Tschentscher, A.A. Romero, H. Li, K. Shih, N. Rajić, Study on the pyrolysis products of two different hardwood lignins in the presence of NiO contained-zeolites, Biomass Bioenergy 103 (2017) 29–34.

[18] M. Saidi, P. Rostami, H.R. Rahimpour, M.A. Roshanfekr Fallah, M.R. Rahimpour, B.C. Gates, S. Raeissi, Kinetics of upgrading of anisole with hydrogen catalyzed by platinum supported on alumina, Energy Fuels 29 (8) (2015) 4990–4997.

[19] S.H. Chang, Bio-oil derived from palm empty fruit bunches: fast pyrolysis, liquefaction and future prospects, Biomass Bioenergy 119 (2018) 263–276.

[20] M. Saidi, M. Yousefi, M. Minbashi, F.A. Ameri, Catalytic upgrading of 4-methylaniosle as a representative of lignin-derived pyrolysis bio-oil: process evaluation and optimization via coupled application of design of experiment and artificial neural networks, Int. J. Hydrog. Energy 46 (12) (2021) 8411–8430.

[21] M. Saidi, F. Samimi, D. Karimipourfard, T. Nimmanwudipong, B.C. Gates, M.R. Rahimpour, Upgrading of lignin-derived bio-oils by catalytic hydrodeoxygenation, Energy Environ. Sci. 7 (1) (2014) 103–129.

[22] F. Cheng, C.E. Brewer, Producing jet fuel from biomass lignin: potential pathways to alkyl-benzenes and cycloalkanes, Renew. Sustain. Energy Rev. 72 (2017) 673–722.

[23] Z. He, X. Wang, Hydrodeoxygenation of model compounds and catalytic systems for pyrolysis bio-oils upgrading, Catal. Sustain. Energy 1 (2012).

[24] Y. Yu, X. Li, L. Su, Y. Zhang, Y. Wang, H. Zhang, The role of shape selectivity in catalytic fast pyrolysis of lignin with zeolite catalysts, Appl. Catal. A Gen. 447–448 (2012) 115–123.

[25] M. Saidi, B. Rahzani, M.R. Rahimpour, Characterization and catalytic properties of molybdenum supported on nano gamma Al2O3 for upgrading of anisole model compound, Chem. Eng. J. 319 (2017) 143–154.

[26] M. Saidi, M. Safaripour, Ni–Mo nanoparticles stabilized by ether functionalized ionic polymer: a novel and efficient catalyst for hydrodeoxygenation of 4-methylanisole as a representative of lignin-derived pyrolysis bio-oils, Int. J. Hydrog. Energy 46 (2) (2021) 2191–2203.

[27] M. Saidi, A. Jahangiri, Refinery approach of bio-oils derived from fast pyrolysis of lignin to jet fuel range hydrocarbons: reaction network development for catalytic conversion of cyclohexanone, Chem. Eng. Res. Des. 121 (2017) 393–406.

[28] S.A. Khromova, A.A. Smirnov, O.A. Bulavchenko, A.A. Saraev, V.V. Kaichev, S.I. Reshetnikov, V.A. Yakovlev, Anisole hydrodeoxygenation over Ni–Cu bimetallic catalysts: the effect of Ni/Cu ratio on selectivity, Appl. Catal. A Gen. 470 (2014) 261–270.

[29] M. Saidi, P. Moradi, Catalytic hydrotreatment of lignin-derived pyrolysis bio-oils using Cu/γ-Al2O3 catalyst: reaction network development and kinetic study of anisole upgrading, Int. J. Energy Res. 45 (6) (2021) 8267–8284.

[30] A.N. Kay Lup, F. Abnisa, W.M.A. Wan Daud, M.K. Aroua, A review on reactivity and stability of heterogeneous metal catalysts for deoxygenation of bio-oil model compounds, J. Ind. Eng. Chem. 56 (2017) 1–34.

[31] L. Nie, D.E. Resasco, Kinetics and mechanism of m-cresol hydrodeoxygenation on a Pt/SiO2 catalyst, J. Catal. 317 (2014) 22–29.

[32] S. Elumalai, B. Arumugam, P. Kundu, S. Kumar, Phenol derivatives of lignin monomers for aromatic compounds and cycloalkane fuels, Biomass, Biofuels, Biochemicals: Recent Advances in Development of Platform Chemicals, Elsevier Inc., 2019, pp. 459–483.

[33] W. Mu, H. Ben, X. Du, X. Zhang, F. Hu, W. Liu, A.J. Ragauskas, Y. Deng, Noble metal catalyzed aqueous phase hydrogenation and hydrodeoxygenation of lignin-derived pyrolysis oil and related model compounds, Bioresour. Technol. 173 (2014) 6–10.

[34] A.S. Ouedraogo, P.R. Bhoi, Recent progress of metals supported catalysts for hydrodeoxygenation of biomass derived pyrolysis oil, J. Clean. Prod. 253 (2020) 119957.

[35] X. Yang, M.V. Pereira, B. Neupane, G.C. Miller, S.R. Poulson, H. Lin, Upgrading biocrude of *Grindelia squarrosa* to jet fuel precursors by aqueous phase hydrodeoxygenation, Energy Technol. 6 (9) (2018) 1832–1843.

[36] M.B. Figueirêdo, Z. Jotic, P.J. Deuss, R.H. Venderbosch, H.J. Heeres, Hydrotreatment of pyrolytic lignins to aromatics and phenolics using heterogeneous catalysts, Fuel Process. Technol. 189 (2019) 28–38.

[37] K. Routray, K.J. Barnett, G.W. Huber, Hydrodeoxygenation of pyrolysis oils, Energy Technol. 5 (1) (2017) 80–93.

[38] M. Hellinger, H.W.P. Carvalho, S. Baier, D. Wang, W. Kleist, J.-D. Grunwaldt, Catalytic hydrodeoxygenation of guaiacol over platinum supported on metal oxides and zeolites, Appl. Catal. A Gen. 490 (2015) 181–192.

[39] G. Zhu, K. Wu, L. Tan, W. Wang, Y. Huang, D. Liu, Y. Yang, Liquid phase conversion of phenols into aromatics over magnetic Pt/NiO–Al2O3@Fe3O4 catalysts via a coupling process of hydrodeoxygenation and dehydrogenation, ACS Sustain. Chem. Eng. 6 (8) (2018) 10078–10086.

[40] M.S. Zanuttini, C.D. Lago, C.A. Querini, M.A. Peralta, Deoxygenation of m-cresol on Pt/γ-Al2O3 catalysts, Catal. Today 213 (2013) 9–17.

[41] X. Zhu, L.L. Lobban, R.G. Mallinson, D.E. Resasco, Bifunctional transalkylation and hydrodeoxygenation of anisole over a Pt/HBeta catalyst, J. Catal. 281 (1) (2011) 21–29.

[42] M. Saidi, S.N.R. Baharan, Kinetic modeling and experimental investigation of hydro-catalytic upgrading of anisole as a model compound of bio-oils derived from fast pyrolysis of lignin over Co/γ-Al2O3, ChemistrySelect 5 (8) (2020) 2379–2387.

[43] C. Liu, Y. Shang, S. Wang, X. Liu, X. Wang, J. Gui, C. Zhang, Y. Zhu, Y. Li, Boron oxide modified bifunctional Cu/Al2O3 catalysts for the selective hydrogenolysis of glucose to 1,2-propanediol, Mol. Catal. 485 (2020) 110514.

[44] Y. Nakagawa, M. Tamura, K. Tomishige, Recent development of production technology of diesel- and jet-fuel-range hydrocarbons from inedible biomass, Fuel Process. Technol. 193 (2019) 404–422.

[45] X. Xu, E. Jiang, Z. Li, Y. Sun, BTX from anisole by hydrodeoxygenation and transalkylation at ambient pressure with zeolite catalysts, Fuel 221 (2018) 440–446.

[46] Z. Yu, Y. Wang, Z. Sun, X. Li, A. Wang, D.M. Camaioni, J.A. Lercher, Ni3P as a high-performance catalytic phase for the hydrodeoxygenation of phenolic compounds, Green Chem. 20 (3) (2018) 609–619.

[47] X. Zhang, W. Tang, Q. Zhang, T. Wang, L. Ma, Hydrodeoxygenation of lignin-derived phenoic compounds to hydrocarbon fuel over supported Ni-based catalysts, Appl. Energy 227 (2018) 73–79.

[48] Y. Yang, C. Ochoa-Hernández, V.A. de la Peña O'Shea, P. Pizarro, J.M. Coronado, D.P. Serrano, Effect of metal–support interaction on the selective hydrodeoxygenation of anisole to aromatics over Ni-based catalysts, Appl. Catal. B Environ. 145 (2014) 91–100.

[49] S. Boullosa-Eiras, R. Lødeng, H. Bergem, M. Stöcker, L. Hannevold, E.A. Blekkan, Catalytic hydrodeoxygenation (HDO) of phenol over supported molybdenum carbide, nitride, phosphide and oxide catalysts, Catal. Today 223 (2014) 44–53.

[50] A.A. Smirnov, S.A. Khromova, D.Y. Ermakov, O.A. Bulavchenko, A.A. Saraev, P.V. Aleksandrov, V.V. Kaichev, V.A. Yakovlev, The composition of Ni-Mo phases obtained by NiMoOx-SiO2 reduction and their catalytic properties in anisole hydrogenation, Appl. Catal. A Gen. 514 (2016) 224–234.

[51] T. Mochizuki, S.-Y. Chen, M. Toba, Y. Yoshimura, Deoxygenation of guaiacol and woody tar over reduced catalysts, Appl. Catal. B Environ. 146 (2014) 237–243.

[52] N.T.T. Tran, Y. Uemura, S. Chowdhury, A. Ramli, Vapor-phase hydrodeoxygenation of guaiacol on Al-MCM-41 supported Ni and Co catalysts, Appl. Catal. A Gen. 512 (2016) 93–100.

[53] K. Leiva, C. Sepulveda, R. Garcia, D. Laurenti, M. Vrinat, C. Geantet, N. Escalona, Kinetic study of the conversion of 2-methoxyphenol over supported Re catalysts: sulfide and oxide state, Appl. Catal. A Gen. 505 (2015) 302–308.

[54] R.N. Olcese, M. Bettahar, D. Petitjean, B. Malaman, F. Giovanella, A. Dufour, Gas-phase hydrodeoxygenation of guaiacol over Fe/SiO2 catalyst, Appl. Catal. B Environ. 115–116 (2012) 63–73.

[55] P. Mäki-Arvela, D. Murzin, Hydrodeoxygenation of lignin-derived phenols: from fundamental studies towards industrial applications, Catalysts 7 (9) (2017) 265.

[56] C. Ranga, V.I. Alexiadis, J. Lauwaert, R. Lødeng, J.W. Thybaut, Effect of Co incorporation and support selection on deoxygenation selectivity and stability of (Co)Mo catalysts in anisole HDO, Appl. Catal. A Gen. 571 (2019) 61–70.

[57] Y. He, Y. Bie, J. Lehtonen, R. Liu, J. Cai, Hydrodeoxygenation of guaiacol as a model compound of lignin-derived pyrolysis bio-oil over zirconia-supported Rh catalyst: process optimization and reaction kinetics, Fuel 239 (2019) 1015–1027.

[58] K.A. Resende, P.M. de Souza, F.B. Noronha, C.E. Hori, Thermodynamic analysis of phenol hydrodeoxygenation reaction system in gas phase, Renew. Energy 136 (2019) 365–372.

[59] X. Liu, W. An, C.H. Turner, D.E. Resasco, Hydrodeoxygenation of m-cresol over bimetallic NiFe alloys: kinetics and thermodynamics insight into reaction mechanism, J. Catal. 359 (2018) 272–286.

[60] F. Yang, H. Wang, J. Han, Q. Ge, X. Zhu, Influence of Re addition to Ni/SiO2 catalyst on the reaction network and deactivation during hydrodeoxygenation of m-cresol, Catal. Today 347 (2020) 79–86.

[61] K. Venkatesan, J.V.J. Krishna, S. Anjana, P. Selvam, R. Vinu, Hydrodeoxygenation kinetics of syringol, guaiacol and phenol over H-ZSM-5, Catal. Commun. 148 (2021) 106164.

[62] K.L. Deutsch, B.H. Shanks, Hydrodeoxygenation of lignin model compounds over a copper chromite catalyst, Appl. Catal. A Gen. 447–448 (2012) 144–150.

[63] C. Ranga, R. Lødeng, V. Alexiadis, J. Thybaut, Hydrodeoxygenation of anisole as bio-oil model compound over supported non-sulphided CoMo catalysts: effect of Co/Mo ratio and support, AIChE Annual Meeting, AIChE, 2017, pp. 1–4.

[64] M. Shetty, E.M. Anderson, W.H. Green, Y. Román-Leshkov, Kinetic analysis and reaction mechanism for anisole conversion over zirconia-supported molybdenum oxide, J. Catal. 376 (2019) 248–257.

CHAPTER 4

Main feedstock for sustainable alternative fuels for aviation

Vânya Marcia Duarte Pasa[a], Cristiane Almeida Scaldadaferri[b], and Henrique dos Santos Oliveira[a]
[a]Department of Chemistry, Institute of Exact Sciences, Federal University of Minas Gerais, Belo Horizonte, Minas Gerais, Brazil
[b]College of Engineering and Physical Sciences, Energy and Bioproducts Research Institute, Aston University, Birmingham, United Kingdom

1. Introduction

The planet has been driven toward an energy transition that leads to maximum decarbonization. Therefore, it is undeniable that we have to find new sources of energy and carbon matrix to support the development of our civilization. In the short and medium term, the use of biofuels is the best alternative, due to the higher degree of maturity, already consolidated in the automotive industry, using ethanol (109.9 million liters) [1], and biodiesel (35–45 million tons in 2019) [2].

The main challenge to produce biofuels is the availability of raw material which guarantees sustainability and economic competitiveness with fossil derivatives, since feedstock is one of the main components of the final cost (40%–80%) [2]. The importance of raw material cost is even greater in the case of Sustainable Aviation Fuels (SAF) production because the conversion processes to obtain drop-in biofuels are complex, involve many steps, high-cost catalysts and high hydrogen pressures.

Raw materials must be available in high quantities and low price as well as to be sustainable, which implies low impact on soil, water, and biodiversity. Feedstocks with low water demand, low consumption of fertilizers and pesticides, which are not competitive with food and do not demand expansion of agricultural frontiers will be more suitable and desirable. This avoids deforestation, with destruction of native forests, which are essential for human live. In order to guarantee sustainability, it is desirable to have certifications by international and independent bodies, which guarantee the observance of environmental and social legislations.

In order to accelerate the production of SAFs, we should use the established biomass production chains and, above all, waste processing to convert fatty materials, and forest, agriculture, industrial, and municipal wastes into biokerosene. Since there are many routes homologated by ASTM D 7566 for SAFs production, different industrial arranges

Sustainable Alternatives for Aviation Fuels
https://doi.org/10.1016/B978-0-323-85715-4.00005-7
Copyright © 2022 Elsevier Inc.
All rights reserved.
69

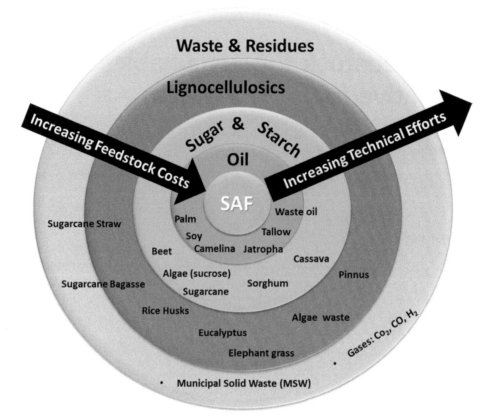

Fig. 1 Different feedstocks for SAF production: Waste and residues, lignocelluloses, sugar and starch, and fatty material. *(Reproduced with permission from Boeing/Embraer/FAPESP and UNICAMP.)*

can be possible. There is no one best option, but the best alternative for each region or country, which is suitable for that local reality [3].

This chapter presents a range of materials that can be used as feedstock for different production processes of SAF. Fig. 1 shows the different feedstock alternatives and shows that the low-cost and higher sustainable feedstocks require a more challenger conversion process, consequently, with a lower Technology Readiness Level (TRL).

At present, the main pathways to produce drop-in biofuels which are homologated by the Standard Specification for Alternative Aviation Turbine Fuel Containing Synthesized Hydrocarbons (ASTM D 7566) [3] are presented in Fig. 2, which also shows the key intermediates and the main feedstocks used in each pathway, all of them are solar energy converted into biomass. The different feedstock kinds were separated into (a) lipids or triglycerides (TG) derivatives; (b) sugars and starch; and (c) lignocellulosic materials.

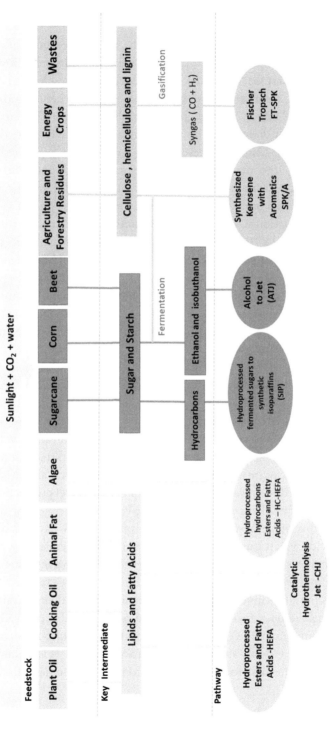

Fig. 2 Main kinds of feedstock to produce Sustainable Aviation Fuels (SAF) by different pathways.

2. Biofuel generations

Biofuels from renewable raw materials are generally categorized into different generations, considering the feedstock used to produce them, as explained below:

(a) **First generation (1st G)**: these fuels are produced from edible food crops, such as oil palm oil, soy oil, sugar cane, corn, wheat, and beets. The benefits observed are the easy conversion to biofuels and the great availability of crops. This kind of fuels can affect the food supply and present limited cultivation area, which are disadvantages;

(b) **Second generation (2nd G)**: these fuels are synthesized from inedible biomass, which is classified into main groups—energy crops and lignocellulosic residues. The main advantage is no competing with edible croplands, ensuring food safety. They can be obtained on a lower scale, as a native plant or on a larger scale when a commercial plantation is available. Some plants that can be used to produce 2nd G biofuels are jatropha (*Jatropha curcas*), macauba (*Acrocomia aculeata*), oilseed radish (*Raphanus sativus*), crambe (*Crambe abyssinica*), cottonseed (*Gossypium* sp.); karanja (*Millettia pinnata*), linseed (*Linum usitatissimum*); Mahua (*Madhuca longifolia*), Neem (*Azadirachta indica*), and others [4]. Cellulosic ethanol is one of the most important second-generation biofuels, produced by the treatment of sugar cane bagasse and straw which are converted into ethanol, which can also be transformed into SAF by the alcohol-to-jet pathway.

(c) **Third-generation (3rd G)**: For this fuel category, the main feedstock is co-products of the fishing and animal fat industry, in addition, algae. Algae are very interesting because they have no food value and no land requirement. The used water for algae growth can be seawater or wastewater, which increases sustainability. However, they have high energy consumption for cultivation and low lipid content; in addition, they present expensive technology in the oil extraction process.

Animal fatty feedstock has great importance, due to sustainability and low costs. These materials have been used worldwide for biodiesel and green diesel production and can be very important for SAF markets, too. However, the volume and collection limitations are challenges to use this animal raw material. Biofuels from tallow have a large volume of saturated fatty acid, which can affect some cold properties, demanding cracking, and isomerization additional step of processing to correct this characteristic for using in low temperatures, as requested in aviation.

Other interesting alternative feedstock in this category are other industrial residual as the cashew nut shell liquid, which has been studied as a potential waste feedstock to produce biohydrocarbons for biofuels [5]. This waste has the cardanol as the main compound and can be used to produce linear and cyclic hydrocarbons [6]. It is a phenolic oil produced as cashew nut industrial process in different countries as Vietnam, India, Indonesia, Philippines, and Brazil.

(d) **Fourth-generation (4th G)**: These fuels are produced from nonbiological resources and those from genetically modified organisms, e.g., microalgae, cyanobacteria, fungi, and yeasts, which have artificially increased the production of oil and/or sugar. There are benefits as a rapid growth rate, more lipid content, and more CO_2 absorbing ability. The electro-fuels synthesized by the reaction of CO_2 with H_2 from water hydrolysis, producing syngas to be converted into hydrocarbons are also highlighted in this category of drop-in biofuels. High initial investments and the initial research stage, which means low TRLs, are the disadvantages of these electrofuels.

In the next section, the main important raw materials that can be used for alternative drop-in biofuels for aviation are discussed and are separated in the following groups: lipids or fatty acids, sugars or starches, and lignocellulosic materials, as follow:

3. Aviation biofuel feedstocks—Lipids

There are many kinds of molecules that can be classified as lipids and have been used to produce biofuels. Lipids include fats (solid at room temperature) and oils (liquid at room temperature) obtained from vegetable or animal sources. Oils and fats are present in oilseeds (30%–58%), fruit pulps (30%–58%), animal tissues (60%–90%) including fish (10%–20%) [7]. They encompass molecules such as TGs, free fatty acids (FFAs) and their derivatives (diglycerides, monoglycerides, and phospholipids), as well as other sterol-containing metabolites such as cholesterol, as shown in Fig. 3 [8].

Crude vegetable oils are mainly made up of triacylglycerol (95%–97%) and when refined this content rises to 99%. TGs are obtained from the reaction of long-chain fatty acids with glycerol. The unsaponifiable fraction corresponds to sterols, hydrocarbons, fatty alcohols, tocopherols, and dyes. When saturated fatty acids predominate, the fatty material tends to be solid and when unsaturated chains predominate, the material is liquid. The carbon chains vegetable oils are normally linear and when there are double bonds they are preferably in the cis conformation.

Vegetable oil has one of the highest trade shares (41%) of the production of all agricultural commodities. This share is expected to remain stable with global vegetable oil exports reaching 96 Mt by 2027. Vegetable oil exports will to be dominated by Malaysia and Indonesia through palm oil. About 80% of Malaysian and more than 70% of Indonesian vegetable oil production is sold in the international market. The share of exports is expected to slightly decline in both countries due to their uses as feedstock for biofuels [8,9].

Vegetable oils and fatty materials have been used for the production of alternative fuels as biodiesel, green diesel, or biokerosene, due to their long aliphatic chains similar to the fossil fuel chains. The oil deoxygenation is performed by hydroprocessed esters and fatty acids (HEFA) pathway, considered the more widespread technology for SAF

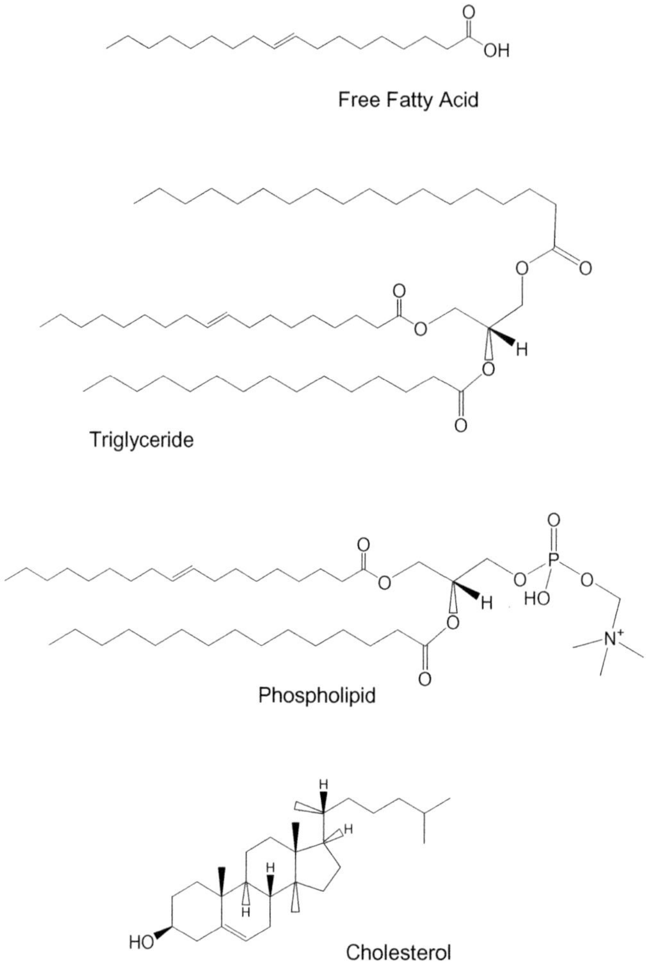

Fig. 3 Main compounds of lipids.

production in this initial phase. Recently, these oils can be also converted into biojet by catalytic hydrothermal pathway.

The pressure has increasing to have more and more sustainable and low-cost lipids for fuel markets. As some oilseeds are also used for food, it has stimulated studies via genetic engineering to increase the oil content in the seeds, avoiding the increase of the area to be cultivated. Another alternative is to find new nonedible species of plants, with high oil productivity, which can be used as feedstock for sustainable biofuels.

There are more than 350 oil-bearing plants identified, with thousands of subspecies that could be used to produce biofuels every year. The productivity of perennial plants is higher; they avoid erosion and can also be cultivated in mountain areas. Some species can be harvested more than once a year. The potentialities of many vegetable oils for fuels were considered as early as 1930, including the oils from castor, grape seed, maize, camelina, beechnut, rapeseed, lupin, pea, poppy seed, peanut, hemp, linseed, chestnut, sunflower seed, palm, olive, soybean, cottonseed, and shea butter.

Palm trees highlights as the vegetable with the highest oil production per area, about 3000–8000 L/ha/year, being the most promising alternatives for SAF production. These vegetables normally have two kinds of oils, kernel and pulp oils. The first oil has short and saturated chains and the second, long and unsaturated chains. There are many different species as buriti (*Mauritia flexuosa*), palm *(Elaeis guineensis)*, and coco (*Cocos nucifera*). Other important palm tree is macauba (*A. aculeata*) that is available in South America as native plants and has been pointed out as a promising source of biofuels due to their high productivity, rusticity, low water demand, and good oil quality.

Although there are many plants rich in oils, it is necessary to focus attention on species whose production already occurs on a large scale or which can be spread without great efforts, in the medium and short term, in order to reduce time, investments and risks [10,11].

The quality of the vegetable oils is also relevant, since the composition of the different fatty materials determines what kind of reactions is necessary to be performed in order to produce the best hydrocarbon chains for the biofuel formulation. For feedstock with higher number of carbons in their structure, it is useful to perform other reactions besides deoxygenation, as for example, hydrocracking and isomerization. These reactions permit to modulate the biofuel physical-chemical properties aiming to meet the quality requirements for biokerosene, especially the freezing point and the distillation curve. Plant oils with a large number of unsaturated chains are more reactive and are easily modified by these reactions. However, they also expend more hydrogen to guarantee high thermal-oxidative stability after decarbonization reactions.

Some noble and light vegetable oils are mainly constituted by saturated TGs with a short carbon chain (C10–C16), as for example the kernel oils produced by some palm trees, which has high thermal stability. These oils are very suitable to produce SAF by HEFA and catalytic hydrothermolysis jet fuel (CHJ) pathways, but it also has high prices in the pharmaceutical and cosmetic markets, which reduce their economic competitiveness for the fuel uses. Kernel oils for SAFs purpose is only economically feasible if there are efforts for palm tree plantation on a large scale, which can be done in tropical areas in different countries of the world. The palm trees have also second oil, the pulp oil, which has long carbon chains (C16–C18). They have lower thermal stability due to the high content of unsaturated molecules in their composition. Consequently, they are cheaper than kernel oil and more suitable for SAF feedstock.

Generally, degraded lipids have high acidity and low prices, and are not used for biofuels as biodiesel production. However, acid oil can be easily converted into aviation fuel by HEFA or CHJ pathways. It can be explained because the FFAs are intermediates in the

TGs deoxygenation processes. In order to produce them from TGs, there is a high consumption of hydrogen to generate the FFA. Consequently, acid oils are more inexpensive and have a lower cost to be converted into hydrocarbons, reducing the final production costs of SAF. The used cooking oil is one example of acid oil, but is difficult to be collected in large scale and provided with a homogeneous and constant quality. Many fruits as macauba or palm have enzymes in their pulp, capable to quickly hydrolyze the TGs into FFA, which is avoided with processing immediately after the fruit collection, which increase the oil cost. This tendency of oil acidity evolution can be a problem for biodiesel industry but a positive point for SAF manufacture.

Table 1 presents the promising vegetable fatty materials for biokerosene production with their more important characteristics, and their main world producers.

3.1 Main vegetable oils

Some vegetable oils are more prominent for fuel production and will be highlight, as following:

3.1.1 Palm oil

Palm oil is extracted from the palm trees (*E. guineensis* and *Elaeis oleifera*) and is the most used oil in the world for different applications. There are two different kind of palm oils, kernel, and pulp oil, as already mentioned. Both are also processed with the aim to partially separate the saturated than the unsaturated TGs chains. Palm olein is the liquid fraction obtained from the physical process of cooling, followed by filtration (winterization process). Its typical composition is: C16:0 (38%–43%), C18:1 (39.8%–43.9%), and C18:2 (10%–12.7%). The solid fraction is named palm stearin and has the main compounds: C16:0 (49.8%–68%), C18:1 (20%–34%), C18:2 (5%–9%), and C18:0 (3.9%–5.6%).

Worldwide palm oil is by far the most important oil in quantity; their plantations are mostly located in tropical and subtropical countries. There is a great and a growing interest in oil palm as SAF, due to this high productivity and availability, but the food completion is a disadvantage. Originated from West Africa, oil palm is currently one of the leading perennial oleaginous food crops grown widely in many tropical regions of Southeast Asia. Malaysia and Indonesia are leaders in palm oil production that supply more than 80% of the global demand, followed by Thailand, Colombia, Nigeria, Guatemala, Papua New Guinea, Ecuador, and Honduras [28,29]. A total palm biomass of 2.3 t (approximately 1 t of empty fruit bunches (EFB), 0.7 t of palm fibers, 0.3 t of palm kernels, and 0.3 t of palm shells) can yield a tonne crude oil palm [30]. Other palm tree that had been pointed as a promising feedstock for SAFs is Macauba, due to its especial productivity and characteristics.

Table 1 Different feedstock for lipids production, main countries which are producers, the height of the tree, type of culture, and data of oil composition.

Feedstock	Main world producers	Plant type/height (m)	Main Free Fatty Acid (Kernel) %	Main free fatty acid (pulp) %	Oil composition (%) Saturated fatty acid	Mono unsaturated fatty acid	Polyunsaturated fatty acid	Reference
Acrocomia aculeata (Macauba)	Brazil, Mexico, Paraguay, Venezuela, and Bolivia	Palm tree/perennial 10–15	C12:0–37 C14:0–9.7 C16:0–8.0 C18:1–28	C16:0–15.0 C16:1–3.0 C18:1–68.4 C18:2–10.1	Kernel (64.5–77) Pulp (19–21)	Kernel (19–32) Pulp (61–71)	Kernel (3–3.5) Pulp (10–19)	[12,13]
Elaeis guineensis (Palm)	Indonesia, Malaysia, Thailand, Ecuador, Colombia, Guatemala, Brazil, and Nigeria	Palm tree/perennial 6–16	C16:0–45 C18:0–4 C18:1–40 C18:2–10	C16:0–44.0 C18:1–40.0 C18:2–10.1 C18:0–4.0	Kernel 82 Pulp 40–50	Kernel 15.3 Pulp 35–44	Kernel 2.7 Pulp 9–12	[4,14]
Cocos nucifera (Coconut)	India, Indonesia, Brazil, Philippines, Thailand, and Sri Lanka	Palm tree/perennial 15–18	C12:0–41 C14:0–24 C16:0–16 C18:1–9.0		92–94	5–6	1–2	[4,15]
Orbignya phalerata (Babassu)	Colombia and Brazil	Palm tree/perennial 15–30	C12:0–47 C14:0–15 C16:0–8.0 C18:1–11.0		78	15	2	[16,17]

Continued

Table 1 Different feedstock for lipids production, main countries which are producers, the height of the tree, type of culture, and data of oil composition—cont'd

Feedstock	Main world producers	Plant type/height (m)	Main Free Fatty Acid (Kernel) %	Main free fatty acid (pulp) %	Oil composition (%)			Reference
					Saturated fatty acid	Mono unsaturated fatty acid	Polyunsaturated fatty acid	
Astrocaryum vulgare (Tucumã)	Bolivia, Colombia, Venezuela, Guyana, Suriname, and Brazil	Palm tree/perennial 10–25	C12:0–47.9 C14:0–23.8 C16:0–6.3 C18:1–12.0		89	8	3	[18,19]
Syagrus coronata (Licuri)	South America	Palm tree/Perennial 5–20	C12:0–36.0 C14:0–16.5 C16:0–8.9 C18:1–14.2		Kernel (72) Pulp (49)	Kernel (24) Pulp (17)	Kernel (4) Pulp (35)	[20,21]
Brassica napus (Rapeseed)	Canada, China, Europe	Herb/Seasonal 1.2	C18:1–14.6 C18:2–28.2 C18:3–21.9 C20:1–19.5		11	34	50	[22,23]
Crambe abyssinica (Crambe)	United States, the Netherlands, and Mexico, and Brazil	Herb/Perennial 0.2–1.5	C18:1–19.2 C18:3–7.9 C20:1–7.1 C22:1–57.3		3–5	79–83	13–16	[24,25]
Helianthus annus (Sunflower)	Europe	Herb/Seasonal 3	C18:0–4.0 C18:1–27.5 C18:2–50.0		13	68	18	[4]
Gossypium herbaceum (Cotton seed)	Europe, China and United States	Herb/Perennial 1.2	C16:0–15.8 C18:0–2.5 C18:1–21.2 C18:2–54.2		24	19	55	[4]

Species	Origin	Type/Lifespan (years)	Fatty acid composition				References
Jatropha curcas (Jatropha)	India, Argentina, United States, Paraguay, Brazil, Africa, and Mexico	Tree/Semi-Perennial 5–8	C16:0–16.0 C18:0–6.5 C18:1–43.5 C18:2–34.4	22	44	34	[4,26]
Zea mays (Corn)	United States, China, Brazil, Argentina, Ukraine, and Indonesia	Grass/Seasonal 1–2.5	C16:0–10.2 C18:0–2.1 C18:1–31.1 C18:2–55.3	12	86	2	[27]
Linum usitatissimum (Linseed)	Canada, Europe, India, and China	Herb/Annuals 1.2	C16:0–6.0 C18:1–22.1 C18:2–14.1 C18:3–48.6	10–11	20–24	46–51	[4]
Glycine max (Soybean)	Brazil, United States, Argentina, and China	Herb/Seasonal 0.5–1.3	C16:0–8.0 C18:1–25.0 C18:2–55.0 C18:3–8.0	14	28	57	[4]
Millettia pinnata (Karanja)	Asia, Australia, China, and United States	Tree/Perennial 15–25	C16:0–11.7 C18:0–7.5 C18:1–51.6 C18:2–16.5	19	52	17	[4,26]

3.1.2 Macauba oil

The *A. aculeata*, known as Macauba, is a palmaceae found in tropical and subtropical America from southern Mexico and the Antilles to the southern regions of Brazil, reaching Argentina and Paraguay [31]. It has high adaptability, thereby enabling its cultivation in different seasons and contributes to income generation for families besides being intercropped with other crops and livestock. Targeting the cultivation of Macauba as an energy crop for the energy sector is attractive because it does not compete with food, and can produce 5000–6000 kg/ha, about 10 times more oil than soybean (375–500 kg/ha), which is the main raw material used in the production of biodiesel, and a major crop used by the food industry [32]. Macauba kernel oil is the raw material with the greatest potential for use in aviation biokerosene because its almond is rich in lauric acid (33.3%) besides still classified as nonedible feedstock [33]. Its pulp oil can be also converted into SAF but cracking is necessary to reduce the resultant chain carbon sizes. This palm tree has been planted in Brazil to recover degraded areas and water natural sources, besides the aim to be used for biofuels production. There is a SAF industrial plant being constructed in Paraguay for green diesel and biokerosene production, which will use macauba oil from native trees as feedstock. This oil has been used in Brazil for biodiesel production, using fruits from native trees, but efforts have been done for macauba plantation in large scale. It is considered a promising alternative for SAF production in this country.

3.1.3 Soybean oil

Soy oil is the second more consumed lipid feedstock for biofuels production, after palm oils. It is consumed by cooking process and it is a by-product of protein industry. It is used by a source of biodiesel in Brazil and the United States, the main world producers of this alternative diesel. The increasing of soy seed plantations for SAF production does not seem to be the best option because its productivity is low and there is a relevant food competition [9]. Otherwise, the plantation of palm species seems to be more promising, as already discussed. The soy oil has as the main component the oleic acid and it has low thermal stability due to the high content of unsaturated molecules. Therefore, hydrogenation and cracking reactions are also necessary, besides deoxygenation step to synthesize biokerosene from this oil.

3.1.4 Carinata (Brassica carinata)

Carinata is cultivated as an oilseed crop in Ethiopia, the United States, Italy, and other countries and recently, it has been used to develop aviation biofuels. Carinata meal is low fiber, high protein source for the livestock industry. In addition, carinata produces a lot of biomasses, so it useful for preventing erosion and improving the soil. It requires no changes in land use. It is a new profitable alternative plant, its oil presents the following fat acids: stearic (1%), oleic (9%); linoleic (16%), linolenic (14%), eicosenoic (7%), erucic (40%), and nervonic (3%) [34].

3.1.5 Algae oil

Algae is another sustainable alternative for oil production, although the technologies involved are yet immature. A challenge in the production of algae-based biofuels is the selection of suitable microalgae strains, which there are many types of microalgae. Algae can adapt to variable water sources like brackish water and wastewater and do not require arable land. It is possible to culture the microalgae near power plants using the flue gas (carbon dioxide) emitted from boilers. These types have substantially different properties, including content of lipids, proteins, vitamins, chlorophyll, and adaptation to the environment, life cycle, among others. Studies report that there are more than 40,000 strains of microalgae that have been categorized into two groups based on whether the strain has a nucleus or not. A large number of microalgae strains contain high lipid content, up to 50% of the dry cell weight [35,36]. The Chlorella genre represents a category with good potential for SAFs due to high yields in carbon (73.2%) a value near to petroleum (83%–87%). In addition, algae have a capacity for photosynthesis efficiency 10 to 20 times greater than terrestrial plants, resulting in the generation of biomass and the sequestration of CO_2 much faster. When analyzed in terms of biomass productivity, i.e., studies with *Chlorella vulgaris*, shows values around 1.5–2.6 g L^{-1} [35].

From the data found in the literature, there is a small fluctuation in values around 1%–3% of the composition of fatty acids, which is expected due to water stress, availability of nutrients, seasonal, and geographical influences for the same species of raw material. Table 2 presented the main compounds present in this oil. Process to extract the algae oils is expensive, but this technology probably will be representative for fuel market in the next years, especially regarding the possibility to have sea farms.

3.1.6 Waste oil

A wide range of waste oil can be used for SAFs production, similar to what happens with biodiesel production chains. These oils are cheap and are an ecological problem because have to be disposed of without environmental impact. Its use is an extra ecological force, but also a logistic challenge because most of the time their generation occurs in a decentralized way, and in a small scale. This oil is classified regarding its origin into three categories: waste oil from restaurants and households; from the food industry and, from the nonfood industry [39]. These oils from cooking or food can be a mixture of different seed oils (soy, palm, rapeseed, sunflower, etc.), water, and food residues, demanding initial pretreatment. Due to the cooking process performed at high temperatures, the oils can have trans chains and large chains due to the degraded products generated by thermo-oxidative processes.

Table 2 Fatty acid composition of beef tallow, pork lard, poultry fat, mutton tallow, anchovy, cod liver, and sardine pilchard.

Fatty acid		Beef tallow [37]	Pork lard [38]	White grease [39]	Yellow grease [39]	Poultry fat [40]	Mutton tallow [41]	Anchovy [39]	Cod liver [39]	Sardine pilchard [42]
Myristic	C14:0	1.6	1.6	1.6	0.7	0.4	2.2	7.4	3.2	8.0
Palmitic	C16:0	21.6	25.1	22.0	14.3	21.6	21.1	17.4	13.5	18.0
Stearic	C18:0	17.7	12.6	10.0	8.2	6.3	11.6	4.0	2.7	–
Palmitoleic	C16:1	2.5	2.8	5.1	1.4	3.2	2.1	10.5	9.8	10.0
Oleic	C18:1	31.5	36.5	42.5	43.3	30.0	38.7	11.6	23.7	13.0
Paullinic	C20:1	–	–	0.6	0.5	–	–	–	–	4.0
Erucic	C22:1	–	–	–	–	–	–	–	–	3.0
Linoleic	C18:2	3.3	16.5	13.2	26.3	28.4	10.2	1.2	1.4	–
Linolenic	C18:3	1.3	1.1	0.9	2.5	2.4	0.6	–	–	–
Arachidonic	C20:4	–	0.3	0.3	–	3.4	–	–	–	–
Eicosapentaenoic	C20:5	–	–	0.1	–	–	–	17.0	11.2	18.0
Docosapentaenoic	C22:5	–	0.2	–	–	0.3	–	–	–	–
Docosahexaenoic	C22:6	–	–	–	–	0.8	–	8.8	12.6	9.0
Total saturated	SFA	49.1	39.4	33.6	23.2	29.1	40.4	28.8	19.4	26.0
Total monounsaturated	MUFA	41.0	39.7	48.2	45.2	33.2	47.1	22.1	33.5	30.0
Total polyunsaturated	PUFA	10.0	20.9	14.5	28.8	37.6	12.5	27.0	25.2	27.0

Chhetri and collaborators reported that the average per capita waste cooking oil is about 9 pounds in the United States and Canada [43]. Similarly, as the current population is 7.8 billion, the world cooking oil production can be estimated as 70.2 billion pounds or 280 million tons, an expressive value to be considered as a potential feedstock, despite the uncertainty associated with the selective collection.

The oils produced by tire and plastic pyrolysis are also considered waste oil from the nonfood industry. They are not renewable when their feedstocks are fossils but are alternative oils with important sustainability and availability [4].

3.1.7 Animal oils and fatty materials

Fats from animal slaughter are an economic attraction for the production of biofuels, since the large amount of fatty residues produced implies low cost and immediate availability of the raw material in agro-industrial areas. In addition, the use of animal fats helps to reduce environmental impacts, avoiding the improper destination of waste. Residues such as beef tallow, mutton tallow, pork lard chicken fat, turkey fat, fish oil have been used for biodiesel production and can be also as SAF feedstock.

Animal fats have different composition, but generally, most of them have 16 to 20 carbons. The main fatty acids are: palmitic (16:0), stearic (18:0), oleic (18:1), linoleic (18:2), and arachidonic (20:4). Ruminant and pig fats are predominantly solids due to their high content of saturated fatty acids (about 40%). Chicken fats are semisolid or almost liquid and have about 30%–33% of saturated fatty acids. The typical compositions of the main animal fats are presented in Table 2.

The use of animal fat in the preparation of biofuels is hampered by its solidification at room temperature and by the percentage of sulfur higher than that found in vegetable oils, general characteristics of this type of material. The content of phosphorous can be also higher due to the presence of phospholipids, which can reduce the biofuels quality. Moreover, the presence of protein as contaminants can also be a problem, increasing the nitrogen content. All these undesirable characteristics can be corrected during the fatty acid pretreatment or by the conversion process to produce hydrocarbons. On the other hand, animal fat normally has a high number of FFAs, which can be an advantage because can reduce the H_2 in the deoxygenation step. For example, the animal's carcass contains 12% fatty acids, although it may contain 15% protein [39].

When faced with data on the production of fats and oils of animal and marine origin through slaughtered animals and through fishing, some countries and regions such as Argentina, Australia, Brazil, Peru, Japan, United States, Scandinavian Peninsula, European Union lead this complex market. In 2019, 800 thousand tons of animal fats and their derivatives were produced in the EU, an amount that has remained constant since 2014 [44]. In the United States, the animal fats represented 8.4% of total biodiesel feedstock through poultry fat, tallow, and white grease with amounts of 74, 132, and 243 thousand tons, respectively [2]. Brazil, one of the most important biodiesel producers, has also used animal fat to produce biodiesel. In 2018 this country produced 5.35 million m^3 of biodiesel, 70% from soybean, 16% from animal fat, 1% cottonseed oil, and 13% of other vegetable lipids, including macauba oil [45].

Fats from the meat processing industry and industrial cooking business are recycled greases and are classified as yellow grease and brown grease. The first one has FFAs <15%, and the second one has the FFA >15%. Regarding the definitions, white grease classified as inedible fat derived from swine. Yellow grease is a product that has its vegetable and animal origin based on the residues of oils and fats during the processing of restaurant fries. It is a very misunderstood category regarding its source and use. Due to the preference for animal fat in food preparation, a new market niche was created due to the residues of this chain and brought a new category of fats: feed-grade animal fat or yellow grease. Yellow grease is best defined as a fat product that does not meet the definitions of animal fat, vegetable fat or oil, hydrolyzed fat, or fat ester. Like any other type of fat, it must be sold in its specifications, which include a minimum percentage of total fatty acids >90%, the maximum percentage of unsaponifiable matter <1%, the maximum percentage of insoluble impurities <0.5%, the maximum percentage of FFAs, and the amount of moisture <15%. In addition, it must meet the Food and Drugs Administration rules.

Lipids of animal origin, such as beef tallow, pork lard, mutton tallow, poultry fat, fish oil, as well white grease as yellow grease also make up the matrix for the production of SAFs. The most common fatty acids are shown in Table 2.

It should be noted that lipids of animal origin have been increasing in value in the biofuels market due to increased demand. They are usually of nonhomogeneous quality and have the challenge of collection logistics, but they can be very important for SAF production, especially if generated on a large scale, centrally in industrial processes.

4. Aviation biofuel feedstocks: Sugar and starchy crops

Regarding the feedstocks for the first-generation biojet fuel production (from edible food crops), sugar, and starches crops are the feedstocks for the biochemical routes such as alcohol to jet process (ATJ) and synthesized isoparaffins (SIP) produced from hydroprocessed fermented sugars. SIP route uses genetically modified strains of *Saccharomyces cerevisiae* yeast to process sugars and subsequently to produce hydrocarbons, where the fermentation of sugars results to farnesene production. The next stages are the hydroprocessing of farnesene into farnesane ($C_{15}H_{32}$) and fractionation to produce SIP biojet fuel [3,46,47].

In ATJ route, fermentable sugars are processed to obtain alcohols, mainly ethanol and *n*-butanol. This biojet fuel pathway starts from alcohol and it is processed through the following reactions: dehydration, oligomerization, hydrogenation, and fractionation. Before these steps, the alcohol production involves three major steps (1) obtaining solution that contains fermentable sugars, (2) converting sugars to alcohol by fermentation, and (3) alcohol separation and purification [3]. For ethanol production, *S. cerevisiae* is the most common employed yeast in industrial scale due to its high ethanol productivity, high ethanol tolerance, and ability of fermenting wide range of sugars. Biobutanol also is a chemical precursor for ATJ aviation fuel and it can be produced by fermentation of sugars through the acetone-butanol-ethanol (ABE) process. Traditionally, the most common strains of bacteria used in ABE fermentation process is *Clostridia* [48,49].

4.1 Sugar and starchy crops composition

Feedstocks for these biochemical routes include raw materials such as sugar cane, sugar beet, sweet sorghum, corn (maize), wheat, rice, potato, cassava, sweet potato, barley, etc., which are feedstocks rich in sugars, commonly named carbohydrates. Monosaccharides are the simplest carbohydrates since they cannot be hydrolyzed to smaller carbohydrates [50]. The most important monosaccharides are glucose and fructose, followed by other minor monosaccharides such as mannose, galactose, xylose, and arabinoses (Fig. 4A). Agriculture products such as corn and sweet potatoes contain high content of these types of sugars [50,52,53]. Oligosaccharides are carbohydrates formed by 2–10 units of monosaccharides which are classified as disaccharides, trisaccharides, tetrasaccharides, and pentasaccharides are. Disaccharides are composed of two units of monosaccharides and are one of the major oligosaccharides present in fruits and vegetables. Sucrose is the most important of the disaccharides; consist of glucose and fructose units (Fig. 4B) [53]. Among the agriculture crops used as feedstock in biochemical routes, sugarcane, and sugar beet have the highest sucrose content [54]. Many of oligosaccharides are components of fiber from plant tissue, which are found in some grains and cereals, such as wheat and barley. Another class of carbohydrates are polysaccharides. Starch is the main reserve of polysaccharide and it is composed of two polymers of glucose named amylose and amylopectin (Fig. 4C). Amylose is a linear molecule of about 200–1000 units of glucose units while amylopectin is a branched molecule of about 2000–2,000,000 units of glucose. Starch is present in various cereal and tuber crops, such as corn (maize), wheat, rice, potato, cassava, sweet potato, barley, etc. [53,55,56].

4.2 Sugar crops for ethanol and butanol production

On a global scale, sugar crops and molasses, which is a by-product of the sugar production process, have a share of approximately 30% in total ethanol production, and corn has the highest share, around 53% (Table 3) [56]. In this scenario, Brazil is the world's leading producer of ethanol based on sugarcane, using approximately 50% of its sugarcane production to produce ethanol. Other raw materials such as wheat and lignocellulosic biomass have a minor contribution in the sector (Table 3) [56].

In terms of sugar crops production, the cultivation of cane and beet sugar differs depending on the climatic conditions. Sugarcane is the main feedstock for sugar

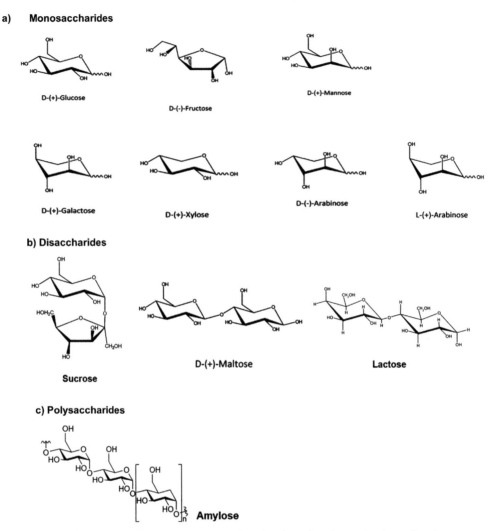

Fig. 4 Chemical structures of the major monosaccharides, disaccharides, and polysaccharides present in sugar and starch crops used as feedstock in biochemical route to produce biofuels [51].

production in tropical countries, whereas sugar beet is grown mostly in the temperate zones of the Northern hemisphere. More than 100 countries grow sugar beets, sugarcane, or both, mainly to produce sugar for food consumption [56]. Approximately 1.85 billion tonnes of sugarcane and 270 million tonnes of sugar beet are produced annually, for direct consumption and for conversion to ethanol and chemical feedstocks as reported by the Food and Agriculture Organization of the United Nations (FAO) [57].

Among these countries, Brazil and India are the major producers of sugar cane. In United States, the size of the beet sugar and cane sugar sector is comparable, while

Table 3 Share of ethanol production worldwide from each feedstock.

Crops	Share of ethanol production
Maize (corn)	53.7%
Sugar crops	25.1%
Molasses	6.7%
Wheat	2.6%
Lignocellulosic biomass	0.4%
Other	11.5%

Data correspond to the average of production from 2015 to 2018 [56].

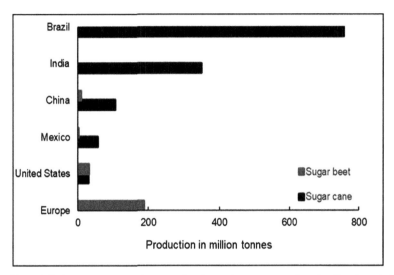

Fig. 5 Leading global sugarcane and sugar beet producers according to Food and Agriculture Organization of the United Nations database (FAO) [57].

the EU produces mainly beet sugar [56,57]. Fig. 5 shows the main producers of sugar crops (with respect to an average from 2015 to 2018) in the world. Both sugar crops are used to produce ethanol for usage as fuel and/or feedstock for ATJ and SIP processes.

Sugar cane (*Saccharum* genus) presents a higher photosynthetic rate in the CO_2 capture and it is a perennial plant cultivated in more than 100 countries [58,59]. Brazil has the largest sugar cane planted area (40%), which represents only about 1% of the total Brazil area. In terms of production Brazil produces 35.48 million tonnes of sugar and 28.16 billion liters of ethanol. From straw and bagasse is generated 25,482 GWh of electricity. It is the crop with the highest agricultural productivity on the planet, and its potential of energy production can increase [58,59].

There are many studies to develop different sugar cane varieties to increase its performance for sugar, ethanol or biomass production. Recently, it is observed that there are varieties which produce more biomass instead of sucrose, as the case of energy cane, resulting from the backcrossing between commercial and ancestral species [59]. The energy cane has been used for second generation ethanol production and it presents higher resistance to water and nutritional stress because its roots are more abundant, with a denser and deeper rhizosphere. [59]. It can produce about 180 t/ha, which means about two to three times more than sugarcane variety, with a cost of production that is a half that is considered for sugarcane [59].

Ethanol production may increase in Brazil in the next years due to the implementation of the RenovaBio, a Brazilian program to stimulate biofuel production. Comparing with other similar programs in other countries, its main innovation is the CBIO, a kind of currency, applied for different biofuel production, and it can be valued like the virtual currencies, but has the advantage to have the ballast the mitigation of greenhouse gases in the atmosphere. According this scenery, the production of biojet from bioethanol can be a promissory for this country using the ATJ or SIP pathways [59].

In terms of bio-butanol production, bio-butanol market is in the nascent stage and China has been leading the worldwide production and might achieve self-sufficiency shortly. Some major companies such as Gevo, DuPont, BP, Cobalt Green Biologics LTD, Syntec Biofuel, Butalco, Russian Technologies (Russian State-Owned Company) and new research platforms such as Plantaonix W2, Energy, ZeaChem, Energy Quest, Metabolic Explorer, OptinolM, Abengoa, Celtic Renewables have also interested in the biobutanol production to be used as chemical precursor and biofuel, and as an alternative to an oil-based economy [49,60].

4.3 Ethanol and bioethanol production from starch-based feedstock

Cereals such as maize/corn, wheat, sorghum, rye, and barley as well as tubers, e.g., potatoes and cassava are starch-based feedstocks used to produce ethanol. The main cereals produced worldwide are corn, wheat, rice, and barley with a starch content of about 70% [56,61]. They account for over 90% of total cereal production, however, the main use of cereals and tubers is direct food and feed use. Approximately, 36% and 43% of cereals were used as food and feed, respectively [56]. Table 4 shows the cereal and tubers production in the major regions and countries in the world, and represents the main cereal crops including maize, wheat, barley, rye, and others [57].

The United States is the largest producer of starches crops in the world, at approximately 440 million tonnes a year, and corn is the major raw material cultivated in this country, accounting for more than 90% of the ethanol production (Table 4). In the EU, the main starch crop for bioethanol production is wheat, where is produced approximately 280 million tonnes of starches including wheat, maize, and barley. The production of bio-ethanol represents 0.7% of EU agricultural land and 2% of Europe's grain

Table 4 Cereal and tubers production in the major regions and countries in the world.

	Cereal and tuber production (in million tonnes)				
	Most regions and countries in the world				
Crops	EU	United States	Brazil	China	India
Wheat	149.06	54.33	5.53	132.93	94.26
Maize	64.14	386.95	82.42	261.39	26.26
Sorghum	0.73	11.46	1.95	2.42	4.77
Barley	59.35	3.87	3.00	1.64	1.65
Rye	7.26	0.26	0.01	1.15	0.00
Sweet potato	0.09	1.42	0.70	52.75	1.39
Cassava	0.00	0.00	20.06	4.85	4.38

Data correspond to the average of production from 2015 to 2018 [57].

supply. In addition, EU has the largest sugar beet-based ethanol industry and sugar beet and molasses had a share of approximately 20% in total EU ethanol production [56,61].

Other crops such as barley (a winter crop), rye (the rather robust grain that also grows on poorer soils), and sorghum are potential biofuel feedstocks and can play important roles in certain regions, especially in Europe where they can be cultivated in rotation with crops such as corn and soybean. Sweet potatoes and cassava, with 70% and 40% of starch content, respectively, can be cultivated in tropical or warm regions and they provide relatively high ethanol yields. As shown in Table 4, sweet potatoes production occurs predominantly in China, where both crops are exploited as a biofuel feedstock [61].

4.4 Sugar from industrial wastes

It is important to mention that the use of dedicated energy crops for aviation fuels, such as sugar or starch-based feedstocks might have potential competitive tensions with biofuels for road transport and agricultural crops as well as have a negative impact in terms of sustainability [62]. In this scenario, research and development have focused attention on the use of industrial wastes. Food industry wastes such as cheese-whey, high sugar content beverage, and food wastes are potential feedstocks for butanol and ethanol production by biotechnological route.

Cheese-whey is a subproduct from the dairy industries. According to Procentese et al. [63] processing of 1 kg of milk produces 0.2 kg of cheese and 0.8 kg of wastewater, known as cheese-whey. This industrial waste, with high concentrations of lactose, is an option as feedstock for the fermentation processes to produce butanol and ethanol or SIP jet fuel. High sugar content beverages such as fruit juices, syrups, soft drinks, and sport drinks area also sugar sources and feedstock for producing alcohols. They are composed of sucrose and fructose, at concentrations ranging from a few grams per liter to hundreds of grams per liter [63]. The main advantages of cheese-whey and high sugar content beverages as feedstock for ATJ and SIP biojet fuel are their continuous availability over the year at an almost constant rate and a negative cost of supply. Food waste is another resource of carbohydrate-based biomass that may be potentially used as feedstock in biochemical routes. Annual global food waste is estimated to be about 1.3 billion tons, about one third of the total food production for human consumption. The household fraction (42%) is the main contributor to food waste; the fraction of food waste from the food processing (39%) and the catering and restaurant services (14%) are also remarkable. This resource is available at quite a high annual rate and at a negative cost of supply [63].

In addition to the feedstocks based on sugars and starch, and food waste, fermentable sugars can be produced by nonedible biomass resources, such as feedstocks that come from lignocellulosic biomass such as wood, straw, grasses, etc., as presented in the next section.

5. Lignocellulosic biomass

Lignocellulosic biomass is considered the most abundant and bio-renewable widespread form of biomass on earth. Lignocellulosic materials are nonedible biomass particularly attractive as feedstocks for biofuel production due to their relatively low cost, great abundance and availability, and sustainable supply [64–66]. It is estimated that 181.5 billion tonnes of lignocellulose biomass are produced annually. Of the 8.2 billion tonnes that are currently used, about 7 billion tonnes are produced from agricultural, grass, and forests and another 1.2 billion tonnes stem from agricultural residues [67].

This inedible biomass does not compete in the food market, differently than the feedstock for first generation biofuels, which is an important advantage, but requires conversion process with many steps due to its high oxygen content, which impacts in the fuel costs and reduction technology readiness levels [68]. Despite these points, several studies have shown that lignocellulosic biomass has enormous potential for sustainable production of chemicals and fuels due to the great availability, without demand of efforts for its cultivation since, most of them are considered wastes.

Lignocellulosic feedstocks can be divided into a few groups which include agricultural residues (cereal straws, stovers, husks, stalks, and bagasse), forest biomass and residues (softwood and hardwood, sawdust, pruning and bark thinning residues), energy crops (perennial grasses and other dedicated energy crops), industrial, and municipal solid wastes (MSWs) [68–72]. In Fig. 6, the different classes of lignocellulosic biomass are shown.

5.1 Lignocellulosic composition

Typically, lignocellulosic biomass is composed primarily of cellulose, hemicellulose, lignin, and small amounts of structural proteins, extractives, and ash. Cellulose and hemicellulose are sources of C6 and C5 sugars and lignin is a source of phenolic compounds.

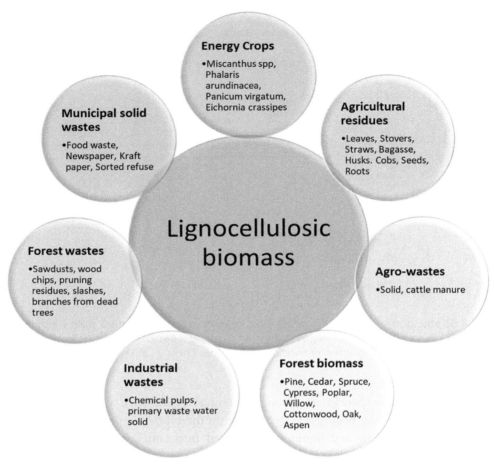

Fig. 6 Sources of lignocellulosic biomass [68–72].

Cellulose is the major component of lignocellulosic biomass and it is composed of glucose units with intra- and intermolecular hydrogen bonding networks that forms robust crystal structures, with high chemical stability. Hemicellulose is also a polysaccharide as cellulose; however, it contains different 5- and 6-carbon monosaccharide units of pentoses and hexoses, which results in amorphous structure. The components of hemicelluloses are varied depending on plants and their part, and it includes pentoses (xylose, arabinose), hexoses (mannose, glucose, galactose), and acetylated sugars. Lignin is a three-dimensional polymer of phenylpropanoid units, and it consists of aromatic *p*-hydroxyphenyl (H), guaiacyl (G), and syringyl (S) units varying in amount depending on biomass type [68–70,73]. Fig. 7 shows the layout and main constituents of lignocellulosic biomass.

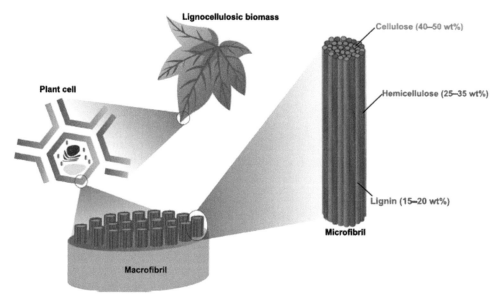

Fig. 7 Scheme of the structure of lignocellulosic biomass indicating the three main constituents: Cellulose, hemicellulose, and lignin. *(Adapted from S. Bertella, J.S. Luterbacher, Lignin functionalization for the production of novel materials, Trends Chem. 2 (2020) 440–453.)*

Lignocellulosic feedstocks are found in almost all plants, such as softwoods, hardwoods, and herbaceous plants, etc. Depending on the type of lignocellulosic biomass, cellulose, hemicellulose and lignin are organized into complex nonuniform three-dimensional structures to different degrees and its composition can vary greatly due to the heterogeneous nature of biomass [68,69,74]. Generally, lignocellulosic biomass consists of 35%–50% cellulose, 20%–35% hemicellulose, and 10%–25% lignin, by dry weight [69–71,73]. Table 5 summarizes types of lignocellulosic biomass and their chemical composition.

5.2 Lignocellulosic agricultural residues

Agricultural residues are by-products from harvesting and processing of agricultural crops. There are two categories of agricultural residues, primary and secondary residues. Primary residues or field-based residues are generated in the field at the time of harvest. Examples of primary residues are: paddy straw, sugarcane top, maize stalks, palm oil frond and bunches, etc. Secondary residues or processing based residues are those co-produced during processing, such as paddy husk, bagasse, maize cob, coconut shell, coconut husk, coir dust, saw dust, palm oil shell, fibers, and empty bunches [70–72].

Table 5 Cellulose, hemicellulose, and lignin content from biomass feedstock.

Biomass feedstocks	Cellulose (%)	Hemicellulose (%)	Lignin (%)	Reference
Woods				
Hardwood	20–25	45–50	20–25	[75]
Hardwood stem	40–55	24–40	18–25	[76]
Hardwood bark	20–40	20–38	30–55	[77]
Softwood	27–30	35–40	25–30	[75]
Softwood stem	45–50	25–35	25–35	[76]
Softwood bark	18–38	15–33	30–60	[77]
Poplar	47–50	27–28	18–19	[78,79]
Spruce	44	6	28	[68]
Eucalyptus	54	18	21	[69]
Pine	44–46	9–26	29–30	[78,79]
Agro-wastes and agricultural residues				
Rice straw	32	24	18	[74]
Corn cobs	42–45	35–39	14–15	[71]
Corn stover	38–40	24–26	7–19	[71]
Nut shells	25–30	25–30	30–40	[74]
Wheat straw	33–38	26–32	17–19	[77]
Barley straw	31–45	27–38	14–19	[77]
Oat straw	31–37	27–38	16–19	[77]
Coffee husk	43	7	9	[71]
Rye straw	33–35	27–30	16–19	[71,77]
Sugarcane bagasse	42–48	19–25	20–42	[71]
Banana waste	13	15	14	[74]
Bamboo	26–43	15–26	21–31	[71]
Solid cattle manure	5–16	1.3–1.4	2.7–5.7	[76]
Sweet sorghum	45	27	21	[74]
Leaf Fiber Sisal (Agave)	43–56	21–24	7–9	[76]
Leaf Fiber Abaca (Manila)	61	17	9	[76]
Energy crops				
Switch grass	45	31	12	[74]
Grasses	25–40	25–50	10–30	[69]
Miscanthus	38–40	18–24	24–25	[80]
Elephant grass	22	24	24	[76]
Sugarcane whole	25	17	12	[81]
Grass Esparto	33–38	27–32	17–19	[76]
Napier grass	32	20	9	[81]
Municipal Solid Wastes (MSW) and Industrial Wastes				
General (MSW)	33–49	9–16	10–14	[71]
Office paper	69	12	11	[82]
Food waste	55	7	11	[82]
Sorted refuse	60	20	20	[74]
Primary wastewater solids	8–15	0	24–29	[76]

Continued

Table 5 Cellulose, hemicellulose, and lignin content from biomass feedstock—cont'd

Biomass feedstocks	Cellulose (%)	Hemicellulose (%)	Lignin (%)	Reference
Newspaper	40–55	25–40	18–30	[71,83]
Waste papers from chemical pulps	6–10	50–70	12–20	[74]
Paper	85–99	0	0–15	[74]
Chemical pulp	60–80	20–30	2–10	[68]
Kraft paper	57	10	21	[82]

Primary residues for energy use have a relative low availability mainly because they have other applications as fertilizer, animal feed, etc. Owing to the slow rate of digestibility, there is a limitation for animal feeding (2% of body weight). As a result, burning or ploughing the straw into the land directly as a fertilizer or using it as fodder for animals are the main practices. On the other hand, secondary residues are usually available in relatively large quantities at the processing site, and they can be used in energy co-generation involving minimal transportation and handling costs [70–72]. Although agricultural residues have this application, the remaining available part can be potentially utilized in biofuel production.

Majority of the global agricultural residues are obtained from four crops, which are sugarcane, corn, rice and wheat. Countries and some blocks which are considered the main granaries in the world, such as the United States, South America, Asia and Europe, supply large quantities of corn straw, sugarcane bagasse, rice, and wheat straw, respectively. According to Tye et al. [72], it is estimated that the total annual production of cereal straw per year is around 1580.2 million tons, which is produced mainly in Europe (barley and oats), followed by the United States (corn and sorghum) and China (rice and wheat) [72,84].

Among the agricultural crops, corn crops have obtained the highest production yield, with an estimated of 376.8 million tons of corn straw or stover available globally per year, under the 60% ground-cover practice. It is estimated a production rate of about 1.0 kg of stalks, leaves, cobs, and husk per kg corn grain after harvesting [71,72,76,84].

Although rice production is relatively low compared to corn and sugarcane production, all the annual rice straw produced globally can be fully utilized since rice straw is not left in the field to prevent erosion (no ground-cover practice). Rice straw is one of the most abundant and promising biomasses in the world and includes stems, leaf blades, and leaf sheaths which are left over after the harvest of kernels, with a global production of 657.1 million tons per year [71,72,77]. According to Tye et al. [72], rice straw contributes to the highest total annual availability of cellulose (210.4–309 million tons) from agricultural residues, which allows a production of 657.5 million tonnes of bioethanol [72].

Wheat straw also has shown a substantial annual availability, with an estimated availability of 472.2 million tons per year.

Bagasse is an important agro-industrial residue obtained from the processing of sugarcane. It is estimated that the yield of sugarcane bagasse is about 0.6 kg for 1 kg of sugarcane with an estimated global yield of 1044.8 million tons per year [72,76].

5.3 Energy crops

The production of native plants (grasses) is around 739.3 million tons in total, although the cellulose availability remarks, in average 22%–40% that corresponds to 162–295 million tons [72]. Currently, crops dedicated to energy stand out, as biomass is cultivated and harvested in order to obtain greater energy. Native plants or energy crops, as they are known, are promising sources of lignocellulosic biomass, because they have low demand for fertilizers and pesticides and less agricultural land [71,82]. In addition, the energy crops located in warm and temperate regions are among the world's high-yielding biomasses, the production is obtained more than twice with high cellulose content when compared as other types of crops [71,72]. Species such as miscanthus (*Miscanthus* spp.), Switchgrass (*Panicum virgatum*), coastal shorts, canary (*Phalaris arundinacea*), giant cane (*Arundo donax*), and alfalfa (*Medicago sativa*) are the most common energy crops for energy use [72].

Switchgrass is a perennial, warm season native grass that is grown in America and Africa. It is also grown in Canada and the United States, although its growth is strongly related to the climate. Around 282.9 million tons of switchgrass is produced annually [72,85]. Among the species, Miscanthus is a biomass suitable for the production of biofuel in Asia and Europe. An annual global production rate of about 14.8–37.1 dry tons of miscanthus per hectare is estimated [72]. Considering an average production rate of 25.9 dry tons per hectare, approximately 256.4 million tons of miscanthus (yield 40% of cellulose) can be produced with approximately 9.9 million hectares of planted land according to Tye et al. [72].

Another perennial herbaceous variety is Coastal Bermudagrass, a crop widely grown in the United States during the hot season. The estimated cultivation of this herbaceous in the world corresponds to an area of 10.1 million hectares. Bermudagrass has an annual production of 14.8–24.7 dry tons per hectare. It is estimated that 200 million tons of biomass are produced annually in the world, from which only 25% of cellulose is obtained (50 million tons) [72].

5.4 Forest biomass

Forest biomass and residues are based on wood lignocellulosic materials which commonly are physically larger, denser, and structurally stronger than nonwood lignocellulosic biomass such as agricultural residues, native plants, and nonwood plant fibers such as sugarcane bagasse, switchgrass, etc. Chemically, it contains a larger amount of lignin than nonwood lignocellulosic biomasses. Forest biomass includes mainly feedstocks such as hardwoods and softwoods, while forest wastes are the sawdust, pruning, bark thinning residues, wood chips, etc. Hardwoods are predominantly found in the Northern hemisphere and comprise trees such as poplar, willow, oak, cottonwood, and aspen. Softwoods includes evergreen species such as pine, cedar, spruce, cypress, fir, hemlock, and redwood, possessing lower densities and faster growth [71,72].

5.5 Lignocellulosic industrial wastes

A variety of industrial wastes could be applied as feedstock for energy production. Currently, most the existing lignocellulosic industrial wastes is generated in pulp and paper industries, wood processing industries and agro-industry as by-product streams. From wood processing industries is generated wood wastes, including sawdust, off-cuts, and wood chips. From the papermaking industry, paper sludge is a solid waste stream rich in cellulose fibers. Furthermore, pulp and paper industry generate industry waste called black liquor, containing mainly lignin in its composition [86–88].

Lignin is a potential feedstock to produce aromatics and cyclic compounds to formulate SAFs. Approximately 50 million tons of lignin are produced annually worldwide, of which 98%–99% is incinerated to produce steam and energy, and 1%–2%, derived from the sulfite pulp industry, is used in chemical conversion to produce lignosulfonates [88]. Lignin can be used to produce aviation biofuels instead be used for steam generation, due to its low cost, high availability in concentrated points. However, the conversion of lignin into liquid hydrocarbons demands the development of robust technologies.

5.6 Municipal solid wastes

MSW can be considered a complex matrix since it consists of several types of wastes, including organic matter from food waste, paper waste, packaging, plastics, bottles, metals, textiles, yard waste, and other miscellaneous items. Commonly, MSW needs to be in conditions for further treatment, which depends heavily on how the classification and disposal of this type of waste occurs. Data show that, total global production of MSW was estimated in 1.3×10^9 tons in 1990, which almost doubled after 10 years, generating 2.3×10^9 tons [71]. Although MSW are not considered ideal inputs for liquid biofuels production currently, due to their wide diversification in terms of components, and contamination, certain fractions of MSW could become potential feedstocks for SAFs production.

5.7 Lignocellulosic conversion processing to produce drop-in biofuels

Lignocellulosic feedstocks can be used to produce SAFs using two distinctive ways—thermochemical and biochemical route. Biochemical routes (ATJ and SIP pathways) were already mentioned in this chapter, referring to production of SAFs from sugars and starch sources. In this context, lignocellulosic feedstocks can be used to produce second-generation ethanol (cellulosic ethanol) or sugars which are precursors for SAFs production. The cellulosic ethanol can be produced from a variety of lignocellulosic residues since the process comprises the cellulose separation, and subsequently, it is converted into fermentable sugars to produce ethanol. To this application, nonwood lignocellulosic feedstocks such as agricultural residues, energy crops, and natural nonwood plant fibers are receiving the most attention, owing to their widespread availability, ease for processing, and requirement of relatively low process energies since they contain more open chemical structures and lower content of lignin compared to wood lignocellulosic materials such as forest biomass and residues. Based on the chemical composition of nonwood fibers, its low lignin content is a substantial advantage for the delignification process [72,89–91].

Thermochemical routes are related to biomass-to-liquid processes (BTL), which comprises technologies such as gasification with Fischer-Tropsch (FT) synthesis, hydrothermal liquefaction, pyrolysis followed by upgrading and etc. In the former pathway, lignocellulosic biomass is converted into syngas (CO and H_2) via gasification and, subsequently, aviation fuels can be produced through catalytic upgrading FT syngas into hydrocarbons [62,92,93]. Furthermore, it is also possible to produce biojet fuel from syngas fermentation. In this case, syngas can be fermented to alcohols by *acetogenic* bacteria and posteriorly can be upgraded into jet fuel via the ATJ technology [94].

Fast pyrolysis and liquefaction are also biomass thermochemical processes to convert solid biomass to liquid fuels [64,65,94]. The liquid products are known as bio-oil or pyrolysis oils and bio-crude, respectively. Its composition corresponds to a complex mixture of lower molecular weight oxygenates, including sugar and lignin oligomers, along with monomeric compounds such as acids, alcohols, aldehydes, ketones, carbohydrates, and phenolics compounds. The large amounts of these reactive oxygenated components have a negative effect on the product properties and limit its direct application as transportation fuel. Therefore, catalytic hydrotreatment, called as bio-oil upgrading, is required to convert bio-oils to stabilized oil products and it involves, typically, a combination of reactions such as hydrodeoxygenation, (hydro)cracking, isomerization, hydrogenation and etc., and requires high temperatures (250–450°C) and pressures (140–200 bar). Another approach is the catalytic fast pyrolysis which provides an attractive technology to obtain fuels from biomass or waste feedstocks, without hydrogen requirements [64–66].

6. Conclusion

This chapter showed many different options of feedstocks spreader worldwide suitable for sustainable aviation fuels production. It is clear that there is no a unique option, but excellent options for different places.

The motivation for energy transition to the detriment of the appropriate use of natural resources enhances scientific studies and efforts in general to achieve a circular economy in the area of sustainable fuels. The arguments show that the use of biomass is a path of no return, capable of offering the most diverse types of raw materials, which leads to different types of homologated industrial routes for a more efficient and safe transformation of the desired products. The freedom to choose fatty acids, sugars, or lignocellulosic materials will depend on the supply available at that location and strictly depends on access to the technology available for use. It is observed that several factors, direct or indirect, contribute to the consolidation of the productive chain of sustainable aviation fuels. It can occur from governmental actions, from the genetic improvement of plants aiming at increasing the productivity of a certain characteristic of the plant per hectare, or even from the maturation of the industrial processes of transformation of biomass to produce SAFs.

References

[1] RFA—Renewable Fuels Association, Markets & Statistics Annual Ethanol Production, U.S. & World Ethanol Production, 2022. https://ethanolrfa.org/statistics/annual-ethanol-production/. (Accessed 2 July 2020).

[2] F. Toldrá-Reig, L. Mora, F. Toldrá, Trends in biodiesel production from animal fat waste, Appl. Sci. 10 (2020) 3644.

[3] ASTM D7566, Standard Specification for Aviation Turbine Fuel Containing Synthesized Hydrocarbons, 2018, Available at: https://www.astm.org/Standards/D7566.htm.

[4] D. Singh, D. Sharma, S.L. Soni, S. Sharma, P.K. Sharma, A. Jhalani, A review on feedstocks, production processes, and yield for different generations of biodiesel, Fuel 262 (2019) 116553.

[5] A.J. Tsamba, W. Yang, W. Blasiak, Pyrolysis characteristics and global kinetics of coconut and cashew nut shells, Fuel Process. Technol. 87 (2006) 523–530.

[6] P. Das, T. Sreelatha, A. Ganesh, Bio oil from pyrolysis of cashew nut shell-characterisation and related properties, Biomass Bioenergy 27 (2004) 265–275.

[7] F.D. Gunstone, F.A. Norris, Lipids in Foods: Chemistry, Biochemistry and Technology, Pergamon Press, Oxford, 1983.

[8] T. Mielke, World markets for vegetable oils: status and prospects, in: M. Kaltschmitt (Ed.), Energy From Organic Materials (Biomass), Encyclopedia of Sustainability Science and Technology Series, Springer, New York, 2019, pp. 261–298.

[9] FAO, Food Outlook—Biannual Report on Global Food Markets: June 2020. Food Outlook, 1, Rome. Available online at:, 2020, https://doi.org/10.4060/ca9509en.

[10] J.C.J. Bart, N. Palmeri, S. Cavallaro, Biodiesel Science and Technology from Soil to Oil, first ed., Woodhead Publishing Series in Energy, vol. 7, CRC Press, Boca Raton, Boston, New York, Washington, 2010.

[11] J. Walton, The fuel possibilities of vegetable oils, Gas Oil Power 33 (1938) 167.

[12] G.K. Souza, F.B. Scheufele, T.L.B. Pasa, P.A. Arroyo, N.C. Pereira, Synthesis of ethyl esters from crude macauba oil (*Acrocomia aculeata*) for biodiesel production, Fuel 165 (2016) 360–366.

[13] W.L.G. Silva, A.A. Salomão, P.T. Souza, M. Ansolin, M. Tubino, Binary blends of biodiesel from macauba (*Acromia aculeata*) kernel oil with other biodiesels, J. Braz. Chem. Soc. 29 (2018) 240–247.

[14] Malaysian Palm Oil Board Technical Advisory Services, Available online at: http://www.palmoilworld.org/about_palmoil.html. (Accessed 15 December 2020).

[15] M. Mahfud, A. Suryanto, L. Qadariyah, S. Suprapto, H.S. Kusuma, Production of methyl ester from coconut oil using microwave: kinetic of transesterification reaction using heterogeneous CaO catalyst, Korean Chem. Eng. Res. 56 (2018) 275–280.

[16] E. Melo, F. Michels, D. Arakaki, N. Lima, D. Gonçalves, L. Cavalheiro, V. Nascimento, First study on the oxidative stability and elemental analysis of babassu (*Attalea speciosa*) edible oil produced in Brazil using a domestic extraction machine, Molecules 24 (2019) 4235.

[17] M.J.F. da Silva, A.M. Rodrigues, I.R.S. Vieira, G.A. Neves, R.R. Menezes, E.G.R. Gonçalves, M.C.C. Pires, Development and characterization of a babassu nut oil-based moisturizing cosmetic emulsion with a high sun protection factor, RSC Adv. 10 (2020) 26268.

[18] E. Pereira, M.C. Ferreira, K.A. Sampaio, R. Grimaldi, A.J.A. Meirelles, G.J. Maximo, Physical properties of Amazon fats and oils and their blends, Food Chem. 278 (2019) 208–215.

[19] R.C.V. Santos, M.R. Sagrillo, E.E. Ribeiro, I.B.M. Cruz, The Tucumã of Amazonas—(*Astrocaryum aculeatum*), in: Exotic Fruits Reference Guide, Academic Press, Elsevier, 2018, pp. 419–425.

[20] I.M.A.S. Cardoso, A.M. Souza, J.E.S. Pereira, The palm tree *Syagrus oleracea* Mart. (Becc.): a review, Sci. Hortic. 225 (2017) 65–73.

[21] P.H.M. Araújo, A.S. Maia, A.M.T.M. Cordeiro, A.D. Gondim, N.A. Santos, Catalytic deoxygenation of the oil and biodiesel of licuri (*Syagrus coronata*) to obtain n-alkanes with chains in the range of biojet fuels, ACS Omega 4 (2019) 15849–15855.

[22] S. Bai, S. Engelen, P. Denolf, J.G. Wallis, K. Lynch, J.D. Bengtsson, M. Van Thournout, B. Haesendonckx, J. Browse, Identification, characterization and field testing of *Brassica napus* mutants producing high-oleic oils, Plant J. 98 (2019) 33–41.

[23] A. Raza, Eco-physiological and biochemical responses of rapeseed (*Brassica napus* L.) to abiotic stresses: consequences and mitigation strategies, J. Plant. Growth Regul. 40 (2020) 1368–1388.

[24] N. Postaue, C.P. Trentini, B.T.F. Mello, L.C. Filho, C. Silva, Continuous catalyst-free interesterification of crambe oil using methyl acetate under pressurized conditions, Energy Convers. Manag. 187 (2019) 398–406.

[25] B. Onorevoli, M.E. Machado, C. Dariva, E. Franceschi, L.C. Krause, R.A. Jacques, E.B. Caramão, A one-dimensional and comprehensive two-dimensional gas chromatography study of the oil and the bio-oil of the residual cakes from the seeds of *Crambe abyssinica*, Ind. Crop Prod. 52 (2014) 8–16.

[26] P.K. Sahoo, L.M. Das, Process optimization for biodiesel production from Jatropha, Karanja and Polanga oils, Fuel 88 (2009) 1588–1594.

[27] Q. Ali, M. Ashraf, Exogenously applied glycinebetaine enhances seed and seed oil quality of maize (*Zea mays* L.) under water deficit conditions, Environ. Exp. Bot. 71 (2011) 249–259.

[28] FAS, Grain: World Markets and Trade, Available online at: https://apps.fas.usda.gov/psdonline/circulars/oilseeds.pdf.

[29] M.A. Sukiran, F. Abnisa, W.M.A.W. Daud, N.A. Bakar, S.K. Loh, A review of torrefaction of oil palm sold wastes for biofuel production, Energy Convers. Manag. 149 (2017) 101–120.

[30] S.H. Chang, An overview of empty fruit bunch from oil palm as feedstock for bio-oil production, Biomass Bioenergy 62 (2014) 174–181.

[31] A. Henderson, G. Galeano, R. Bernal, Field Guide to the Palms of the Americas, Princeton University Press, New Jersey, 1995.

[32] T.P. Pires, E.S. Souza, K.N. Kuki, S.Y. Motoike, Ecophysiological traits of the macaw palm: a contribution towards the domestication of a novel oil crop, Ind. Crop Prod. 44 (2013) 200–210.

[33] L.S. Noemi, C.C. Cardoso, V.M.D. Pasa, Synthesis and characterization of esters from different alcohols using Macaúba almond oil to substitute diesel oil and jet fuel, Fuel 166 (2016) 453–460.

[34] PGF Biofuels, Available online at: http://www.pgfbiofuels.com/carinata/.

[35] F. Li, S.C. Srivatsa, S. Bhattacharya, A review on catalytic pyrolysis of microalgae to high-quality bio-oil with low oxygeneous and nitrogenous compounds, Renew. Sustain. Energy Rev. 108 (2019) 481–497.

[36] M. Packer, Algal capture of carbon dioxide; biomass generation as a tool for greenhouse gas mitigation with reference to New Zealand energy strategy and policy, Energy Policy 37 (2009) 3428–3437.

[37] C. Realini, S. Duckett, G. Brito, M.D. Rizza, D. Mattos, Effect of pasture vs. concentrate feeding with or without antioxidants on carcass characteristics, fatty acid composition, and quality of Uruguayan beef, Meat Sci. 66 (2004) 567–577.

[38] F. Toldrá, M.A. Rubio, J.L. Navarro, L. Cabrerizo, Quality aspects of pork meat and its nutritional impact, in: F. Shahidi, A.M. Spanier, C.T. Ho, T. Braggins (Eds.), Quality of Fresh and Processed Foods, Advances in Experimental Medicine and Biology, vol. 542, Springer, Boston, MA, 2004.

[39] H.W. Ockerman, L. Basu, By-products—inedible, in: Encyclopedia of Meat Sciences, Academic Press, Elsevier, 2014, pp. 125–136.

[40] Z. Zdunczyk, R. Gruzauskas, A. Semaskaite, J. Juskiewicz, A.R. Stupeliene, M. Wroblewska, Fatty acid profile of breast muscle of broiler chickens fed diets with different levels of selenium and vitamin, Eur. Poult. Sci. 75 (2011) 264–267.

[41] T. Castro, T. Manso, A.R. Mantecon, J. Guirao, V. Jimeno, Fatty acid composition and carcass characteristics of growing lambs fed diets containing palm oil supplements, Meat Sci. 69 (2005) 757–764.

[42] C. Jacobsen, Fish oils: composition and health effects, in: Encyclopedia of Food and Health, Academic Press, Elsevier, 2016, pp. 686–692.

[43] A.B. Chhetri, K.C. Watts, M.R. Islam, Waste cooking oil as an alternate feedstock for biodiesel production, Energies 1 (2008) 3–18.

[44] M. Ramos, A.P.S. Dias, J.F. Puna, J. Gomes, J.C. Bordado, Biodiesel production processes and sustainable raw materials, Energies 12 (2019) 4408.

[45] Empresa Brasileira de Pesquisa Agropecuária, Ministério da Agricultura, Pecuária e Abastecimento, Brazil, Available online at: https://www.embrapa.br/busca-de-noticias/-/noticia/47881589/sebo-bovino-e-segunda-materia-prima-na-producao-de-biodiesel.

[46] J. Yang, Z. Xin, Q.S. He, K. Corscadden, H. Niu, An overview on performance characteristics of bio-jet fuels, Fuel 237 (2019) 916–936.

[47] W. Wang, L. Tao, Bio-jet fuel conversion technologies, Renew. Sustain. Energy Rev. 53 (2016) 801–822.

[48] L.C. Basso, T.O. Basso, S.N. Rocha, Ethanol production in Brazil: the industrial process and its impact on yeast fermentation, in: M.A.D.S. Bernardes (Ed.), Biofuel Production—Recent Developments and Prospects, Intech, 2011, pp. 85–100.

[49] A. Kujawska, J. Kujawski, M. Bryjak, W. Kujaws, ABE fermentation products recovery methods—a review, Renew. Sustain. Energy Rev. 48 (2015) 648–661.

[50] P. Halder, K. Azad, S. Shah, E. Sarker, Prospects and technological advancement of cellulosic bioethanol ecofuel production, in: K. Azad (Ed.), Advances in Eco-Fuels for a Sustainable Environment, Woodhead Publishing Series in Energy, Woodhead Publishing, 2019, pp. 211–236.

[51] M.G. Freire, C.L.S. Louro, L.P.N. Rebelo, J.A.P. Coutinho, Aqueous biphasic systems composed of a water-stable ionic liquid carbohydrate and their applications, Green Chem. 13 (2011) 1536–1545.

[52] R. Höfer, Sugar- and starch-based biorefineries, in: A. Pandey, R. Höfer, M. Taherzadeh, M. Nampoothiri, C. Larroche (Eds.), Industrial Biorefineries and White Biotechnology, Elsevier, 2015, pp. 158–235.

[53] E.M. Yahia, A.C. López, L.A.B. Perez, Carbohydrates, in: E.M. Yahia (Ed.), Postharvest Physiology and Biochemistry of Fruits and Vegetables, Woodhead Publishing, 2019, pp. 175–205.

[54] R. Duraisam, K. Salelgn, A.K. Berekete, Production of beet sugar and bio-ethanol from sugar beet and it bagasse: a review, Int. J. Eng. Trends Technol. 43 (2017) 222–233.

[55] E.A. Acevedo, P.C.F. Silva, L.A.B. Perez, Cereal starch production for food applications, in: M.T.P.S. Clerici, M. Schmiele (Eds.), Starches for Food Application—Chemical, Technological and Health Properties, Academic Press, 2019, pp. 71–102.

[56] V. Wolf, M. Haß, World markets for sugar and starch: status and prospects, in: R.A. Meyers (Ed.), Encyclopedia of Sustainability Science and Technology, Springer, New York, 2017, pp. 1–37.

[57] FAO, Production Quantity of Sugar, Cereal and Tuber Crops that can be used as Biofuel Feedstock, According to the Data, Food and Agriculture Organization of the United Nations (FAO), FAOSTAT, 2022. Available online at: https://www.fao.org/3/cb5332en/cb5332en.pdf.

[58] J.A. da Silva, The importance of the wild cane *Saccharum spontaneum* for bioenergy genetic breeding, Sugar Tech 19 (2017) 229–240.

[59] M.C.B. Grassi, G.A.G. Pereira, Energy-cane and RenovaBio: Brazilian vectors to boost the development of biofuels, Ind. Crop Prod. 129 (2019) 201–205.

[60] Bio-Butanol Market-Growth, Trends, Available online at: https://www.mordorintelligence.com/industry-reports/bio-butanol-market.

[61] ETIP Bioenergy, European Technology and Innovation Platform, ETIP Bioenergy, 2022. Available online at: https://www.etipbioenergy.eu/value-chains/feedstocks/agriculture/starch-crops.

[62] S.S. Doliente, A. Narayan, J.F.D. Tapia, N.J. Samsatli, Y. Zhao, S. Samsatli, Bio-aviation Fuel: a comprehensive review and analysis of the supply chain components, Front. Energy Res. 8 (2020) 1–110.

[63] A. Procentese, F. Raganati, G. Olivieri, M.E. Russo, M. De-la-Feld, A. Marzocchella, Renewable feedstocks for biobutanol production by fermentation, New Biotechnol. 39 (2017) 135–140.

[64] W. Yin, M.V.A. Bykova, R.H. Venderbosch, V.A. Yakovlev, H.J. Heeres, Catalytic hydrotreatment of the pyrolytic sugar and pyrolytic lignin fractions of fast pyrolysis liquids using nickel based catalysts, Energies 13 (2020) 285–310.

[65] M. Sharifzadeh, M. Sadeqzadeh, M. Guo, T.N. Borhani, N.V.S.N.M. Konda, M.C. Garcia, L. Wang, J. Hallett, N. Shah, The multi-scale challenges of biomass fast pyrolysis and bio-oil upgrading: review of the state of art and future research directions, Prog. Energy Combust. Sci. 71 (2019) 1–80.

[66] A.R. Ardiyanti, M.V. Bykova, S.A. Khromova, W. Yin, R.H. Venderbosch, V.A. Yakovlev, H.J. Heeres, Ni-based catalysts for the hydrotreatment of fast pyrolysis oil, Energy Fuel 30 (2016) 1544–1554.

[67] N. Dahmen, I. Lewandowski, S. Zibek, A. Weidtmann, Integrated lignocellulosic value chains in a growing bioeconomy: status quo and perspectives, GCB Bioenergy 11 (2019) 107–117.

[68] M. Raud, T. Kikas, O. Sippula, N.J. Shurpali, Potentials and challenges in lignocellulosic biofuel production technology, Renew. Sustain. Energy Rev. 111 (2019) 44–56.

[69] F.H. Isikgora, C.R. Becer, Lignocellulosic biomass: a sustainable platform for the production of bio-based chemicals and polymers, Polym. Chem. 6 (2015) 4497–4559.

[70] B. Volynets, F. Ein-Mozaffari, Y. Dahman, Biomass processing into ethanol: pretreatment, enzymatic hydrolysis, fermentation, rheology, and mixing, Green Process. Synth. 6 (2017) 1–22.

[71] H. Zabed, J.N. Sahu, A.N. Boyce, G. Faruq, Fuel ethanol production from lignocellulosic biomass: an overview on feedstocks and technological approaches, Renew. Sustain. Energy Rev. 66 (2016) 751–774.

[72] Y.Y. Tye, K.T. Lee, W.N.W. Abdullah, C.P. Leh, The world availability of non-wood lignocellulosic biomass for the production of cellulosic ethanol and potential pretreatments for the enhancement of enzymatic saccharification, Renew. Sustain. Energy Rev. 60 (2016) 155–172.

[73] S. Bertella, J.S. Luterbacher, Lignin functionalization for the production of novel materials, Trends Chem. 2 (2020) 440–453.

[74] A.K. Kumar, S. Sharma, Recent updates on different methods of pretreatment of lignocellulosic feedstocks: a review, Bioresour. Bioprocess. 4 (2017) 1–19.

[75] P. McKendry, Energy production from biomass (part 1): overview of biomass, Bioresour. Technol. 83 (2002) 37–46.

[76] C. Sánchez, Lignocellulosic residues: biodegradation and bioconversion by fungi, Biotechnol. Adv. 27 (2009) 185–194.

[77] J.K. Saini, R. Saini, L. Tewari, Lignocellulosic agriculture wastes as biomass feedstocks for second-generation bioethanol production: concepts and recent developments, 3 Biotech 5 (2015) 337–353.

[78] N. Mosier, C. Wyman, B. Dale, R. Elander, Y.Y. Lee, M. Holtzapple, M. Ladisch, Features of promising technologies for pretreatment of lignocellulosic biomass, Bioresour. Technol. 96 (2005) 673–686.

[79] L. Olsson, B. Hahn-Hägerdal, Fermentation of lignocellulosic hydrolysates for ethanol production, Enzyme Microb. Technol. 18 (1996) 312–333.

[80] N. Brosse, R.E. Hage, P. Sannigrahi, A. Ragauskas, Dilute sulphuric acid and ethanol organosolv pretreatment of Miscanthus × Giganteus, Cellul. Chem. Technol. 44 (2010) 71–78.

[81] R. Saxena, D. Adhikari, H. Goyal, Biomass-based energy fuel through biochemical routes: a review, Renew. Sustain. Energy Rev. 13 (2009) 167–178.

[82] E. Schmitt, R. Bura, R. Gustafson, J. Cooper, A. Vajzovic, Converting lignocellulosic solid waste into ethanol for the State of Washington: an investigation of treatment technologies and environmental impacts, Bioresour. Technol. 104 (2012) 400–409.

[83] R. Howard, E. Abotsi, E.J. Van Rensburg, S. Howard, Lignocellulose biotechnology: issues of bioconversion and enzyme production, Afr. J. Biotechnol. 2 (2004) 602–619.

[84] S. Kim, B.E. Dale, Global potential production from waste crops and crop residues, Biomass Bioenergy 26 (2004) 361–375.

[85] G.T. Timilsina, J.J. Cheng, Advanced biofuel technologies: status and barriers, in: Policy Research Working Paper, 2013. Available at: worldbank.org.

[86] J. Jayamuthunagai, I.A. Selvakumari, S. Varjani, P. Mullai, B. Bharathiraja, Valorization of industrial wastes for biofuel production: challenges and opportunities, in: A. Pandey, R.D. Tyagi, S. Varjani

(Eds.), Biomass, Biofuels, Biochemicals: Circular Bioeconomy—Current Status and Future Outlook, Elsevier, 2021, pp. 231–245.

[87] N. Belyakov, Bioenergy, in: N. Belyakov (Ed.), Sustainable Power Generation—Current Status, Future Challenges, and Perspectives, Academic Press, 2019, pp. 461–474.

[88] I.F. Demuner, J.L. Colodete, A.J. Dumener, C.M. Jardim, Biorefinery review: wide-reaching products through Kraft lignin, BioResources 14 (2019) 7543–7581.

[89] B.O. Ogunsile, G.C. Quintana, Modeling of soda-ethanol pulps from *Carpolobia lutea*, BioResources 5 (2010) 2417–2430.

[90] M. Chandra, Use of Nonwood Plant Fibers for Pulp and Paper Industry in Asia: Potential in China, Wood Science and Forest Products, Blacksburg, VA, 1998, p. 12.

[91] E.K. Pye, M. Rushton, A. Berlin, A biorefinery process suitable for non-wood fibers, in: Proceedings of the Tappi Engineering, Pulping and Environmental Conference, 2007.

[92] C. Gutiérrez-Antonio, F.I. Gómez-Castro, J.A. de Lira-Flores, S. Hernández, A review on the production processes of renewable jet fuel, Renew. Sustain. Energy Rev. 79 (2017) 709–729.

[93] H. Wang, B. Yang, Q. Zhang, W. Zhu, Catalytic routes for the conversion of lignocellulosic biomass to aviation fuel range hydrocarbons, Renew. Sustain. Energy Rev. 120 (2020) 109612–109640.

[94] W.C. Wang, L. Tao, J. Markham, Y. Zhang, E. Tan, L. Batan, E. Warner, M. Biddy, Technical Report: Review of Biojet Fuel Conversion Technologies, National Renewable Energy Laboratory, 2016. Review of Biojet Fuel Conversion Technologies (nrel.gov).

CHAPTER 5

Advanced biorefineries for the production of renewable aviation fuel

Araceli Guadalupe Romero-Izquierdo[a], Claudia Gutiérrez-Antonio[a], Fernando Israel Gómez-Castro[b], and Salvador Hernández[b]

[a]Engineering School, Autonomous University of Querétaro, El Marqués, Querétaro, Mexico
[b]Chemical Engineering Department, Natural and Exact Sciences Division, Campus Guanajuato, University of Guanajuato, Guanajuato, Guanajuato, Mexico

1. Introduction

Until 2019, the aviation sector had outstanding growth. According to the International Air Transport Association (IATA), during the period 2014–19, the flights increased by 22.12%, reaching a value of 40.30 million USD [1]; as a consequence, the income of the sector reached a value of 872 billion USD, a value that represents an increment of 13.68% with respect to 2014 [1]. Moreover, during the same period, the fuel demand augmented by 27%, as well as carbon dioxide emissions; the latter reached a value of 936 million tons in 2019 [1]. Considering this growth rate, it was forecasted that carbon dioxide emissions could contribute to 10% of the total value of anthropogenic emissions [2] if no action is taken to counteract this increase. Therefore, a four-pillar strategy was proposed, which will contribute to the objectives proposed by the aviation sector, which includes a 50% reduction in CO_2 emissions by 2050, relative to 2005 emissions levels, along with a neutral growth in CO_2 emissions from 2020 [3,4]. The four-pillar strategy considers technological advances in engines and aircraft structures, operational improvements through online optimization of flight paths, market-based measurements, and development of renewable aviation fuels; from these strategies, the development of renewable aviation fuels has been identified as the most appropriate to substantially reach significant reductions in the CO_2 emissions at the short and medium term [5]. During the implementation of this strategy, the virus severe acute respiratory syndrome coronavirus 2 (SARS-CoV-2) spread very quickly, causing serious effects in all economic sectors. In the aviation sector, the effect has been significant, and potential loss is estimated at 389 billion USD during 2020 [6]. Despite this bleak scenery, a recent report indicates that the development of alternative energy systems will allow having a sustainable economic recovery [7]. Thus, in this context, the development of renewable aviation fuel remains as one strategy that could contribute to the sustainable recovery of the sector.

Sustainable Alternatives for Aviation Fuels
https://doi.org/10.1016/B978-0-323-85715-4.00008-2

Copyright © 2022 Elsevier Inc.
All rights reserved.

Renewable aviation fuel consists of hydrocarbons, mainly paraffinic and naphthenic compounds; this biofuel can contain aromatic compounds, depending on the raw material and production pathway [8]. Considering that the standard ASTM D7566 established a minimum content of aromatic compounds in the fuel (8.4 vol%), the renewable aviation fuel can be used in mixtures with fossil jet fuel until 50% in volume [9]; the amount of aromatics compounds generated in the production process is determinant in the maximum volume of biofuel that can be used in the mixtures with fossil jet fuel. The aromatic compounds are required in the fuel tank, mainly to avoid leakages since the aromatic compounds interact with the elastomers in the seals of the fuel tank [8].

Renewable aviation fuel can be produced from almost any type of biomass through several processing pathways. In general, the production of renewable aviation fuel is feasible from the technical point of view; the yields and operating costs vary between the different production processes and the composition of the biomass used as raw material. However, the production of renewable aviation fuel is not feasible from an economic point of view. Some studies have evaluated the production cost of renewable aviation fuel, finding that in all the cases the minimum selling price is higher than the price of fossil jet fuel [10–12]. For instance, Wang [10] reported a minimum renewable jet fuel selling price of $5.42 USD/gal, in comparison with 1.43 USD/gal of fossil jet fuel [13]. Thus, the efforts have been focused on the reduction of the production cost, and in consequence the price of aviation biofuel. In this context, the application of energy integration strategies has been reported [14–16]; as result, it can be concluded that the energy integration of the process allows reducing the operating costs, which is almost compensated by the increase in the capital costs; however, there is a significant reduction (up to 86%) of the carbon dioxide emissions. Other works have focused on the application of process intensification strategies for the production of aviation biofuel [17–21]; results indicated that it is possibly decreasing the operating costs and also the carbon dioxide emissions. Despite these efforts, the price of biojet fuel is still greater than the one of its fossil counterparts. Another alternative that has been less explored, which can be suitable for the production of renewable aviation fuel, is the processing of biomass in biorefinery schemes.

A biorefinery is defined as a conversion platform that allows the total conversion of a given biomass into bioproducts and bioenergy. Thus, the portfolio of bioproducts of high value-added (and low volume) and low value-added (and high volume) makes profitable the conversion of biomass. Thus, the production of renewable aviation fuel through a biorefinery scheme can help to get a competitive price for this biofuel, since it is obtained along with other bioproducts, which improve the profitability of the biomass conversion.

Therefore, in this chapter the revision of the literature related to the production of renewable aviation fuel under biorefinery schemes is presented; moreover, the future trends in this topic are discussed. The content of the chapter is as follows: Section 2 includes information on the production processes for renewable aviation fuel. Later, the revision of the literature reported on the production of renewable aviation fuel on a biorefinery scheme is presented in Section 3, whereas the future trends related to the production of aviation biofuel through biorefineries are discussed in Section 4. Finally, conclusions are contained in Section 5.

2. Production processes of renewable aviation fuel

Renewable aviation fuel can be produced from almost any type of biomass through several conversion pathways. The biomass can be classified according to the so-called generations or to its chemical nature; this last classification allows to group the conversion processes regardless of the origin of the biomass (edible, nonedible, or residual). Thus, the biomass can be classified as triglyceride, lignocellulosic, sugar, and starchy. Each one of these feedstocks can be processed to produce renewable aviation fuel through several processing pathways, as can be seen in Fig. 1.

It is important to mention that, whatever the chemical nature of the feedstock is, it is necessary to realize pretreatments. These pretreatments can be thermochemical, biochemical, chemical, mechanical, or biological. The objective of all pretreatments is to modify the original state of the biomass through the application of chemical, biochemical, and thermochemical reactions as well as physical operations to have the biomass in the proper condition for its conversion into biojet fuel. In general, traditional pretreatments are observed as inefficient to meet the demand for an industrial adaptation, their energy requirements are high, and usually, the used agents are not environment friendly [22,23]. Due to this, new pretreatment methods such as microwave, ultrasound, deep eutectic solvent, irradiation, and high force assisted [22] have been reported in the literature as strategies to overcome these limitations. To better understand the performance of

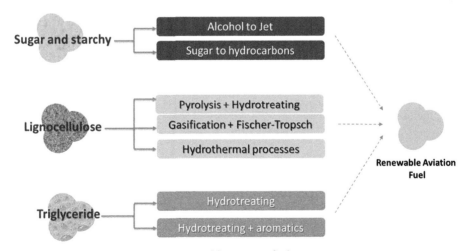

Fig. 1 Conversion routes to produce renewable aviation fuel.

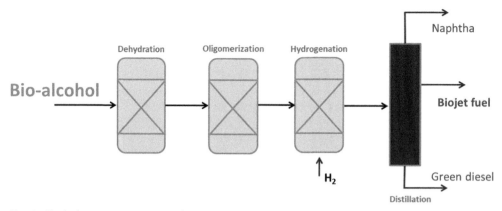

Fig. 2 Alcohol to jet conversion pathway.

pretreatment technologies and develop enhancing strategies, the use of computational methods must be exploited [24].

For sugar and starchy feedstocks, one conversion route certified by ASTM is the alcohol to jet (ATJ) process; this technology was certified in 2016 [25]. In this pathway, a bio-alcohol is dehydrated, oligomerized, and later hydrogenated to obtain renewable hydrocarbons. Fig. 2 shows the block diagram of the ATJ process, which raw material is bio-alcohol that can be produced from sugar and starchy feedstock. According to the Commercial Aviation Alternative Fuels Initiative, the renewable aviation fuel produced by this pathway can be used in mixtures with fossil jet fuel until 50% in volume [26].

In the literature, there are some reports related to the catalytic dehydration of ethanol using lanthanum–phosphorous modified HZSM-5 (0.5%La-2%PHZSM-5), where high selectivity to ethylene is reported [27,28]. Moreover, the dehydration of isobutanol has also been studied with alumina catalysts [29]. Once the alcohols are dehydrated, oligomerization takes place. For the oligomerization reactions, several catalysts have been used such as montmorillonite K-10, and sulfated Zirconia, Ru/Al_2O_3 (Hc), NaOH and Pt/SiO_2-Al_2O_3 (HO), solid phosphoric acid, Group 4 transition-metal with methylaluminoxane, and PtO_2 [30–35]; in these studies, lineal and aromatic hydrocarbons are produced, including not just those in the boiling point range of biojet fuel but also gasoline and green diesel. In 2018, a techno-economic and greenhouse gas emissions assessment was performed for the production of biojet fuel considering alcohol to jet processes as well as direct fermentation [36]; results indicated that significant reduction of the environmental impacts can be assessed, but the production technology must be improved since the minimum selling price in all cases is higher with respect to the fossil jet fuel. Recently, a cofermentation-based process was proposed for the production of renewable aviation fuel from switchgrass [37]; the results indicated that this process has 44% lower global warming potential for 100 years, in comparison with fossil jet fuel.

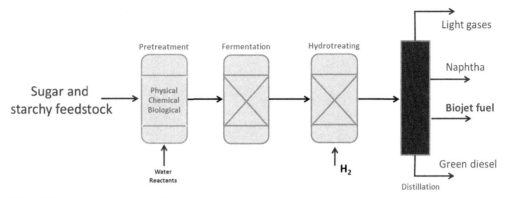

Fig. 3 Direct sugar to hydrocarbons conversion pathway.

On the other hand, the Sugar to Hydrocarbons process, also known as Hydroprocessed Fermented Sugars to Synthetic Isoparaffins (HFS-SIP), uses modified yeasts to convert the sugars, through fermentation, into a hydrocarbon molecule called farnesane, with chemical formula $C_{15}H_{24}$. Then, the farnesane is hydroprocessed to generate renewable hydrocarbons, including aviation biofuel. This conversion pathway was certified by ASTM in 2014, and the produced renewable aviation fuel can be used in mixtures until 10% in volume with fossil jet fuel [26]. The block diagram of the HFS-SIP process is shown in Fig. 3.

Related to this pathway, the reported works in the literature are scarce. Blanch [38] reported that the farnesene is obtained through the mevalonate pathway, where sugars react with water and hydrogen. Later, farnesane must be hydrogenated to obtain farnesene, which later will be converted to biojet fuel through isomerization and cracking reactions [10].

The lignocellulosic feedstock can be processed to renewable aviation fuel through several pathways. The Fischer-Tropsch (F-T) process was the first conversion pathway certified by ASTM in 2009 [25]. In the gasification plus F-T process, the lignocellulosic feedstock is gasified, and the syngas is converted to hydrocarbons through a Fischer-Tropsch synthesis, as can be seen in Fig. 4. Depending on the composition of the lignocellulosic feedstock, the product of this process could require hydrotreating, cracking, and isomerization reactions to generate the renewable aviation fuel. Finally, all the produced hydrocarbons are purified with a distillation train. The renewable fuel produced with gasification plus F-T process can be used in mixtures until 50% in volume with fossil jet fuel [26].

Regarding the gasification plus F-T process, there are some studies reported in the literature. In these works, a yield up to 29% to biojet fuel can be reached using iron catalysts [39–42]. In these processes, the main disadvantage is the high amount of energy required for the gasification stage [8].

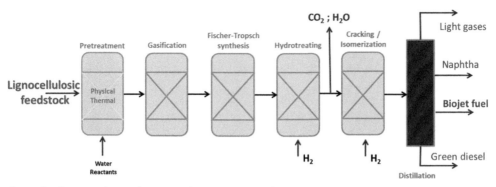

Fig. 4 Gasification plus Fischer-Tropsch conversion pathway.

Another processing alternative to produce renewable aviation fuel considers the pyrolysis of the lignocellulosic biomass; if the pyrolysis is not catalytic, bio-oil is obtained as product, which can be later deoxygenated, cracked, and isomerized to generate renewable hydrocarbons. On the other hand, if the pyrolysis is catalyzed then partial deoxygenation of the bio-oil is performed. In both variants of the process, the produced hydrocarbons are purified through distillation to obtain the renewable aviation fuel, as it is shown in Fig. 5A and B for noncatalyzed and catalyzed pyrolysis. The pretreatment usually includes a size reduction of the feedstock.

With respect to the pyrolysis plus hydroprocessing pathway, the safflower seed press cake was converted to fuel with similar characteristics to kerosene, but with improved ignition properties [43]. Moreover, the conversion of sunflower oil into biojet fuel has been reported, reaching yields up to 30% for hydrocarbons in the range of C7–C12 through the use of ZSM-5 catalyst [44]. Finally, a biojet fuel yield of 69% was obtained through the conversion of stearic acid, waste soybean oil, and palm fatty acid distillate with Pd/beta-zeolite catalyst; an important aspect of this work is that hydrogen is not required [45].

Finally, the lignocellulosic biomass can be pretreated and hydrolyzed, to release the sugars contained in the cellulose and hemicellulose. These sugars are fermented to produce a bio-alcohol, which is dehydrated, oligomerized, and hydrogenated to generate renewable hydrocarbons; these obtained fuels are later purified through distillation, as it is presented in Fig. 6. It is important mentioning that from the alcohol as an intermediate product until the separation of renewable fuels, the processing corresponds to ATJ, as was defined earlier in this chapter.

On the other hand, the conversion of the triglyceride feedstock is the most compact process. The hydrotreating technology, also known as Hydroprocessed Esters and Fatty Acids (HEFA), was certified by ASTM in 2011 [25] and is the most mature of all the conversion pathways. The hydrotreating consists of two consecutive reactors. In the first one, the hydroprocessing takes place through hydrodeoxygenation, hydrodecarboxylation, and hydrodecarbonilation chemical pathways, where long, linear hydrocarbons

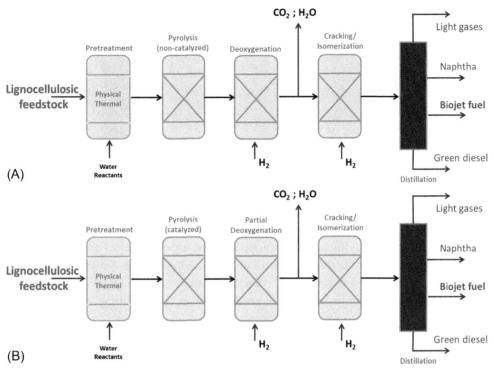

Fig. 5 (A) Noncatalyzed pyrolysis followed by hydroprocessing conversion pathway. (B) Catalyzed pyrolysis followed by hydroprocessing conversion pathway.

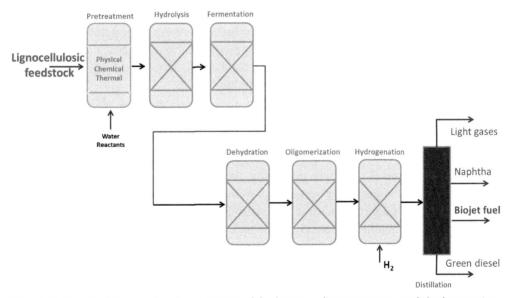

Fig. 6 Hydrolysis followed by fermentation, dehydration, oligomerization, and hydrogenation conversion pathways.

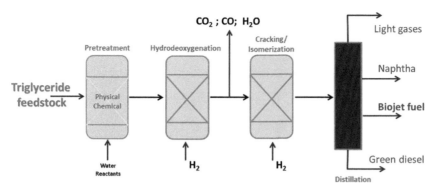

Fig. 7 Hydroprocessing of triglyceride feedstock.

chains are obtained; then, in the second reactor hydrocracking and hydroisomerization reactions allow to obtain linear and branched hydrocarbons. Later, these hydrocarbons are purified through distillation to obtain light gases, naphtha, green diesel, and renewable aviation fuel, as is shown in Fig. 7. According to the Commercial Aviation Alternative Fuels Initiative, the renewable fuel produced from hydrotreating can be used up to 50% in volume with fossil jet fuel [26].

According to Gutiérrez-Antonio et al. [25], the conversion of different oils to biojet fuel has been reported in the literature, such as *Jatropha curcas*, castor, microalgae, soybean, camelina, carinata, coconut, babassu, inedible corn oil, peanut, sunflower, macauba palm, cotton, palm (refined and crude), tung along with chicken fat, vacuum gas oil of vegetable oil mixtures, waste cooking oil, waste vegetable oils, animal fats, microbial oils, and bio-oil from cornstalk and Douglas fir pellets. Indeed, most of the works reported the experimental or modeling studies in two steps: hydrodeoxygenation followed by hydroisomerization and hydrocracking, considering catalysts such as Ni, Mo, Co, Pd, Ro, Pt, and bimetallic sulfide catalysts, such as Ni-Co-Fe, Mo-WU, $NiMoS_2$, Ni-W/SiO_2, $CoMoS_2$, and $NiWS_2$ supported on Al_2O_3 [8,25].

The processing routes have advantages and disadvantages as can be observed in Fig. 8. Considering the cultivated biomasses, the conversion of triglyceride feedstock has low operating costs, but the feedstock is expensive; the lignocellulosic feedstock is a low-cost alternative, but its processing is costly. On the other hand, for sugar and starchy biomass the costs of both, feedstock and their processing, are intermediate, in comparison with the other two types of biomasses; it is important to mention that most of this kind of biomass is edible. Another important factor is the availability of each type of biomass and the creation of a well-established supply chain for each one. On the other hand, the yields to renewable aviation fuel are also important. The hydroprocessing of triglyceride feedstock has yields between 46% and 80%, whereas the lowest values are reported for the transformation of lignocellulosic feedstock, with yields between 29% and 36% [8]. On the other hand, the alcohol-to-jet route has yields between 55% and 90%, which are higher than the

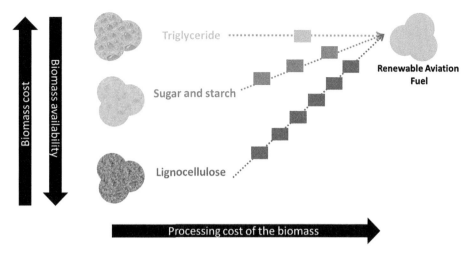

Fig. 8 Advantages and disadvantages of the conversion routes for each type of biomass. *(Based on S.K. Maity, Opportunities, recent trends and challenges of integrated biorefinery: part I, Renew. Sustain. Energy. Rev. 43 (2015) 1427–1445.)*

other processing pathways [8]; however, it is important to notice that this is the yield from alcohol to biojet, not biomass to biojet. If we consider the maximum yield for bioethanol production reported for switchgrass, 29.0% [46], then the total yield from the lignocellulosic feedstock to biojet fuel production ranges between 15.95% and 26.10%. The production of alcohol from sugar, starchy, or lignocellulosic materials is not part of the alcohol to jet process; however, it is important to remark the low yields of the alcohol production process since it would affect the final price of the biojet fuel obtained through this route.

Moreover, for each type of biomass, there are technologies certified by ASTM. At the moment, the certified technologies include Fischer-Tropsch Synthetic Paraffinic Kerosene (FT-SPK), Hydroprocessed Esters and Fatty Acids Synthetic Paraffinic Kerosene (HEFA-SPK), Hydroprocessed Fermented Sugars to Synthetic Isoparaffins (HFS-SIP), Fischer-Tropsch Synthetic Paraffinic Kerosene with Aromatics (FT-SPK/A), Alcohol to Jet Synthetic Paraffinic Kerosene (ATJ-SPK), Catalytic Hydrothermolysis Synthesized Kerosene (CH-SK, or CHJ), Hydroprocessed Hydrocarbons, Esters and Fatty Acids Synthetic Paraffinic Kerosene (HHC-SPK or HC-HEFA-SPK) [26].

3. Biorefineries to produce renewable aviation fuel

3.1 Basic concepts

A biorefinery is defined as a conversion platform that allows the total conversion of given biomass into value-added products, biofuels, and/or bioenergy (heat and/or power). Due to this, the biorefinery includes a wide variety of thermochemical, chemical, biological,

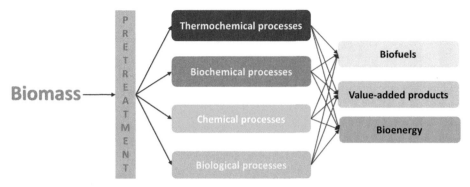

Fig. 9 Conceptual scheme of a biorefinery.

and/or biochemical processes, as can be seen in Fig. 9. An important aspect of this type of conversion platform is that allows using the complete biomass, instead of just a fraction; for instance, the complete fruit of *J. curcas* instead of only the oil derived from their seeds.

The pretreatment is the first step in the processing into the biorefineries, and it allows the separation of biomass in the available fractions, which can include triglyceride, sugar, starch, protein and/or lignocellulose. Thus, there are physical (size reduction, microwave), chemical (hydrolysis, extractions), physicochemical (steam or ammonia fiber explosions), and biological (enzymes, insect cultivation) pretreatments.

According to Basu [47], in the thermochemical conversion the biomass can be transformed into thermal energy, solid products (like fuel pellets), bio-oil, or gases, which can be converted in other products. The main thermochemical conversion processes are combustion, torrefaction, pyrolysis, gasification, and liquefaction. On the other hand, in the biochemical processes large biomass molecules are broken down into smaller ones by the action of bacteria or enzymes [47]. The objective of this type of process is to convert the biomass into biofuels (bioethanol and biogas) and value-added products (digestate) or intermediate raw materials (sugars). The main biochemical processes are digestion, fermentation, and hydrolysis.

The chemical processes focus on the conversion of biomass through chemical reactions (catalytic and noncatalytic) into value-added products, biofuels or chemical products. There is a great amount of chemical reactions that can be carried out; for instance, transesterification for biodiesel production, hydrodeoxygenation for green diesel production, Fischer-Tropsch process for liquid hydrocarbons production, among others.

Finally, the biological process involves the cultivation of different organisms to transform the biomass. The resulting products include value-added products (as fertilizers) or new raw materials (larvae that contains proteins, lipids, and carbohydrates) to generate new bioproducts. Among this type of process, composting and the cultivation of insects (for instance black soldier fly) can be mentioned [25].

Thus, the composition of the biomass gives information about the available fractions, allowing predicting the potential value-added components that can be obtained from it. Based on that information, the products that can be obtained are chosen, and in consequence the required conversion pathways. As it was mentioned before, biojet fuel can be produced from all types of biomass; however, it is important to consider the generation of value-added products to increase the profitability of the conversion processes of biomass. According to Kim et al. [48], there is a tangible need for a renewed paradigm on how aviation fulfills its sustainable development. The production of renewable aviation fuel through biorefinery schemes can be the response. In the next section, the revision of the literature where biorefineries have been proposed for the production of renewable aviation fuel is presented.

3.2 Revision of the literature

In this subsection, a revision of the literature about the advances in biorefineries that include the production of renewable aviation fuel is presented. The discussion of the reported works will be ordered chronologically. These works were selected from a search in the databases of Elsevier (www.sciencedirect.com), American Chemical Society (https://pubs.acs.org/), Springer (https://www.springer.com/gp/), and Wiley (https://onlinelibrary.wiley.com/); the used keywords were "biorefinery" in combination with "biojet fuel," "renewable aviation fuel," "aviation biofuel," "biokerosene," and "synthetic paraffinic kerosene." The articles related to life cycle analysis, supply chain, policies, or individual production processes for biojet fuel were discarded.

Trivedi et al. [49] proposed a conceptual biorefinery for the conversion of microalgae biomass. The biorefinery includes the growth, harvesting, and concentration of the microalgae biomass. The proposed products include carbohydrates, proteins, pigments, and biochemical products, as well as biodiesel, bioethanol, biogas, and biojet fuel. However, there is neither estimation of the amount of microalgae biomass that must be used for the generation of each product, nor the information about the specific conversion technologies and operating conditions. Authors conclude that the "only biofuel" production approach will not be commercially feasible and the economics of other options will play the key role.

Later, in 2016, Vidyashankar and Ravishankar [50] presented a proposal to use micro and macroalgae for wastewater treatment; in addition, they outlined the potential applications of the algae cultivated to produce animal feed, nutraceuticals, bioplastics, biofertilizers, bio-alcohols, light hydrocarbons, biodiesel, and aviation biofuel. They gave general information about the conversion process of these products but separately; there is no estimation of the amount of micro or macroalgae biomass that must be used for the generation of each product.

Wang [10] realized a techno-economic analysis of a biorefinery to process *J. curcas* fruit. The conversion technologies include dehulling, combustion, oil extraction, pyrolysis, hydroprocessing, and product separation. The biorefinery was planned to process 100,000 kg/h of *J. curcas* fruit. The obtained products include biochar, biogas, propane, green diesel, renewable aviation fuel, and electricity. The author estimated a price of 5.42 USD/gal of biojet fuel, concluding that the selling of the coproducts provides a high potential to reduce the bio-jet fuel production cost [10].

Alves et al. [51] realized a techno-economic analysis considering three types of feedstocks to produce biojet fuel, naphtha, diesel, LPG, carbon dioxide, chemicals, lignin, and livestock feed. According to their results, sugarcane and soybean were the most promising feedstock in two regions of Brazil; in spite of this, the scenarios present uncertainties and high financial risk.

Santos et al. [36] realized a techno-economic analysis of the conversion of autarkic sugarcane using the first- and second-generation biorefineries; the studied production scale for biojet fuel was 208 kton/y. Moreover, other products were obtained such as water, furfural, acid acetic, lignin, juice, and grains. Some of the processes of the biorefinery include enzymatic hydrolysis, oligomerization, milling, fast pyrolysis, gasification, Fischer-Tropsch synthesis, fermentation (for ethanol and farnesene) as well as hydroprocessing. They concluded that the price of biojet fuel production from sugarcane is not competitive with its fossil counterpart; however, carbon dioxide emissions for biojet fuel production can be reduced by 50%, in comparison with the production of fossil jet fuel.

Klein et al. [52] proposed several scenarios to couple production processes for biojet fuel into a bioethanol production process from sugarcane. The production processes considered for biojet fuel included hydrotreating, Fischer-Tropsch, and Alcohol to Jet; the raw materials were vegetable oils (palm, macauba, and soybean) and eucalyptus. This coupling allows obtaining hydrous ethanol, green naphtha, renewable jet fuel, green diesel, and electricity; however, value-added products are not generated. Authors conclude that the best scenario, regarding economical performances, included the Fischer-Tropsch process, whereas the greater volumes of biojet fuel were obtained with hydroprocessing technology. In all the cases, there is a reduction of 70% of carbon dioxide emissions derived from the production of biojet fuel, with respect to fossil jet fuel.

In 2019, Tongpun et al. [53] performed the techno-economic analysis of biojet fuel production in Taiwan from Jatropha; covering from the farming processes to the hydroprocessing stage to obtain renewable aviation fuel. The complete processing included fruit dehulling, shell combustion, oil extraction, pyrolysis, oil hydroprocessing, and product separation. Also, the minimum biojet fuel selling price was estimated as 6.25 USD/L. This value is high regarding the processing without including the farming stage.

Romero-Izquierdo et al. [54] proposed a biorefinery scheme to convert waste cooking oil (WCO) to renewable aviation fuel; the biorefinery was planned to process 5,596,032.97 L/y of WCO to produce renewable aviation fuel, green diesel, naphtha, biodiesel, and sodium phosphate. The main processes considered were esterification, transesterification, and hydroprocessing along with the recovery of ethanol. Results show that the total annual cost and carbon dioxide emissions per kilogram of biofuel were 97% and 99% smaller than the values reported for the biodiesel and biojet fuel individual production processes.

Neves et al. [55] proposed scenarios to incorporate the production of renewable aviation fuel into the sugarcane biorefineries. They considered the incorporation of straw as a residue from agricultural activities, and the coupling of pyrolysis, Fischer-Tropsch synthesis, and hydrocracking. The products expected to be obtained include ethanol, sugar, biojet fuel, green diesel, green gasoline, and electricity. However, there is no estimation of the production rates, prices of products, or operating conditions of the conversion processes; also, the amount of each raw material that must be used for the generation of each product is not reported.

In 2020, Michailos and Webb [56] proposed a biorefinery for the conversion of rice straw into ethylene, biojet fuel, gasoline, as well as heat and power. They realized a techno-economic analysis of the biorefinery, considering a biojet fuel production capacity of 11.20 ton/h. The results show that it is possible to produce such an amount of renewable aviation fuel with a selling price of 1.52 USD/L, and the biorefinery has an energy efficiency of 43.7%.

Recently, in 2021, Kim et al. [57] reported an experimental process to produce biojet fuel and omega-3 polyunsaturated fatty acids from heterotrophic microalgae *Schizochytrium* sp. ABC101. The studied processes include lipid extraction, transesterification, distillation, deoxygenation, and hydrocracking. Their results show that 54.60% of omega-3 polyunsaturated fatty acids can be obtained, and 20.40% of renewable aviation fuel. According to the authors, this strategy can provide an economical and unique solution for microalgal biorefinery targeting biofuels in addition to value-added products [57].

Also in 2021, a study for the production of biojet fuel through the integration of the hydroprocessing and alcohol to jet processes along with a transesterification plant to produce glycerol and biodiesel was presented [58]. Moreover, a cogeneration plant that supplies all the thermal and electrical energy to the process is included. The results indicated that the required inversion is elevated, the yield of the biojet fuel production processes is low, and, despite the diversification of the product of the biorefinery, the production chains are not fully prepared to address these new challenges [58].

From all the biorefineries reported in the literature, five of them are conceptual, four of them were simulated and evaluated through a techno-economic analysis, and one of them is an experimental work. Moreover, the explored biomasses are microalgae, macroalgae, *J. curcas*, waste cooking oil, rice straw, sugarcane, and oils from maracuba, palm, and soybean in combination with sugarcane; these raw materials are of high productivity or waste biomass, with exception of the vegetable oils that are coprocessed in a bioethanol process from sugarcane. In all the studies where techno-economic analysis is realized, there is observed a significant reduction of the carbon dioxide emissions, until 70% with respect to the production of fossil jet fuel. However, the prices reported for the aviation biofuel in two of those studies are around 1.50 USD/L, whereas the price of fossil jet fuel was 0.44 USD/L in December 2020 [13]; it is important to mention that in 2019 the price of fossil jet fuel was 0.70 USD/L [13], which evidences the high volatility of fossil fuel prices. The most studied conversion pathways for the production of biojet fuel into the biorefineries are Fischer-Tropsch and Hydroprocessing.

4. Future trends in the production of renewable aviation fuel through biorefineries

According to Shahabuddin et al. [59], the energy efficiency and capital cost of converting biomass and residual wastes to aviation fuels are major barriers to widespread adoption. Therefore, based on the revision of the literature some opportunities areas and challenges for the production of renewable aviation fuel through biorefineries schemes are detected. The future trends in this research area will be discussed next.

In the synthesis of a biorefinery, one of the first and most crucial factors to consider is the choice of renewable feedstock—this is a critical step, as being "renewable" does not necessarily equate to being green and environmentally sustainable [60]. Thus, the most viable raw materials are those of high productivity or low cost. In the case of the low-cost raw materials, all the waste biomasses are an interesting alternative due to its high availability; indeed, in most of the cases, that type of biomass is abandoned in the culture field, burned or it just starts to accumulate and becomes a pollution problem. Thus, the processing of waste biomass through a biorefinery scheme allows solving the problem associated with its disposal and revaluating them, which could benefit the agricultural and agro-industrial producers. For the case of high productivity raw materials, microalgae and macroalgae are the best alternatives due to its high productivity and the CO_2 consumed as part of its growth [61]. It is desirable that the culture of algae is performed using wastewater for greenhouses or water treatment plants; in this way, it is possible to decrease its water footprint. On the other hand, an interesting opportunity area is the use of mixtures of raw materials, to give more strength and flexibility to the operation of the biorefinery [25]; it is desirable to combine waste biomass (that can be found all the year) with those of high productivity (which can be available only in some periods of the year). However, the mixture of raw materials along with the variability in the composition of the biomasses due to the location of the cultivation sites and the seasonality could impact the efficiency or performance of the processes incorporated into the biorefinery; here is a great challenge for the design of robust biorefineries for processing multibiomass mixtures.

For any type of raw materials, it is important to establish appropriate optimal supply chains for the biorefineries. Previous studies have reported that, when the production of aviation biofuel in individual processes is considered, the best alternative is either local or regional supply chains [62,63]; since, the transportation of raw materials and/or products can significantly increase the costs and carbon footprint, becoming infeasible from economic and environmental points of view. However, it is important to mention that, to the authors' knowledge, the determination of the optimal supply chain for the production of renewable aviation fuel through biorefineries schemes is not available in the literature. This challenge is complex since the distribution of several products must be optimized and the variety of locations, transport conditions, as well as the frequency of delivery can be variable among them.

Regarding the processing pathways, it is necessary to develop novel and efficient methods to process biomass, within a biorefinery context, to address the highly complex process operations, the high costs associated with processing, and the low conversion of biomass to products [60]. In this context, the use of energy integration and process intensification strategies may help to reach this target. Energy integration is defined as a strategy to use the available energy into a process to heat or cool the process streams [25]; this allows a reduction in the operating costs, due to the diminishing of the external utilities. However, in most cases, the capital costs are increased since additional pumps and heat exchangers are required to perform the energy exchange. On the other hand, process intensification consists of the development of novel apparatuses and techniques that are compared with those commonly used today, are expected to bring dramatic improvements in manufacturing and processing, substantially decreasing equipment-size/production-capacity ratio, energy consumption, or waste production, and ultimately resulting in cheaper, sustainable technologies [64]. Among intensified processes, multifunctional reactors, microreactors, static mixers, reactive distillation, thermally coupled distillation, and extractive reactions, can be mentioned. The application of process intensification strategies allows decreasing the operating and capital costs, as well as space requirements and piping costs. Thus, the use of process intensification and energy integration strategies for biorefineries could aid to decrease operational and/or capital costs, as well as the associated carbon dioxide emissions.

Another important aspect is given by the pretreatments, which usually have low efficiency and high operating costs; thus, there is an urgent need to enhance the existing pretreatment strategies or develop a universal pretreatment approach for utilization of different biomass and facilitate large scale production of biofuel and value-added products [65]. Some steps in this direction can be found in the literature. For instance, the proposal of self-steam explosion, where the moisture content of the biomass sample is fully utilized as a steam resource to auto-hydrolyze the biomass components, producing fine particles with diameters <1 mm [66]. Another interesting pretreatment for solid waste management is insect cultivation, especially of black soldier fly; the black soldier fly larvae culture is a self-sustained cost-effective method that allows the generation of value-added products [67] and biofuels [68].

On the other hand, the use of alternative solvents is an important tool to develop sustainable biorefineries. Usually, in the oil extraction processes solvents such as hexane, petroleum ether, isopropanol, acetone, chloroform, methanol, and 1-butanol are used [69]. These solvents are not environment friendly, and the use of alternative solvents must be explored. In this context, the green solvents have arisen as low environmental impact entrainers; in this category, ionic liquids, switchable solvents, supercritical fluids, or deep eutectic solvents are found [70,71]. In this topic, the computer-aided molecular design of new solvents is a powerful tool since they are based on reliable property prediction models [72].

Another alternative is the use of metabolic modeling to design cell factories where microorganisms directly produce biojet fuel or a precursor, like alkanes, alkenes, or terpenes [73]. This pathway has been explored for more than a decade; however, the concentrations of hydrocarbons are still in the range of 580 mg/L [74]. On the other hand, the electro-biotechnology has been studied as an alternative process to produce value-added products as well as biofuels; this technology is defined as the use of electricity to improve the conversion of chemical and biochemical processes [75]. According to Zeng [75], another alternative is the microbial production systems, which have achieved great success in the production of large-volume fuels, chemicals and materials from sugar-type substrates or biomass. Moreover, it has been recently reported that *Yarrowia lipolytica* can be used as a cell biorefinery where value-added products, biofuels, oleochemicals, and specialty chemicals can be produced [76]. To the authors' knowledge, electro-biotechnology and microbial production systems have not been reported for the production of renewable aviation fuel. Nevertheless, the use of microorganisms such as bacteria and yeasts as catalysts are a sustainable alternative for the production of liquid biofuels including bioethanol, biomethanol, biobutanol, bio-ammonia, biokerosene, and bioglycerol [77].

Summarizing, no clear winning sustainable jet fuel technology exists, a situation that encourages continued research and development in innovative biojet fuel technologies [78]. In this context, the biorefinery concept offers a unique opportunity to make a major step-change and provide future societies with genuinely green and sustainable products [60]. To achieve this, the complete biomasses, especially wastes, must be converted through highly efficient conversion processes using catalysts, solvents, and reactants with low environmental impact. This could be reached if each biomass fraction (lipids, sugars, and lignocellulose) is processed through the lowest possible number of equipment units with good conversion efficiency. It is important to extract the value-added component first to avoid its degradation, and later the residual biomass can be processed to obtain chemical, solvents, biofuels as well as bioenergy; in other words, the biomass composition must lead to the selection of the most promissory products, and, as a consequence, the required conversion processes that must be integrated into a biorefinery. In this way, it could be possible to ensure the sustainability of the production of renewable aviation fuel, and the other bio-products, in the biorefinery.

It is important to mention that in 2020, there were reported 19 facilities for the production of biojet fuel worldwide [25]; indeed, 15 facilities use the Hydroprocessed Esters and Fatty Acids Synthetic Paraffinic Kerosene technology, 3 of them employs the Fischer-Tropsch synthesized isoparaffinic kerosene, and the last one considers the Hydroprocessed Fermented Sugars to Synthetic Isoparaffins. Based on the information presented in the patents of these technologies all the processes are individual and conventional; to the authors' knowledge, there are no biorefineries at industrial scale to produce biojet fuel as well as value-added products and/or bioenergy (thermal and electrical energy). According to the International Energy Agency, the efficient use of the available biomass will be crucial and the uptake of biorefineries at the industrial level will be required [79]; however, the investment needs are elevated and novel technologies are needed to become profitable the conversion of biomass into biorefinery schemes.

5. Conclusions

In this chapter, the revision of the literature related to the production of renewable aviation fuel through biorefineries schemes has been presented. In this topic, there are few works devoted to the production of biojet fuel through biorefinery schemes. The main opportunity areas identified are the use of waste or high productivity biomass, to achieve either reduced or competitive costs for the feedstocks, respectively; also, the use of biomass mixtures could guarantee the supply of raw materials for the biorefinery. Moreover, the proposal of high efficiency process must be guaranteed; to achieve this, the application of process intensification and energy integration tools constitute promissory strategies. In addition, innovation is mandatory for the development of more efficient and environmentally friendly pretreatments for the biomass; these new processes can help to reduce operating and capital costs while decreasing its environmental impact and increasing the profitability of the processing of the biomass. Indeed, simpler pretreatments can avoid the destruction of valuable components, which can be instead extracted and commercialized. Likewise, the development of green solvents is an important aspect to decrease the environmental impact of oil extraction.

On the other hand, additional challenges are related to the determination of the optimal supply chain for the production of renewable aviation fuels through biorefinery schemes; the complexity of this problem will be increased considering the transportation needs of each product, as well as the frequency and sites of delivery. In addition, more research must be done on the use of cell factories, electro-biotechnological, and microbial production systems for the generation of renewable aviation fuel.

The main challenge is allowing the biomass composition to guide the selection of the most feasible value-added products and biofuels; as consequence, the conversion processes will have a minimum of equipment, also its energetic consumption and environmental impact can be improved through the application of several strategies, such as process intensification and energy integration tools. In this way, the production of renewable aviation fuel through advanced and efficient biorefineries will play a key role in the sustainable recovery and development of the aviation sector.

Acknowledgments

Financial support provided by CONACyT, grant 239765, for the development of this project is gratefully acknowledged. Also, Araceli Guadalupe Romero-Izquierdo was supported by a scholarship from CONA-CYT for the realization of her graduate studies.

References

[1] International Air Transport Association, Industry Statistics, Fact Sheet, 2019, Available from: https://www.iata.org/en/iata-repository/publications/economic-reports/airline-industry-economic-performance- - -december-2019- - -data-tables/. Consulted on December 14th 2020.

[2] C. Chuck, Biofuels for Aviation, Feedstocks, Technology and Implementation, first ed., Press Publication, United Kingdom, 2016, pp. 3–14.

[3] International Air Transport Association, The IATA Technology Roadmap Report, 2019, Issued June 2009. Available from: https://www.iata.org/en/programs/environment/technology-roadmap/. Consulted on December 14th 2020.

[4] International Air Transport Association, Carbon-Neutral Growth by 2020, 2018, Available from: https://www.iata.org/en/pressroom/pr/2009-06-08-03/. Consulted on December 14th 2020.

[5] IRENA, Biofuels for Aviation: Technology Brief, International Renewable Energy Agency, Abu Dhabi, 2017.

[6] International Civil Aviation Organization, Economic Impacts of COVID-19 on Civil Aviation, 2020, Available from: https://www.icao.int/sustainability/Pages/Economic-Impacts-of-COVID-19.aspx. Consulted on December 14th 2020.

[7] International Energy Agency, Sustainable Recovery, 2020, Available from: https://webstore.iea.org/sustainable-recovery-weo-special-report. Consulted on December 14th 2020.

[8] C. Gutiérrez-Antonio, F.I. Gómez-Castro, J.A. de Lira-Flores, S. Hernández, A review on the production processes of renewable jet fuel, Renew. Sustain. Energy Rev. 79 (2017) 709–729.

[9] American Standards of Tests and Materials, ASTM Standard D7566, Standard Specification for Aviation Turbine Fuel Containing Synthesized Hydrocarbons, ASTM International, 2019. Available from: https://www.astm.org/Standards/D7566.html. Consulted on December 15th 2020.

[10] W. Wang, Techno-economic analysis of a bio-refinery process for producing Hydro-processed Renewable Jet fuel from Jatropha, Renew. Energy 95 (2016) 63–73.

[11] M. Pearlson, C. Wollersheim, J. Hileman, A techno-economic review of hydroprocessed renewable esters and fatty acids for jet fuel production, Biofuels Bioprod. Biorefin. 7 (2013) 89–96.

[12] D. Klein-Marcuschamer, C. Turner, M. Allen, P. Gray, R.G. Dietzgen, P.M. Gresshoff, B. Hankamer, K. Heimann, P.T. Scott, E. Stephens, R. Speight, L.K. Nielsen, Techno-economic analysis of renewable aviation fuel from microalgae, *Pongamia pinnata*, and sugarcane, Biofuels Bioprod. Biorefin. 7 (2013) 416–428.

[13] Petróleos Mexicanos, Precio al público de productos petrolíferos, Estadísticas Petroleras, 2019, Available from: https://www.pemex.com/ri/Publicaciones/Indicadores%20Petroleros/epublico_esp.pdf. Consulted on May 14th 2020.

[14] C. Gutiérrez-Antonio, A.G. Romero-Izquierdo, F.I. Gómez-Castro, S. Hernández, A. Briones-Ramírez, Simultaneous energy integration and intensification of the hydrotreating process to produce biojet fuel from *Jatropha curcas*, Chem. Eng. Process. Process Intensif. 110 (2016) 134–145.

[15] A.G. Romero-Izquierdo, C. Gutiérrez-Antonio, F.I. Gómez-Castro, S. Hernández, Energy integration and optimization of the separation section in a hydrotreating process for the production of biojet fuel, in: A. Espuña, M. Graells, L. Puigjaner (Eds.), Proceedings of the 27th European Symposium on

Computer Aided Chemical Engineering, vol. 40, Computer Aided Chemical Engineering, Barcelona, Spain, 2017, pp. 661–666. ISSN 1570-7946, ISBN 9780444639653. October 1st–5th, 2017.

[16] C. Gutiérrez-Antonio, F.I. Gómez-Castro, A.G. Romero-Izquierdo, S. Hernández, Energy integration of a hydrotreating process for the production of biojet fuel, in: Z. Kravanja, M. Bogataj (Eds.), Proceedings of the 26th European Symposium on Computer Aided Chemical Engineering, vol. 38, Computer Aided Chemical Engineering, Portoroz, Slovenia, 2016, pp. 127–132. ISSN 1570-7946, ISBN 9780444634283. June 12th–15th, 2016.

[17] A.L. Moreno-Gómez, C. Gutiérrez-Antonio, F.I. Gómez-Castro, S. Hernández, Modelling, simulation and intensification of the hydroprocessing of chicken fat to produce renewable aviation fuel, Chem. Eng. Process. Process Intensif. 159 (2021) 108250.

[18] A.G. Romero-Izquierdo, F.I. Gómez-Castro, C. Gutiérrez-Antonio, S. Hernández, M. Errico, Intensification of the alcohol-to-jet process to produce renewable aviation fuel, Chem. Eng. Process. Process Intensif. 160 (2021) 108270.

[19] C. Gutiérrez-Antonio, M.L. Soria Ornelas, F.I. Gómez-Castro, S. Hernández, Intensification of the hydrotreating process to produce renewable aviation fuel through reactive distillation, Chem. Eng. Process. Process Intensif. 124 (2018) 122–130.

[20] C. Gutiérrez-Antonio, F.I. Gómez-Castro, S. Hernández, A. Briones-Ramírez, Intensification of a hydrotreating process to produce biojet fuel using thermally coupled distillation, Chem. Eng. Process. Process Intensif. 88 (2015) 29–36.

[21] C. Gutiérrez-Antonio, A. Gómez-De la Cruz, A.G. Romero-Izquierdo, F.I. Gómez-Castro, S. Hernández, Modeling, simulation and intensification of hydroprocessing of micro-algae oil to produce renewable aviation fuel, Clean Technol. Environ. Policy 20 (2018) 1589–1598.

[22] D. Haldar, M.K. Purkait, A review on the environment-friendly emerging techniques for pretreatment of lignocellulosic biomass: mechanistic insight and advancements, Chemosphere 264 (2021) 128523.

[23] V.B. Agbor, N. Cicek, R. Sparling, A. Berlin, D.B. Levin, Biomass pretreatment: fundamentals toward application, Biotechnol. Adv. 29 (2011) 675–685.

[24] P.R. Seidl, A.K. Goulart, Application of computational methods for pretreatment processes of different biomass feedstocks, Curr. Opin. Green Sustain. Chem. 26 (2020) 100366.

[25] C. Gutiérrez-Antonio, A.G. Romero-Izquierdo, F.I. Gómez-Castro, S. Hernández, Production Processes of Renewable Aviation Fuel, Elsevier, 2021.

[26] Commercial Aviation Alternative Fuels Initiative, Fuel Qualification, 2021, Available from: http://www.caafi.org/. Consulted on February 9th 2021.

[27] N. Zhan, Y. Hu, H. Li, D. Yu, Y. Han, H. Huang, Lanthanum–phosphorous modified HZSM-5 catalysts in dehydration of ethanol to ethylene: a comparative analysis, Catal. Commun. 11 (2010) 633–637.

[28] Y. Hu, N. Zhan, C. Dou, H. Huang, Y. Han, D. Yu, Y. Hu, Selective dehydration of bioethanol to ethylene catalyzed by lanthanum-phosphorous modified HZSM-5, influence of the fusel, Biotechnol. J. 5 (2010) 1186–1191.

[29] J.D. Taylor, M.M. Jenni, M.W. Peters, Dehydration of fermented isobutanol for the production of renewable chemicals and fuels, Top. Catal. 53 (2010) 1224–1230.

[30] B.G. Harvey, R.L. Quintana, Synthesis of renewable jet and diesel fuels from 2-ethyl-1-hexene, Energy Environ. Sci. 3 (2010) 352–357.

[31] H. Olcay, A.V. Subrahmanyam, R. Xing, J. Lajoie, J.A. Dumesic, G.W. Huber, Production of renewable petroleum refinery diesel and jet fuel feedstocks from hemicellulose sugar streams, Energy Environ. Sci. 6 (2013) 205–216.

[32] T.M. Sakuneka, A. Klerk, R.J.J. Nel, A.D. Pienaar, Synthetic jet fuel production by combined propene oligomerization and aromatic alkylation over solid phosphoric acid, Ind. Eng. Chem. Res. 47 (2008) 1828–1834.

[33] M.E. Wright, B.G. Harvey, R.L. Quintana, inventors; Wright Michael E., Harvey Benjamin G., Quintana Roxanne L., assignee, Diesel and Jet Fuels Based on the Oligomerization of Butene, United States patent US 20120209045, 2012.

[34] M.E. Wright, B.G. Harvey, R.L. Quintana, inventors, Diesel and jet fuels based on the oligomerization of butene, The United States of America as represented by the secretary of the navy, assignee, United States patent US 8395007, 2013.

[35] M.E. Wright, B.G. Harvey, R.L. Quintana, inventors, Diesel and jet fuels based on the oligomerization of butene, US Government As represented by the secretary of department of the navy, assignee, United States patent US 20140051898, 2014.

[36] C.I. Santos, C.C. Silva, S.I. Mussatto, P. Osseweijer, L.A.M. van der Wielen, J.A. Posada, Integrated 1st and 2nd generation sugarcane bio-refinery for jet fuel production in Brazil: techno-economic and greenhouse gas emissions assessment, Renew. Energy 129 (Part B) (2018) 733–747.

[37] A.S. Prasad Pamula, D.J. Lampert, H.K. Atiyeh, Well-to-wake analysis of switchgrass to jet fuel via a novel co-fermentation of sugars and CO2, Sci. Total Environ. (2021) 146770.

[38] H.W. Blanch, Bioprocessing for biofuels, Curr. Opin. Biotechnol. 23 (2011) 390–395.

[39] M.J.A. Tijmensen, A.P.C. Faaij, C.N. Hamelinck, M.R.M. van Hardeveld, Exploration of the possibilities for production of Fischer Tropsch liquids and power via biomass gasification, Biomass Bioenergy 23 (2002) 129–152.

[40] R.L. Espinoza, A.P. Steynberg, B. Jager, A.C. Vosloo, Low temperature Fischer–Tropsch synthesis from a Sasol perspective, Appl. Catal. A Gen. 186 (1999) 13–26.

[41] A.P. Steynberg, R.L. Espinoza, B. Jage, A.C. Vosloo, High temperature Fischer–Tropsch synthesis in commercial practice, Appl. Catal. A Gen. 186 (1999) 41–54.

[42] J.C. Viguié, N. Ullrich, P. Porot, L. Bournay, M. Hecquet, J. Rousseau, BioTfueL project: targeting the development of second-generation biodiesel and biojet fuels, Oil Gas Sci. Technol. 68 (2013) 935–946.

[43] S. Şensöz, D. Angın, Pyrolysis of safflower (*Charthamus tinctorius* L.) seed press cake in a fixed-bed reactor: part 2. Structural characterization of pyrolysis bio-oils, Bioresour. Technol. 99 (2008) 5498–5504.

[44] X. Zhao, L. Wei, J. Julson, Q. Qiao, A. Dubey, G. Anderson, Catalytic cracking of nonedible sunflower oil over ZSM-5 for hydrocarbon bio-jet fuel, New Biotechnol. 32 (2015) 300–312.

[45] I.H. Choi, K.R. Hwang, J.S. Han, K.H. Lee, J.S. Yun, J.S. Lee, The direct production of jet-fuel from non-edible oil in a single-step process, Fuel 158 (2015) 98–104.

[46] M. Morales, A. Arvesen, F. Cherubini, Integrated process simulation for bioethanol production: effects of varying lignocellulosic feedstocks on technical performance, Bioresour. Technol. 328 (2021) 124833.

[47] P. Basu, Biomass Gasification, Pyrolysis and Torrefaction, Elsevier, 2018.

[48] Y. Kim, J. Lee, J. Ahn, Innovation towards sustainable technologies: a socio-technical perspective on accelerating transition to aviation biofuel, Technol. Forecast. Soc. Change 145 (2019) 317–329.

[49] J. Trivedi, M. Aila, D.P. Bangwal, S. Kaul, M.O. Garg, Algae based biorefinery—how to make sense? Renew. Sustain. Energy Rev. 47 (2015) 295–307.

[50] S. Vidyashankar, G.A. Ravishankar, Algae-based bioremediation: bioproducts and biofuels for biobusiness, in: M.N.V. Prasad (Ed.), Bioremediation and Bioeconomy, vol. 1, Elsevier, India, 2016, pp. 457–493 (Chapter 18).

[51] C.M. Alves, M. Valk, S. Jong, A. Bonomi, L.A.M. van der Wielen, S.I. Mussatto, Techno-economic assessment of biorefinery technologies for aviation biofuels supply chains in Brazil, Biofuels Bioprod. Biorefin. 11 (2017) 67–91.

[52] B.C. Klein, M. Ferreira Chagas, T. Lopes Junqueira, M.C.A. Ferreira Rezende, T.F. de Cardoso, O. Cavalett, A. Bonomi, Techno-economic and environmental assessment of renewable jet fuel production in integrated Brazilian sugarcane biorefineries, Appl. Energy 209 (2018) 290–305.

[53] P. Tongpun, W.-C. Wang, P. Srinophakun, Techno-economic analysis of renewable aviation fuel production: from farming to refinery processes, J. Clean. Prod. 226 (2019) 6–17.

[54] A.G. Romero-Izquierdo, F.I. Gómez-Castro, C. Gutiérrez-Antonio, R. Cruz Barajas, S. Hernández, Development of a biorefinery scheme to produce biofuels from waste cooking oil, in: A.A. Kiss, E. Zondervan, R. Lakerveld, L. Özkan (Eds.), 29th European Symposium on Computer Aided Chemical Engineering, vol. 46, Computer Aided Chemical Engineering, Eindhoven, The Netherlands, 2019, pp. 289–294. June 16th–19th, 2019.

[55] R.C. Neves, B.C. Klein, R. Justino da Silva, M.C.A. Ferreira Rezende, A. Funke, E. Olivarez-Gómez, A. Bonomi, R. Maciel-Filho, A vision on biomass-to-liquids (BTL) thermochemical routes in integrated sugarcane biorefineries for biojet fuel production, Renew. Sustain. Energy Rev. 119 (2020) 109607.

[56] S. Michailos, C. Webb, Valorization of rice straw for ethylene and jet fuel production: a technoeconomic assessment, in: M.R. Kosseva, C. Webb (Eds.), Food Industry Wastes, second ed., vol. 1, Academic Press, 2020, pp. 201–221 (Chapter 10).

[57] T.-H. Kim, K. Lee, B.-R. Oh, M.-E. Lee, M. Seo, S. Li, J.-K. Kim, M. Choi, Y.K. Chang, A novel process for the coproduction of biojet fuel and high-value polyunsaturated fatty acid esters from heterotrophic microalgae *Schizochytrium* sp. ABC101, Renew. Energy 165 (Part 1) (2021) 481–490.

[58] A.A.V. Julio, E.A.O. Batlle, C.J.C. Rodriguez, J.C.E. Palacio, Exergoeconomic and environmental analysis of a palm oil biorefinery for the production of bio-jet fuel, Waste Biomass Valoriz. 12 (2021) 5611–5637, https://doi.org/10.1007/s12649-021-01404-2.

[59] M. Shahabuddin, M.T. Alam, B.B. Krishna, T. Bhaskar, G. Perkins, A review on the production of renewable aviation fuels from the gasification of biomass and residual wastes, Bioresour. Technol. 312 (2020) 123596.

[60] T.M. Attard, J.H. Clark, C.R. McElroy, Recent developments in key biorefinery areas, Curr. Opin. Green Sustain. Chem. 21 (2020) 64–74.

[61] M. Bhattacharya, S. Goswami, Microalgae—a green multi-product biorefinery for future industrial prospects, Biocatal. Agric. Biotechnol. 25 (2020) 101580.

[62] S. Domínguez-García, C. Gutiérrez-Antonio, J.A. De Lira-Flores, J.M. Ponce-Ortega, Optimal planning for the supply chain of biofuels for aviation in Mexico, Clean Technol. Environ. Policy 19 (2017) 1387–1402.

[63] S. Domínguez-García, C. Gutiérrez-Antonio, J.A. De Lira-Flores, J.M. Ponce-Ortega, M.M. El-Halwagi, Strategic planning for the supply chain of aviation biofuel with consideration of hydrogen production, Ind. Eng. Chem. Res. 56 (46) (2017) 13812–13830.

[64] A.I. Stankiewicz, J.A. Moulijn, Process intensification: transforming chemical engineering, Chem. Eng. Prog. 1 (2000) 22–34.

[65] B. Kumar, N. Bhardwaj, K. Agrawal, V. Chaturvedi, P. Verma, Current perspective on pretreatment technologies using lignocellulosic biomass: an emerging biorefinery concept, Fuel Process. Technol. 199 (2020) 106244.

[66] D.E. Priyanto, S. Ueno, H. Kasai, K. Mae, Rethinking the inherent moisture content of biomass: its ability for milling and upgrading, ACS Sustain. Chem. Eng. 6 (3) (2018) 2905–2910.

[67] A. Singh, K. Kumari, An inclusive approach for organic waste treatment and valorisation using Black Soldier Fly larvae: a review, J. Environ. Manag. 251 (2019) 109569.

[68] R. Raksasat, J. Wei Lim, W. Kiatkittipong, K. Kiatkittipong, Y. Chia Ho, M. Kee Lam, C. Font-Palma, H.F. Mohd Zaid, C. Kui Cheng, A review of organic waste enrichment for inducing palatability of black soldier fly larvae: wastes to valuable resources, Environ. Pollut. 267 (2020) 115488.

[69] Y.G. Keneni, L.A. Bahiru, J.M. Marchetti, Effects of different extraction solvents on oil extracted from Jatropha seeds and the potential of seed residues as a heat provider, Bioenergy Res. 14 (2021) 1207–1222, https://doi.org/10.1007/s12155-020-10217-5.

[70] L. Soh, M.J. Eckelman, Green solvents in biomass processing, ACS Sustain. Chem. Eng. 4 (11) (2016) 5281–5837.

[71] Y. Chen, T. Mu, Revisiting greenness of ionic liquids and deep eutectic solvents, Green Chem. Eng. 1 (2021).

[72] N.G. Chemmangattuvalappil, Development of solvent design methodologies using computer-aided molecular design tools, Curr. Opin. Chem. Eng. 27 (2020) 51–59.

[73] B. de Jong, V. Siewers, J. Nielsen, Systems biology of yeast: enabling technology for development of cell factories for production of advanced biofuels, Curr. Opin. Biotechnol. 23 (4) (2012) 624–630.

[74] M. Das, P. Patra, A. Ghosh, Metabolic engineering for enhancing microbial biosynthesis of advanced biofuels, Renew. Sustain. Energy Rev. 119 (2020) 109562.

[75] A.-P. Zeng, New bioproduction systems for chemicals and fuels: needs and new development, Biotechnol. Adv. 37 (4) (2019) 508–518.

[76] M. Jingbo, G. Yang, M. Marsafari, P. Xu, Synthetic biology, systems biology, and metabolic engineering of *Yarrowia lipolytica* toward a sustainable biorefinery platform, J. Ind. Microbiol. Biotechnol. 47 (2020) 845–862.

[77] H.K. Shariat Panahi, M. Dehhaghi, J.E. Kinder, T.C. Ezeji, A review on green liquid fuels for the transportation sector: a prospect of microbial solutions to climate change, Biofuel Res. J. 6 (2019) 995–1024.

[78] L. Zhang, T.L. Butler, B. Yang, Recent trends, opportunities and challenges of sustainable aviation fuel, in: A.A. Vertès, N. Qureshi, H.P. Blaschek, H. Yukawa (Eds.), Green Energy to Sustainability: Strategies for Global Industries, Wiley, 2020 (Chapter 5).

[79] International Energy Agency, The Role of Industrial Biorefineries in a Low-Carbon Economy Summary and Conclusions From the IEA Bioenergy/IEA IETS Workshop, 2017, Available from: https://iea-industry.org/app/uploads/ieabioenergy-iets-industrial-biorefineries-workshop-report.pdf. Consulted on March 31st 2021.

CHAPTER 6

Role of catalysts in sustainable production of biojet fuel from renewable feedstocks

Abu Yousuf[a], Md. Anisur Rahman[a], Mohammad Jalilur Rahman[b], and Md. Shahadat Hossain[a]

[a]Department of Chemical Engineering and Polymer Science, Shahjalal University of Science and Technology, Sylhet, Bangladesh
[b]Department of Chemistry, Shahjalal University of Science and Technology, Sylhet, Bangladesh

1. Introduction

In this era of globalization and socialization, the tourism, business, and education sector have crossed the national border. Due to advancement in aviation industries, these sectors are expanding more than ever, contributing significantly to the development of national and international economy, culture, and education. Aviation's global economic impact has been estimated to be $2.7 trillion which is equivalent to 3.6% of the world's gross domestic product (GDP) [1]. In 2017, 4.1 billion passengers were carried by airlines, which is around 8% increase from 2016, and it is forecasted that the number of air passengers will be doubled over the next 20 years [1,2]. Each year the aviation industry consumes approximately 341 billion liters of conventional petroleum derived jet fuel, which is roughly 10% of global liquid fuel usage [1]. Moreover, it is expected that the air travel demand will increase at a rate of 3%–5% for passengers and 10% for goods over the next 30 years, doubling the fuel consumption and increasing the CO_2 emissions six times by 2050 [3]. At the same time, aviation fuel derived from fossil-based source is predicted to experience a fall each year by 2026 [4]. The price of petroleum derived jet fuel is dependent on that of the crude oil, and any fluctuation in the crude oil price is directly reflected in the price of jet fuel. The U.S. Energy Information Administration (EIA) estimated that the annual growth rate of global consumption of jet fuel is 1.5% leading to a 2.7% annual growth rate in average jet fuel price from 2016 to 2050 [5]. Aviation industry also contribute to global warming by releasing greenhouse gases (GHGs). It has been estimated that aviation sector has increased the global carbon emission by some 2%–6% [6]. Annually, 859 million tons of CO_2 are released into the Earth's atmosphere by the aviation transportation system [1]. Due to the depleting source and the increasing demand of crude oil, instability in crude oil price, and the associated environmental impact, the aviation industry might face an unsustainable future in terms of long-term planning and

Copyright © 2022 Elsevier Inc.
All rights reserved.

operating expenses. Therefore, the increasing urge to mitigate the energy crisis and the environmental pollution has driven the researchers to find alternative jet fuel sources that are clean, affordable, environmentally friendly, and reproducible. Among the proposed alternatives, biomass has emerged as one of the most prospective sources for jet fuel production due to their easy availability, high productivity, low pollution, and carbon neutrality, paving a route to sustainable energy production ensuring energy security [7]. Liquid jet fuel derived from biomass, such as plants, algae, organic wastes is known as biojet fuel. Major sources of feedstocks for obtaining biojet fuel come from sugar sources (sugar cane, starch, etc.), lipids (vegetable oils, animal fat, and algae) and lignocellulosic biomass (wood, agricultural, forest residues, etc.) [8]. Biojet fuel is carbon neutral as they have a closed carbon cycle where the plants reabsorb the CO_2 released from burning of the fuel during photosynthesis and starts the cycle by being converted to biofuel again. It has been proved that burning of biojet fuel can significantly reduce the GHG emission compared to the conventional one. Open pond algal oil derived biojet fuel demonstrated a GHG emission of 1.5 g CO_2e/MJ, which is a 98% reduction relative to petroleum-derived jet fuel [9].

Based on the type of feedstocks, there are many processes available for the conversion of biomass to biojet fuel of which some are at commercial stage and others are in the research and development stage. Among them, oils-to-jet, syngas-to-jet, alcohols-to-jet, and sugars-to-jet have been widely studied, of which biojet fuel from oils-to-jet and syngas-to-jet route have received approval by ASTM international method (D7566) to be blended with conventional aviation fuels up to 50% [10,11]. Triglycerides (TG)-based feedstocks, such as vegetable oil, animal fat, and microbial oil can be hydrotreated to aviation fuel. Lignocellulosic biomass can be converted to biooil and syngas through fast pyrolysis and gasification, respectively, which can further be hydroprocessed to jet fuel. Lignocellulosic biomass can also be depolymerized and subsequently transformed to various platform molecules that can undergo catalytic upgradation to jet fuel. Sugar-based feedstocks can either be catalytically converted to platform molecules or biochemically converted to alcohols, followed by conversion to biojet fuel. The jet fuel is designed specifically to run gas-turbine engines and has stricter quality requirements compared to other transportation fuels. Whatever the feedstock or intermediate products, the overreaching aim of the upgradation of the biomass to jet fuel is to reduce the oxygen content and unsaturation, increase the cracking or coupling of C—C to obtain chain length in the range of C_8–C_{17}, branching, isomerization, and aromatization. The upgraded fuel will have the desired jet fuel properties including high heating value, high energy density, low viscosity, improved clod flow properties, low freezing point, and high flash point. Catalyst plays an important role in all of these upgradation steps by providing suitable reaction pathway to obtain the desired properties.

Considerable attention has been focused on biojet fuel as an alternative to traditional fossil-based fuels because of the rapid depletion of finite fossil fuel sources and the environmental pollution caused by their combustion. Many governments and international organizations are issuing directives to promote the use of biomass-derived jet fuels. International aviation industry has committed to cut the GHG emissions by 50% before 2050. In 2011, a memorandum of understanding for USD 500 million investment was signed between the U.S. Department of Agriculture and the U.S. Department of Energy to produce drop-in aviation fuels for military and commercial applications [3]. International Civil Aviation Organization is promoting Carbon Offsetting Reduction Scheme for International Aviation that is aimed at mitigating around 2.5 billion metric tons of CO_2 in the airline sector between 2021 and 2035 along with cutting the net carbon emission in half by 2050 compared to that of 2005 [12]. To achieve this goal by fully exploiting the benefit of biojet fuel, commercialization of biojet fuel is necessary which largely relies on the feasible conversion technologies based on the type of feedstock and catalyst used. Therefore, development of low cost and easily available catalyst with high activity and selectivity to targeted product is crucial to the industrial scale upgradation of biomass to biojet fuel. For this purpose, comprehensive knowledge on the types of catalyst and their specific role on different steps of biomass conversion to biojet fuel is necessary, and the aim of this chapter is to address them with asserting associated limitations and suggesting future outlooks.

2. Roles of catalyst on different feedstock

The biomass that are typically used for biojet fuel production include triglycerides, lignocellulosic biomass, and sugar and starchy materials. Their conversion pathway to biojet fuel depends on the nature of the feedstock and generally comprises hydroprocessing of triglyceride feedstock, thermochemical processing of biomass to gas or bio-oil and subsequent hydroprocessing, and sugar derived alcohol to jet [13]. Catalyst plays an important role in the transformation processes of biomass to aviation fuel, by providing reaction pathway with moderate conditions. However, among the processes, only the hydroprocessing of triglyceride and the thermochemical conversion of biomass by gasification and the subsequent Fischer-Tropsch conversion of the gases to liquid hydrocarbons are ASTM-certified pathways for commercial production of biojet fuel [14].

2.1 Roles on vegetable oil

Vegetable oils are a suitable alternative to conventional fossil fuels as they come from renewable sources. Moreover, they are environment friendly in nature and can be converted to fuel to be used in transportation facilities. Vegetable oils consist of triacylglycerols (TAGs) or triglycerides (TGs) which are mainly triesters of glycerol and possesses the similar number of carbons in their backbone as the fossil source derived liquid fuel [15,16]. Vegetable oils possess a high viscosity ranging from 25 to 50 cP, due to its long carbon chain and large molecules containing oxygen [17]. The H/C molar ratio of vegetable oil typically ranges from 1.64 to 2.37 and the oxygen content varies between 10.5% and 14.5% [18]. They differ in the length of their fatty acid chains on which the quality and yield of the produced hydrocarbon fuel depends. Most vegetable oils contain C_{16} and C_{18} long fatty acids [19], which can be saturated and/or unsaturated. The saturated fatty acids contain mainly palmitic acid ($C_{16}H_{32}O_2$), stearic acid ($C_{18}H_{36}O_2$), arachidic acid ($C_{20}H_{40}O_2$), and docosanoic acid ($C_{22}H_{44}O_2$), whereas the unsaturated one contain mainly oleic acid ($C_{18}H_{34}O_2$) and linoleic acid ($C_{18}H_{32}O_2$) [17]. The higher heating value (HHV) of TGs based oil varies from 37.1 to 40.6 MJ/kg, which are significantly higher than the HHV of biomass derived pyrolysis oils which ranges from 16.00 to 20.00 MJ/kg. Because of the presence of high paraffinic components, TGs show good potentiality as a renewable jet fuel in commercial airlines and jet fuel industries [20].

Though the H/C mole ratio, heat of combustion, and the molecular formation of TGs based vegetable oils nearly match with that of the jet fuel [21], they need to be further processed to attain the applicability in aviation sector. Aviation or jet fuel should contain a mixture of different types of hydrocarbons, such as paraffins, cycloparaffins or napthenes, aromatics, and olefin compounds, having the carbon chain length ranging from C_8 to C_{16} with a low level of oxygen content [21–23]. Among them, paraffins, isoparaffins, and cycloparaffins are the dominating one and typically remain in the range of 70%–85%. Owing to high hydrogen to carbon ratio, normal paraffins are needed for high specific heat, whereas isoparaffins and cycloparaffins are required to reduce the freezing point and increase the flash point of jet fuel suitable for operation in higher altitude. Aromatic hydrocarbons typically present in the range of 8%–25% and contribute to the enhancement of energy density and foster the swelling of elastomeric seal to a certain extent and thus prevent potential fuel leakage issues [23]. However, their presence in excess may lead to excessive smoke formation that can reduce the lifespan of engine, damaging the combustion chamber, or the turbine blades [24]. Therefore, to remove the oxygen content from TGs and convert them into straight chain paraffins, cycloparaffins, aromatics, and olefin compounds, hydroprocessing or catalytic hydrotreating of vegetable oil in the presence of hydrogen and an appropriate catalyst is mandatory.

In hydroprocessing pathway, the vegetable oil feedstocks go through several consecutive and/or simultaneous processes, such as hydrogenation/hydrogenolysis, hydrodeoxygenation (HDO), hydroisomerization, and hydrocracking to produce biojet fuel [3,25]. The produced hydrocarbon is also known as hydroprocessed renewable jet (HRJ). With hydrogen and solid catalyst at high pressure and temperature, the TGs undergo hydrogenation reaction to get saturated if there is any unsaturation in the carbon backbone of the fatty acid and subsequently the hydrogenolysis reaction frees the fatty acids and propane gas from the TGs. Here, the propane gas is produced from the glycerol portion of the TGs through the saturation of C=C bond followed by the cleavage of C—O in the presence of hydrogen. Decomposition of TGs also follows other reaction pathways, such as β-elimination, γ-H migration, carbon-carbon scission, and direct deoxygenation that led to mainly formation of fatty acids with some hydrocarbons [26–28]. In particular, direct deoxygenation occurs over noble metal and metal sulfide catalysts. Later, these fatty acids get converted to long chain hydrocarbons through an oxygen removal process in the presence or absence of hydrogen. In the presence of hydrogen, carbonyl oxygens get removed through hydrodehydration (HDH) and decarbonylation (DCO), leaving water and CO as byproduct, respectively. In general, three times hydrogen is required for HDH process than the DCO process [29]. On the other hand, release of the carboxyl groups in fatty acids and esters through decarboxylation (DCO$_2$) process produces CO$_2$ as a byproduct and does not require the presence of hydrogen. HDH pathway produces hydrocarbon with the same carbon length as in the fatty acid, while DCO and DCO$_2$ pathway produce hydrocarbon that are one carbon atom less than the fatty acid chain. Fig. 1 demonstrates the effect of catalyst on different deoxygenation routes to alkane production from TGs. The deoxygenation process produces the linear alkanes and do not contribute to the formation of other essential components of jet fuel such as isoalkane, cycloalkane, and aromatics that are required to lower the pour and cloud points in their linear counterparts and thus improve the cold flow properties of the jet fuel. The produced long chain hydrocarbons are then subjected to hydroisomerization and hydrocracking to attain the jet fuel characteristics of having hydrocarbon chain length ranging from C$_8$ to C$_{16}$. Being an exothermic reaction, HDH is favorable at low reaction temperature, whereas high reaction temperature favors the DCO and DCO$_2$, as they are endothermic in nature [30]. Hydroprocessing of the vegetable oil generally produces hydrocarbons in the range of C$_{15}$–C$_{18}$ since the most common vegetable oils (e.g., soybean, palm, canola, rapeseed, corn, camelina, jatropha) are composed of C$_{16}$ or C$_{18}$. Hence, a certain degree of hydrocracking (C—C bond breaking) is also desired to adjust the carbon chain length within the range of jet fuel, i.e., C$_8$–C$_{16}$. Isomerization and hydrocracking generally occurs over acidic catalyst supports. An alternative route to the production of TGs is thermal hydrolysis or catalytic hydrothermolysis where the TG feedstock is processed with 3 mol of water which dissociates to produce hydrogen ion and hydroxyl ion; the hydrogen ions attach to the glycerol backbone to form 1 mol of glycerol and the hydroxyl ions attach to the ester group to yield 3 mol of fatty acids [31,32]. This process of extracting fatty acid from vegetable oil is energy intensive and, hence, is not well practiced [33,34]. However, the produced fatty acids with oxygen content and unsaturation have to go through the steps of hydrotreatment pathway, such as hydrogenation, HDO, hydroisomerization, and hydrocracking process to become compatible for aviation transports. Fig. 2 shows the catalytic conversion of alkanes to other jet fuel components.

In all of the steps, catalyst plays a significant role and the best candidates are those who have high stability, surface area, moderate porosity, and activity and selectivity toward the jet fuel range hydrocarbons. Usually heterogeneous catalyst is employed for hydrotreatment owing to its easy separation, reusability, low corrosivity compared to the

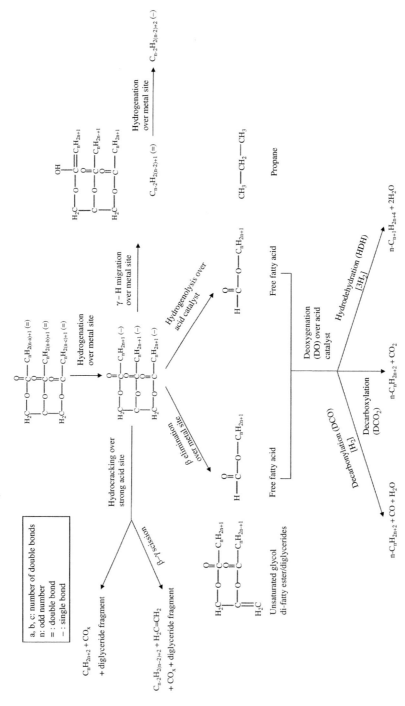

Fig. 1 Effect of catalyst on different deoxygenation routes to alkane production from TGs.

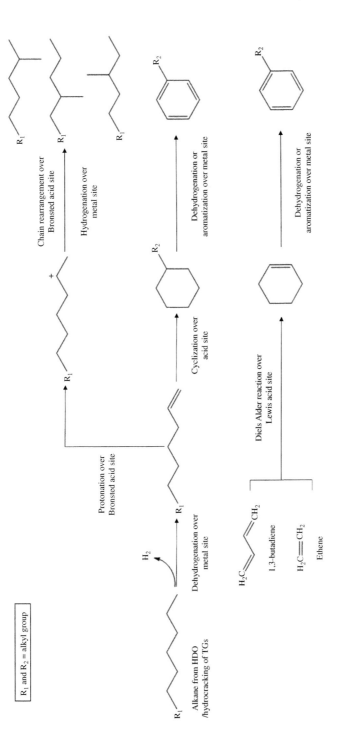

Fig. 2 Catalytic conversion of alkanes to other jet fuel components.

homogeneous one, and high selectivity. Among them, noble metals (e.g., Pt, Pd, Ru, etc.) and transition metal composite catalysts like NiMo, CoMo, NiCo have been found to facilitate the HDO process, such as HDH, DCO, DCO_2 reactions [35]. Acidity of the catalyst is also a crucial factor and the catalysts with moderate acidity are desirable as they facilitate the HDO reaction and produce alkanes with C_7–C_{14} selectivity. However, high acidity may lead to excessive hydrocracking and thus reduce the jet range hydrocarbon yield and cause undesired gas and coke formation [35]. However, noble metals come with some limitations as they are susceptible to sintering at elevated temperature that cause their activity to decay with time. Water produced during HDO reaction can also poison the active sites of noble metal catalysts. Moreover, they are costly enough to be economically viable for large-scale application [35]. Transition metals are a better choice as catalyst and/or catalyst support because of their high surface-to-volume ratio, low cost, and high surface energy, which makes their surface atoms very active [36].

Transition metal niobium-based materials are promising catalysts for deoxygenation reaction of vegetable oil due to their acidic and redox properties. Moreover, they also perform isomerization, cyclization, and aromatization of the deoxygenates produced from the deoxygenation reaction of TGs. During the hydroprocessing of soybean oil in the presence of hydrogen at 10 bar, niobium phosphate ($NbOPO_4$) catalyst showed a great efficiency in converting TGs into jet range fuel that included linear and branched alkanes [26]. Apart from the isomerization reaction, $NbOPO_4$ catalyst also took part in the formation of aromatic compounds from olefines and cycloalkane through its acid sites. The pyridine adsorption-desorption study showed the predominance of Brønsted acid sites and the NH_3 TPD (temperature programmed desorption) confirmed the catalyst was of high acid strength with moderate and strong acid sites. The decomposition of TGs followed the β-elimination pathway as evidenced by the formation of fatty acids. The fatty acids then followed HDH and DCO_x to yield hydrocarbons. Some portion favored the carbon-carbon scission between the β and γ carbon atoms that produced C_{n-3} hydrocarbons (pentadecane, C_{15}). However, pentadecane could also be generated form DCO_x (DCO and DCO_2) of C_{16} fatty acid such as palmitic acid. Moreover, a considerable amount of thermal cracking occurred on $NbOPO_4$ that led to alkanes and olefins.

Incorporation of a second metal has synergistic effect as demonstrated by Kubička and Kaluža [37] during the upgradation of rapeseed oil over Ni, Mo, and NiMo supported on Al_2O_3 catalyst. NiMo catalyst showed 100% conversion efficiency compared to 80% with each of the Ni and Mo catalyst. Moreover, the selectivity to C_5–C_{15} hydrocarbon was nearly 100% with NiMo system, whereas with Ni and Mo the selectivity was 60% and 50%, respectively. The synergistic effect improved the fuel grade alkane selectivity while minimizing the drawbacks of high undesired oxygenated intermediates with Ni and low alkane yield with Mo. Ni can be used as a promoter to increase the activity of Mo catalyst [38]. The synergistic effect of NiMo was pronounced even without the sulfidation, with n-C_{17} and n-C_{18} as the predominant product produced through HDH and DCO$_x$ [39]. The use of Mo (or Co) as a support could improve the hydrogenation sites of the bimetallic Co-Mo catalysts [36]. Bimetallic catalyst can reduce the production of CO, a catalyst poison, by weakening the interaction of CO with catalyst and converting it to methane and water. Lee et al. [40] observed that PtRe on ultra-stable Y (USY) zeolite catalyst showed better CO tolerance compared to the conventional Pt/USY. Bimetallic catalysts such as Pt-Re and Pt-Sn on γ-Al_2O_3 were tested in deoxygenation of stearic acid (a lipid model compound) in the absence of external hydrogen [41]. In-situ hydrogen was supplied from the dehydrogenation of ethyl-cyclohexane on the bimetallic sites. However, excessive cracking of C_{17} or C_{18} alkanes occurred over the Lewis acid sites possibly originated from Re and Sn oxides. Hence, metal site should be controlled accordingly. Moreover, metal loading can alter the reaction pathway preference. Zhang et al. [42] investigated the effect of MoS_2 modified with Ni and Co in upgradation of canola oil and found that HDH was more active on NiMoS and DCO$_x$ was the preferred pathway over CoMoS. However, the synergistic effect of Ni and Co showed more than 90% selectivity toward the n-C_7 to n-C_{18} alkanes. In another study, Ni doping in MoS_2 favored the substitution of S edge with Ni that enhanced the deoxygenation activity in Ni-MoS_2 compared to MoS_2 [43]. The overall effect was a reduction in activation energy for C=O hydrogenation and C—OH bond breaking steps induced by Ni-Mo-S active phase [44]. Addition of W with Ni favored DCO$_x$ pathway, whereas addition of Mo with Ni chose HDH route while supported over either SAPO-11 or silica-alumina catalyst [45,46].

Nonmetallic inclusion in catalyst such as sulfur and phosphorous has impact on the catalytic conversion of feedstock to biojet fuel. Incorporation of sulfur improves the catalyst stability and increase the availability of active metal sites for reaction [47]. However, during hydroprocessing, considerable amount of hydrogen in the feed can react with sulfur to produce H_2S that can promote the decay of catalyst activity. Moreover, the produced H_2S and sulfur containing residues can get released into the environment and cause contamination and global warming. Therefore, nonsulfide catalyst, especially nonsulfide zeolite catalyst is favorable.

Catalyst support enhances the activity of a catalyst by increasing the dispersion and accessibility to the reactant materials and thus, they play an important role in heterogeneous catalysis. If the catalyst support also contains active sites, then they can participate in the reaction by forming bifunctional catalyst. Catalyst supports are mainly composed of oxides, zeolites, carbon, heteropoly acids, etc. Usually, acidic catalyst supports are sought for upgradation of TGs to biojet fuel. Among them, the widely used catalyst supports are ZSM-5, SAPO-11, Al_2O_3, ZrO_2, SiO_2. Neutral activated carbon (AC) is also used as catalyst support. It has been shown that the catalyst support can act as the main source of catalyst acidity, i.e., Brønsted and/or Lewis acidity that contribute to isomerization and cracking reaction [43]. However, catalyst support with moderate acidity is always favorable as high acidity can increase the cracking of generated liquid (both n- and i-) alkanes, reducing the biojet fuel selectivity. During the upgrading of rapeseed oil over zeolites HZSM-5 and H-Y supported Pt, HZSM-5 produced higher (21%) light C_1–C_4 cracking species than that of H-Y (<4%) due to higher strong acid sites of HZSM-5 than H-Y [48]. Supports with narrow pore size are not desirable as the large TGs and fatty acids cannot come into contact with the active sites of catalyst. Thus, mesoporous supports are favorable for hydrotreating reaction as they help to avoid the diffusion limitation and minimize pore blocking by large hydrocarbons [49].

Alumina is a promising catalyst support for the catalytic hydroprocessing of vegetable oil due to their high surface area and acidity [17]. The BET surface area of alumina-based catalysts usually ranges from 148 to 358 m^2/g. [50]. Modification with noble metal catalysts such as Pd, Pt can enhance the hydrogen activation and also increases the catalyst lifetime as they are less likely to be deactivated. They have a high hydroprocessing activity, but their application on commercial scale is limited due to their high cost [51]. Metal loading on alumina can promote electron pair acceptors or Lewis acid sites. Noble metal sulfide or transition metal (e.g., Co, Mo, Ni, W) sulfide on alumina support are widely used for hydroprocessing of vegetable oil and they have been in use in petrochemical industry for the removal of heteroatoms such as nitrogen or sulfur from crude petroleum [52,53]. However, alumina support has a drawback as it becomes instable in water environment during hydroprocessing [54]. In the presence of water, the acidic sites of alumina are susceptible to poisoning along with crystalline rupture. Zeolite, silica, and AC are more stable in acidic environment than alumina. Modification of alumina support with silica further enhanced the hydrotreatment of jatropha oil over NiMo catalyst with 100% conversion and >90% selectivity [55]. AC is a promising support because of its textural properties, thermal stability, low catalyst deactivation due to its neutrality [17]. Sousa et al. have studied the conversion of palm oil to biojet fuel over Pd/C at 10 bar H_2 and 300°C and obtained a biofuel yield higher than 85% [56]. Morgan et al. investigated the performance of three types of catalysts, such as Pd/C (5 wt%), Ni/C (20 wt%), and Pt/C (1 wt%) in the deoxygenation of triglycerides (tristearin, triolein, and soybean oil) under 7 bars of N_2 pressure at 350°C in a batch reactor [57]. In this condition, high yields of cracking products were observed with increasing unsaturation of the triglyceride. Pd/C and Pt/C showed lower activity toward the triglyceride deoxygenation compared to Ni/C.

Zeolite-based catalysts are also popular for vegetable oil upgrading due to their high BET surface area ranging from 56 to 1126 m^2/g. Various metal elements such as Zn, Ni, Mo are used to modify the activity of the zeolite catalysts [58]. Zeolite shows both Brønsted and Lewis acidity required for hydrocracking and hydroisomerization. The acid sites originate from the Al atoms: when they are located into zeolite framework, Brønsted acidity generates, and when the dealumination process occurs, Lewis acidity is encountered in zeolites [59]. Acid sites can be classified as strong acid sites, medium acid sites, and weak acid sites, and control of their density and strength can affect the product selectivity [17]. Moderate acid site is required for vegetable oil conversion to biojet fuel. With increasing Si/Al ratio, the acid site decreases that results in increased higher hydrocarbon in the final product. Wang et al. [60] investigated the role of acidity of zeolite supports on the conversion of soybean oil to liquid hydrocarbons using Ni as the active metal site. Among the studied catalysts (Ni/SAPO-11, Ni/ZSM-22, Ni/ZSM-23, Ni/ZSM-5, Ni/Beta), Ni/ZSM-5 yielded the lowest conversion (80.1%) and alkane selectivity (67.4%) with excessive cracking (52.3%) due to its higher acidic sites (2.920 mmol (NH_3) g^{-1}). However, the catalyst with moderate acidity (Ni/SAPO-11, 1.842 mmol (NH_3) g^{-1}) was more selective to the formation of liquid alkanes (100% alkane selectivity and 86.5% isomerization selectivity) with limited cracking (16%). Crystal size of zeolite also plays an important role on the processing of vegetable oils. In spite of having high acid sites, with large crystal size that is responsible for small pore structure, the vegetable oils may not approach the acid site and cannot take the advantage of the catalytic effect. On the other hand, larger pore structure can release the heavy molecules without secondary cracking which may contribute to diesel like fuel selectivity, impairing the jet fuel characteristics. Mesoporous material can promote the diffusion of reactants to the active site of the catalyst. For instance, mesoporous zeolites (e.g., Ni/mesoporous-Y and Ni/mesoporous-HZSM-5) facilitated the transport of bulky vegetable oil molecules to the catalyst active sites through their larger mesopores [61]. Therefore, the introduction of mesopores into a microporous zeolite can enhance the diffusion of large molecules while harnessing the benefit of crystalline structure and acidity of the zeolite. Example of such catalyst includes hierarchical mesoporous crystalline β zeolite synthesized by using a dual-functional template that contains both micro- and mesopores [62]. Moreover, catalysts with medium-sized porosity are better for improved cracking process [63]. Table 1 lists some important catalysts that are suitable for the conversion of TGs to aviation fuel.

Kim et al. investigated the conversion of palm oil derived TGs to aviation fuel in a two-step process: firstly, the TGs were converted to alkane through hydrogenation of C=C, hydrogenolysis into fatty acids, and deoxygenation (DCO_x was prevalent over HDH) over Pt/γ-Al_2O_3 catalyst; then the alkanes were hydrocracked and hydroisomerized over acidic Pt/nano-Beta catalyst [80]. Therefore, it is evident that the hydrotreatment catalyst mainly produces linear and slightly branched alkane, but cannot produce isomers, cycloalkanes, and aromatics that are desired in a jet fuel. To get the hydrogenation/dehydrogenation and isomerization reaction with a single catalyst, bifunctional catalyst with metallic sites could be a promising option. Here, the metal site is responsible for hydrogenation/dehydrogenation reaction and the acid sites are responsible for isomerization reaction. Normal paraffins get dehydrogenated over the metal sites to produce olefin that protonates on the acid site with formation of the alkylcarbenium ion. The produced carbenium ion get rearranged to branched ions on the acid site.

Table 1 Successful catalysts for the conversion of TGs to jet range fuel.

Catalyst	Feed	Reaction condition	Product	Reference
PtRe/USY	Palm oil	$P = 5$ MPa H_2, $T = 220–310°C$, WHSV $= 2\ h^{-1}$	41 wt% biojet fuel [C_{15}–C_{18} (70%), C_8–C_{14} (20%)]	[40]
NiMo/γ-Al$_2$O$_3$	Soybean oil	$P = 9.5$ MPa H_2, $T = 400°C$, $t = 2$ h	92.9% hydroprocessing conversion	[64]
CoMo/γ-Al$_2$O$_3$	Soybean oil	$P = 9.5$ MPa H_2, $T = 400°C$, $t = 2$ h	78.9% hydroprocessing conversion	[64]
Ni-W/ SiO$_2$-Al$_2$O$_3$	Jatropha oil	$P = 0.2–0.9$ MPa H_2, $T = 340–420°C$	25%–30% yield of aviation bio-kerosene	[65]
WO$_x$/Pt/TiO$_2$	Jatropha oil	$T = 240°C$, $t = 6$ h	86% conversion to C_{17} alkane	[66]
NiMo/SAPO-11	Jatropha oil	$P = 6–8$ MPa H_2 $T = 375–450°C$ LHSV $= 1\ h^{-1}$	C_9–C_{14} yield 22%–37.5%	[45]
NiMoS$_2$/γ-Al$_2$O$_3$	Palm oil	$P = 5$ MPa H_2, $T = 270°C$, WHSV $= 1\ h^{-1}$	28% biojet yield	[67]
Ni$_2$P/silica	Soybean oil	$P = 3$ MPa H_2, $T = 330°C$, LHSV $= 3\ h^{-1}$	82.1–82.6 wt% alkane yield	[68]
Ni/SAPO-11	Palm oil	$P = 4$ MPa H_2, $T = 360°C$, LHSV $= 1\ h^{-1}$	70% alkane yield	[69]
NiP/SAPO-11	FAME	$P = 2$ MPa H_2, $T = 340°C$, WHSV $= 2.5\ h^{-1}$	84.5% C_{15}–C_{18} yield	[70]
CoMo/Al$_2$O$_3$	Cotton seed oil	$P = 3$ MPa H_2, $T = 320°C$, WHSV $= 5\ h^{-1}$	90% conversion to liquid alkane	[71]
NiMoCe/Al$_2$O$_3$	Jatropha oil	$P = 3.5$ MPa H_2, $T = 370°C$, LHSV $= 0.9\ h^{-1}$	80% yield and 90% selectivity to C_{15}–C_{18}	[72]
NiMo/Al$_2$O$_3$	Jatropha oil	$P = 2.8$ MPa H_2, $T = 390°C$, LHSV $= 32.28\ h^{-1}$	25% jet fuel	[73]
CoMO/Al$_2$O$_3$	Jatropha oil	$P = 2.8$ MPa H_2, $T = 390°C$, LHSV $= 32.28\ h^{-1}$	25% jet fuel	[73]
7 wt% Ni/nano-sized SAPO-11	Palm oil	$P = 4$ MPa H_2, $T = 280°C$, $t = 6$ h, LHSV $= 2\ h^{-1}$	79 wt% liquid alkane yield	[74]

Table 1 Successful catalysts for the conversion of TGs to jet range fuel—cont'd

Catalyst	Feed	Reaction condition	Product	Reference
Ni/SiO$_2$-Al$_2$O$_3$	Soybean oil	$P = 9.5$ MPa H$_2$, $T = 400°C$, $t = 2$ h	n-Alkane content (46.3 wt%)	[64]
Pd/γ-Al$_2$O$_3$	Soybean oil	$P = 9.5$ MPa H$_2$, $T = 400°C$, $t = 2$ h	n-Alkane content (85.7 wt%)	[64]
Ni-MoS$_2$/γ-Al$_2$O$_3$	Palm kernel oil	$P = 5$ MPa H$_2$, $T = 333°C$, LHSV $= 1$ h^{-1}	Jet fuel-like hydrocarbon (>90%)	[43]
Zn/ZSM-5	Camelina oil	In absence of H$_2$, $T = 500°C$, LHSV $= 0.6$ h^{-1}	C$_7$–C$_{15}$ content 77.48%	[75]
Ni/mesoporous-Y	Palm oil	$P = 3$ MPa H$_2$, $T = 333°C$	31% yield	[61]
NiMo/Al$_2$O$_3$	Coconut oil	$P = 3$ MPa H$_2$, $T = 360°C$, LHSV $= 1$–3 h^{-1}	60% yield of jet fuel	[76]
MoNi/γ-Al$_2$O$_3$	Coconut oil	$P = 0.8$ MPa H$_2$, $T = 350°C$, LHSV $= 1$ h^{-1}	71% yield of C$_8$–C$_{16}$	[77]
Sulfided NiW/HZSM-5	Jatropha oil	$P = 6$ MPa H$_2$, $T = 400°C$, LHSV $= 1$ h^{-1}	99% conversion and 39.6% yield of C$_9$–C$_{15}$	[78]
Sulfided NiMo/HZSM-5	Jatropha oil	$P = 8$ MPa H$_2$, $T = 400°C$, LHSV $= 1$ h^{-1}	99% conversion and 38.3% yield of C$_9$–C$_{15}$	[78]
Ni/MCM-41-APTES (7.5%)-USY	Castor oil	$P = 3$ MPa H$_2$, $T = 300°C$, WHSV $= 2$ h^{-1}	99% conversion and 80.3% yield of C$_5$–C$_{18}$	[79]

Finally, the alkylcarbenium ions get deprotonated and produce isoparaffins [81]. Liu et al. reported an 80% yield of C$_8$–C$_{15}$ alkanes (iso/n-paraffin ratio = 4.4) from castor oil, over a (3-aminopropyl)-triethoxysilane (APTES)-modified MCM-41/USY composite-supported Ni bifunctional catalyst [79]. In this system, the acidity of the catalyst can be moderated by changing the metal concentration to tune the degree of hydrocracking. However, these bifunctional catalysts can experience catalyst deactivation through coke deposition. Alkenes are considered as carbon-deposit precursors that can be generated over the active metal sites via dehydrogenation of naphthenes and then get converted to polymeric aromatics or coke on the adjacent acid sites [82]. HY zeolite is more acidic (1.71 mmol NH$_3$ g^{-1}) than HZSM-5 (0.27 mmol NH$_3$ g^{-1}) and produce more polycyclic hydrocarbons, coke, and tar during jet fuel production from TGs [21]. The high hydrogen pressure help decrease the coke formation. However, due to its high cost, the target is to minimize the amount of hydrogen during hydroprocessing. At elevated temperature, acidic bifunctional catalyst can increase the cracking reaction by breaking C—C bond, decreasing the yield of jet fuel range hydrocarbon. It has been reported that during the biojet fuel conversion of sunflower oil over acidic 5 wt% Pd/Al-SBA-15, the yield of C_{15}–C_{18} alkanes decreased from 73% to 53% due to an increase in temperature by 50°C [83]. Dominion of DCO_x over HDH increased the odd/even carbon chains that contributed to more fission and unwanted lighter hydrocarbon production.

Catalyst loading has an impact on the product selectivity in hydroprocessing reaction. Low amount of catalyst can yield low acidity, whereas high catalyst loading can cause severe hydrocracking leading to increased gas and coke formation and decreased jet fuel yield [79]. Overcracking results in the formation of light species with carbon chain length ranging from C_1 to C_4 and naphtha ranging from C_5 to C_8 [10]. In one study, Liu et al. [69] employed Ni/SAPO-11 catalysts with different Ni loadings (2, 5, 7, and 9 wt%) to investigate the impact of catalyst loading on the HDO product of palm oil. The yield of liquid alkane and the isomerization selectivity were significantly increased from 60% to 67.4% and 46% to 61.5%, respectively, with the increase in Ni loading from 2 to 7 wt%. Further increase in Ni loading to 9 wt% did not alter the alkane yield, but improved the isomerization selectivity to >83%. Wang et al. [60] also observed the same phenomena with increasing the Ni loading from 0 to 8 wt% over SAPO-11 support during the conversion of soybean oil. The improve in isomer selectivity was due to the balancing of strong acid sites on the catalyst support and the increase of active metal sites due to the increase in catalyst loading. However, a higher loading is not encouraged due to the chance of excessive cracking from pore blockage [35]. Therefore, an optimized catalyst loading is crucial for ensuring the desired selectivity and yield of the product.

2.2 Roles on microbial oil

Among the potential biomass feedstocks for biojet fuel production, microalgae have emerged as a promising source as they do not compete with edible food and crops and arable land, and can grow in aquatic environment and produce higher amount of lipid content than terrestrial biomass. Microalgae can hold lipid up to 50% of their dry biomass [84]. Moreover, continuous removal of algal biomass can improve the native ecology of the aquatic system by ensuring nutrient balance [85]. Hence, algal biomass (micro- and macro-algae) is a potential source of biojet fuel production and is classified as third-generation biofuel. Microalgae have gained favor over macroalgae as the microalgae have faster growth rates, higher oil contents, and less complex structures than macroalgae. Moreover, macroalgae are more difficult to grow in bioreactors which limited their acceptance in scientific community [86]. Carbon numbers of algae derived fatty acids (mainly varies between C_{16} and C_{18}) generally exceed those of aviation fuel hydrocarbons (C_8–C_{16}) and hence, they need to be further processed [87]. Phyla of Cyanobacteria, Haptophyta, and Euglenozoa have been reported as the most suitable species for biojet production [88].

The oil in most algae is stored in the form of TGs which need further processing to be used as transportation fuel [89]. Transesterification of the extracted oil/lipid to biodiesel is not suitable for transportation fuel as it contains considerable amount of oxygen as ester, fatty acid, and alcohol, which reduces the energy content of the fuel [90]. One viable option is hydroprocessing of esters and fatty acids (HEFA) for the upgradation of microbial oil to biojet fuel. HEFA mainly involves deoxygenation reaction, however it includes some other catalytic reactions, such as cracking, isomerization, hydrogenation, dehydrogenation, alkylation, oligomerization, etc. in the presence of catalyst [29]. Generally, the reaction process occurs as follows: hydrogenation of carbon-carbon double bonds; hydrogenolysis of TGs to fatty acids and propane; and HDO (HDH/DCO_x) of fatty acids and their derivatives to n-alkanes. In order to achieve the desired physical properties (e.g., freezing point and flash point) of aviation fuel, the resulting n-alkanes must then be isomerized and/or cracked to appropriate branched i-paraffins. This process requires a lot of hydrogen because of oxygen removal as well as hydrogenation of olefin structures. For the hydroprocessing of lipid, a wide range of catalyst have been employed, such as metal sulfide, noble metal, transition metal, bimetallic metals, zeolites, oxides, and acid catalysts. Different catalysts have been tested including Ru/C, Pd/C, Ni-Cu/Al_2O_3, Ni-Cu/SiO_2, Ru/TiO_2, Ru/Al_2O_3, CoMo/γ-Al_2O_3, and NiMo/Al_2O_3 [91]. The most common HDO catalysts are CoMo-based and NiMo-based catalysts which are also industrially used for removal of sulfur, nitrogen, and oxygen [92]. Metals activate the hydrogen for deoxygenation reaction. However, they are suitable for deoxygenation but not for isomerization or branching and, therefore, need another step. For instance, Robota et al. reported conversion of algal TGs over 3% Pd/C to alkane that went through a hydroisomerization step over 0.5% Pt/US-Y zeolite catalyst to enhance the quantity of branched portion [93]. Among the catalysts, metal-based catalysts are used for deoxygenation purpose and acid catalysts are used for cracking, isomerization, cyclization, and aromatization reaction [90,94]. Hence, the catalysts that can perform both purposes are desired the most as jet fuel requires cyclic and aromatic hydrocarbons along with normal liquid hydrocarbon in the range of C_8–C_{16}. Verma et al. reported a bifunctional sulfided Ni-Mo catalyst on ZSM-5 zeolite support for the one-step conversion of algal oil to jet fuel and obtained a 77% yield of C_9–C_{15} hydrocarbons with iso/n-paraffin ratio of 2.5 [78]. Several other research studies have been done to assess the potentiality of bifunctional catalyst for one-pot/single conversion of TGs to biojet fuel, but for algal TGs it is very limited, hence this area seeks further attention. In order to meet ASTM jet fuel specifications, some cracking of fatty acids is required which involves both C—C bond scission and rearrangements that result in chain branching.

Hydrocracking can be done using acid site containing zeolite as catalyst support to get the heavier fraction in the jet fuel range hydrocarbon. The extent of cracking can be controlled by adjusting the acid sites and operating temperature. The shape selectivity of the zeolite micropore has influence on the isomer to alkane ratio and the extent of lighter alkane production. Zeolite with smaller pore (ZSM-5) yield lighter alkane and lower iso/n-paraffin ratio since the formation of highly branched isomers prior to cracking is inhibited due to small pore size. Therefore, larger pore zeolite Betas are suggested as they can accommodate multibranched alkanes, leading to higher iso/n-paraffin ratios which is highly desirable [80]. Lipid deoxygenation can also be done in absence of hydrogen. Lu et al. suggested an integrated process where dehydrogenation of lignin-derived cycloalkanes provided the necessary hydrogen for the deoxygenation of lipid over Pt-Ir/γ-Al$_2$O$_3$ at 400°C in the absence of external hydrogen [41]. The cycloalkane also provided some aromatic hydrocarbons necessary for good jet fuel. The in-situ glycerol and water also hold potentiality as hydrogen donor during hydrogen free deoxygenation reaction.

Microbes can also be converted into energy intensive crude bio-oil through thermochemical conversion such as hydrothermal liquefaction (HTL) and pyrolysis, of which HTL enhances the energy density of crude bio-oil more than the pyrolysis [95]. Pyrolysis requires high energy input to dry the feedstock, whereas HTL occurs in aquatic environment which is an excellent option for aquatic and wet biomass. Pyrolysis is typically done at medium to high temperatures (350–700°C) and 1–5 atm in the absence of oxygen. Catalysts can be employed in order to increase the yield or quality of the bio-oil. Catalysts such as Co/Al$_2$O$_3$, Ni/Al$_2$O$_3$, γ-Al$_2$O$_3$, ZSM-5, HZSM-5, and nickel phosphide have been used during pyrolysis and Na$_2$CO$_3$, KOH, CH$_3$COOH, HCOOH, NiO, Ca$_3$(PO$_4$)$_2$, H$_2$SO$_4$, and zeolite have been used during HTL [91]. In thermochemical conversion, apart from lipid, the protein and carbohydrates in the algae can be transformed into bio-oil and its yield and quantity depend on the feedstock biochemical properties, thermochemical process, and operating conditions. HTL-derived bio-oil has lower oxygen content and consequently higher energy content than pyrolysis derived oils since hydrogenolysis, DCO$_x$, and hydrogenation occur to some extent during HTL [96]. However, the crude bio-oil in general has several undesired properties, such as high viscosity, high oxygen and acid content, and low HHV, and thus it needs to be further processed before gaining applicability as transportation fuel. Bio-oil consists of several hundreds of organic compounds, such as acids, alcohols, aldehydes, esters, ketones, phenols, and guaiacols, to name a few [92,97]. The crude bio-oil can be upgraded to biojet fuel through hydrotreating which is done to increase the HHV and reduce the viscosity and decrease the O, N, and S contents of crude bio-oils. During the upgradation of algae derived bio-oil over Ru/C catalyst, a significant reduction in O, N, and S and subsequent increase in H$_2$O, NH$_3$, and H$_2$S production through HDH, hydrodenitrogenation, hydrodesulfurization reactions were observed and hence the hydrogen content in the final product decreased. Therefore, the authors suggested that the presence of hydrogen is necessary to get hydrocarbon with high hydrogen content [98]. Pentadecane was the predominant alkane in the upgraded bio-oils. Guo et al. derived bio-oil from the pyrolysis of *Chlorella* and *Nannochloropsis* in the presence of MCM-41 catalyst and the subsequent crude bio-oils were then further upgraded via HDO over a bimetallic Ni-Cu/ZrO$_2$ catalyst [99]. Zhao et al. used HTL process to derive bio-oil from *Nannochloropsis oceanica* and hydroprocessed the resulting crude bio-oil over a sulfided Ni/Mo/γAl$_2$O$_3$ catalyst [100]. Crude algal oils may contain considerable amount of phosphorus from phospholipids, nitrogen from proteins and metals (mostly magnesium) from chlorophyll which may contaminate the catalyst [90]. Hence, lipid purification as well as development of contaminant tolerant catalysts are necessary to achieve the most cost-effective process.

Some algal species produce hydrocarbon without heteroatom such as terpenes that are gaining increasing interest due to their similarity to petroleum [101]. Moreover, processing of these hydrocarbons does not require costly hydrogen for the removal of oxygen atoms. Squalene (2,6,10,14,18,22-hexaen-2,6,10,15,19,23-hexamethyltetracosane) from *Aurantiochytrium mangrovei* [102] and botryococcene (polymethylated triterpenes C$_n$H$_{2n-10}$ ($n = 30$–37)) from *Botryococcus braunii* [103] are some of the representative algae generated hydrocarbons. *B. braunii* oil (also known as Bot-oil) is composed of non-oxygenated triterpenic hydrocarbons (mainly C$_{34}$H$_{58}$) [104]. These hydrocarbons are basically heavy oil fraction which needs to be lightened by cleavage of some C—C bonds before their employment as transportation fuels. The C—C cleavage can be done over bifunctional catalysts, such as Pt/zeolite and Pt/SiO$_2$-Al$_2$O$_3$, and monofunctional catalysts, such as Ir/SiO$_2$, Ru/CeO$_2$, Ru/SiO$_2$ [105]. Bifunctional catalyst containing both solid acid and noble metal can lead to deoxygenation or isomerization and hence they are not desired for the processing of branched hydrocarbon, such as squalene and Bot-oil [106]. Monofunctional catalyst such as Ru/CeO$_2$ was found to convert squalane (2,6,10,15,19,23-hexamethyltetracosane) to jet range fuel through regioselective breaking of CH$_2$—CH$_2$ bonds located between the branches that maintained the branching in the final products. However, other monofunctional catalysts such as Rh/C, Ru/SiO$_2$, and Ru/C were not that much selective to jet range fuel due to terminal C—C bond breaking that removed the methyl branch as methane [107]. With bifunctional catalyst, catalytic cracking of Bot-oil using zeolite supported NiMo or CoMo catalysts in absence of hydrogen produced gasoline (C$_5$–C$_9$) or diesel range (C$_{16}$–C$_{20}$) fuel, whereas hydrocracking over Re-modified Pt/SiO$_2$-Al$_2$O$_3$ (SA) system produced biojet (C$_{10}$–C$_{15}$) fuel [108]. The optimum catalyst was pt-3%Re/SA for which the yield of biojet fuel was 47% and 40.3% from squalane (C$_{30}$H$_{62}$) and bot-oil, respectively. The C=C double bond in the oil were protonated on SA to form carbocations that subsequently broken down into lower hydrocarbons which then gone through hydrocracking over Pt-Re metal sites and partial aromatization over SA to produce jet fuel [108].

Biodiesel produced from algae or vegetable derived oil through transesterification process can also be upgraded to aviation range biofuel. Biodiesel is a mixture of fatty acid methyl esters (FAMEs) with carbon chain length of C_{14}–C_{24} [109]. Due to their high viscosity and oxygen content, they are not suitable for aviation turbine and need hydro-processing (HDO, hydrocracking, and hydroisomerization). Usually, noble metal-based bifunctional catalysts are employed, but noble metal is costly and susceptible to oxygen and heavy metal poisoning. Therefore, in one study, transition metal Ni-based bifunctional catalyst Ni/Hβ and Ni/HZSM-5 were used for the upgradation of FAME, however, they mostly produced diesel range ($>C_{16}$) hydrocarbon due to the shape selectivity of the zeolite support [110,111]. To overcome this, Ni-based hierarchical mesoporous zeolite (Ni/meso-Y zeolite) was tested for the conversion of microalgae biodiesel to jet fuel with 91.5% conversion and 56.2% selectivity [109]. Conversion of FAME to aviation fuel is still in nascent stage. Therefore, further studies are required in terms of finding appropriate catalyst for selective and efficient conversion.

2.3 Roles on lignocellulosic biomass

The biofuel versus food controversy of first-generation biofuels, mainly biodiesel and bioethanol, has contributed to a growing interest for second-generation biofuel form lignocellulosic biomass that are abundantly available in the continental earth mostly as low-cost waste material. The global production of lignocellulosic biomass is over 170 billion metric tons per year of which only 5% are being utilized by human [112]. Hence, lignocellulosic biomass holds a great potential to alleviate the growing energy demand as well as the environmental crisis. Generally, lignocellulosic biomass is composed of 35%–50% cellulose, 20%–35% hemicellulose, and 10%–25% lignin [113]. Cellulose is a linear polysaccharide consisting of D-glucose molecules bound together by β-1,4-glycoside linkages; hemicellulose is a branched polysaccharide comprising of copolymer of any of the monomer-glucose, galactose, mannose, xylose, arabinose, and glucuronic acid forming branched structure between cellulose and lignin; and lignin is a highly complexed three-dimensional polymer of three types of phenylpropane units bound together by ether and carbon–carbon bonds [114]. Cellulose and hemicellulose give the plant cell structural and mechanical strength, whereas lignin maintains the stability of these structures. Due to the complex, three dimensional, and insoluble nature of these components, conversion of lignocellulosic biomass is more difficult than vegetable oils or sugars [115].

The conventional way of lignocellulosic biomass conversion to biofuel occurs through fast pyrolysis and hydrothermal liquefaction (HTL), producing mainly a black oxygenated organic liquid called bio-oil [116]. These bio-oils are composed of more than 400 different kinds of oxygenated hydrocarbons including acids, aldehydes, alcohols, esters, ethers, furans, ketones, phenols, and carbohydrates [117,118]. Due to the presence of high oxygen content, bio-oil possess some undesirable characteristics, such as low heating value and low energy density, high density and viscosity, poor thermal and chemical stability, immiscibility with other hydrocarbon fuel, and high corrosivity [119]. Generally, the bio-oils derived from fast pyrolysis contain higher oxygen content, lower heating value and higher acidity than HTL derived one [120]. Therefore, bio-oils need significant modification in terms of oxygen content and chain structure to be used widely as transportation fuel. Due to the simplicity of the process, fast pyrolysis is widely practiced for biofuel production from lignocellulosic biomass which involves a rapid heating of the biomass in absence of oxygen at around 500°C and the subsequent quenching the produced vapors to generate the bio-oil [119]. As different portion of the biomass degrade to different intermediate products at different temperature, a relatively new concept-multistage thermal decomposition is also in practice instead of fast pyrolysis to ease the selective catalytic conversion of the bio-oil [121]. Typically, the bio-oil upgradation involves oxygen removal through either HDO process over noble metal supported or metal sulfide supported catalysts in presence of hydrogen or catalytic cracking over acidic zeolite catalyst at atmospheric pressure. However, the hydrocarbons produced through these routes are of low carbon chain length and hence are not suitable for aviation purpose [122]. Therefore, a route that comprises C—C coupling reaction followed by HDO is preferred. During the thermal degradation, cellulose yields furanics and anhydrous sugars (e.g., levoglucosan), hemicellulose depolymerizes into light oxygenates (e.g., acetic acid and acetol) and furanics (e.g., furfural and furan), and the lignin decomposes into single- and multimethoxylated phenolics (e.g., guaiacol and syringol) [123]. Carbon-carbon bond coupling of these oxygenates includes ketonization, alkylation, acylation, hydroxyl alkylation, and aldol condensation. During ketonization, two carboxylic acids are coupled to form a larger ketone with $2n - 1$ carbon chain length, expelling one molecule of H_2O and CO_2. Metal coated oxide catalysts, such as TiO_2, CeO_2, ZrO_2, and Ru/TiO_2, are generally used for this reaction [124,125]. Alkylation, acylation, and hydroxyl alkylation reactions combine light oxygenates with either furanics or phenolics acting as carbon acceptors, without loss of any carbon. Heterogeneous acidic zeolite-based catalysts (e.g., HZSM-5) and homogeneous acid catalysts (e.g., sulfuric acid and hydrofluoric acid) play a good catalytic role during this reaction [123]. Aldol condensation reaction occurs between the oxygenates containing carbonyl group (aldehydes and/or ketones) over basic catalyst without losing any carbon [126,127]. During the aldol condensation, α-H abstraction from a carbonyl molecule at a basic site forms enolate that attacks another electrophilic carbonyl molecule generated over acid site due to polarizing of the C=O group and thus C—C coupling is formed [128]. Moreover, hydrogenation and oxidation of the produced compounds over the metallic catalyst site can lead to further formation of acids and alcohols which can undergo C—C coupling through the above-mentioned reactions to larger hydrocarbon chain. Finally, hydrogenation and HDO over bifunctional catalysts make the hydrocarbon suitable for aviation application. Overall, this process of biojet fuel production form lignocellulosic biomass is known as hydro-treated depolymerized cellulosic jet (HDCJ) which has not been approved yet by ASTM [10]. Wang et al. proposed a new transformation route consisting of three steps for the conversion of pyrolysis bio-oil to jet range hydrocarbon [122]. Bio-oil derived from straw stalk was first converted to C_6–C_8 low-carbon aromatics and C_2–C_4 light olefins through catalytic cracking, deoxygenation (HDH, DCO_x), and aromatization over the acidic sites of the HZSM-5 zeolite. A [bmim] Cl-2AlCl$_3$ (1-butyl-3-methylimidazolium chloroaluminate) ionic liquid catalyst performed the alkylation of low-carbon aromatics with light olefins to produce C_8–C_{15} aromatics under the low temperature. Brønsted acid and Lewis acid of the ionic liquid promotes the protonation of light olefins to form active electrophilic species which enhances the alkyl addition to the aromatics. Finally, in the presence of hydrogen over Pd/C catalyst, the C_8–C_{15} aromatics get converted to C_8–C_{15} cyclic alkanes which is the second most abundant components found in the commercial and military jet fuels with the mass percent up to 50 wt% [122].

Biomass to liquid (BTL) is another process for the transformation of lignocellulosic biomass to aviation fuel where syngas (CO and H_2) produced form the biomass gasification is converted to liquid alkane through Fischer-Tropsch synthesis (FTS) reaction over metal catalysts. The liquid alkane is then hydrotreated (hydro deoxygenated, cracked, and isomerized) to achieve the molecular weight and structure of jet fuel range hydrocarbons [129–131]. The commonly employed catalysts for the FTS process are group VIII to X metals, such as Fe, Co, Ni, and Ru [132]. Among them, Ru has a good catalytic activity and selectivity, though it is not widely used due to its high cost [133]. Fe and Co are the most widely used catalysts for industrial production. Li et al. investigated the conversion of syngas to jet fuel over Co catalyst, which is an excellent catalyst for CO hydrogenation, on ZrO_2-SiO_2 bimodal support with both large and small pores [134]. The large pores contributed to speedy molecular diffusion to get high C_{5+} selectivity and low CH_4 selectivity and the small pores were responsible for large active surface area to increase the dispersion of the metal catalyst. The BET surface area as well as the metal dispersion was higher in case of bimodal support compared to the unimodal support (SiO_2), providing better performance in FTS reaction. Normal paraffins and 1-olefins are produced as the intermediate reaction products of FTS reaction. The 1-olefins get desorbed form one site and undergo various secondary reactions, such as hydrogenation to n-alkane, hydrogenolysis, and reinsertion into growing chains (mostly effective for low chain olefins C_2H_4 and C_3H_6), over another site [135]. The C—C chain propagation occurs on the CH_2 species over the catalyst surface where the olefins add quickly to produce alkanes in the range of jet range. The Co/ZrO_2-SiO_2 bimodal catalyst exhibited a CO conversion of 51.6% and a jet fuel selectivity (C_8–C_{16}) of 29.0%. However, the authors also conducted the same reaction in presence of pure olefin and mixed olefin additives and observed a significant increase in the jet fuel selectivity with marked decrease in olefin/paraffin ratio. For instance, with the addition of mixed 1-olefins of

1-decene and 1-tetradecene, the C_8–C_{16} selectivity reached up to 83.3% [134]. The activity and selectivity of the catalysts can also be adjusted by the addition of some promoters, such as alkali metals, alkaline earth metals, and transition metals [136]. They can increase the reducibility and decrease the deactivation of the catalyst. Alkali promoters especially K is widely used as catalyst promoter which decrease the selectivity toward CH_4 and preserves the carbide phase of the catalyst which is active for the FTS and the water gas shift (WGS) reaction to adjust H_2/CO ratio [137,138]. Moreover, alkali promoters facilitate chain growth reaction by improving CO dissociative adsorption, enhancing selectivity to heavier hydrocarbons in the jet range [139]. Monte et al. investigated the effect of the K addition and the coaddition of K and Co on the FTS catalyst Fe and observed a 25% and 30% selectivity toward jet range hydrocarbon (C_9–C_{16}), whereas for pure Fe catalyst, it was 20% [140].

A promising route for the synthesis of aviation fuel from the lignocellulosic biomass is direct catalytic conversion of lignocellulose where they are catalytically depolymerized into small monomers or platform molecules that undergo re-oligomerization through aldol condensation, hydroxyalkylation, and ketonization reactions to proper carbon chain length, and finally are deoxygenated to jet range fuel via HDO reaction [141]. Because of the complex structure of the lignocellulosic materials, several pretreatment methods are applied to separate lignocellulose in its main constituents and depolymerize them to soluble fractions (pentose, hexose, phenol, etc.) suitable for further efficient chemical or biological conversion. Mostly, acid or base hydrolysis and steam explosion are used for this purpose, however, some alternative pretreatment processes have also been developed, such as milling, organosolv processes, wet oxidation, ozonolysis, application of supercritical CO_2, ionic-liquid-assisted deconstruction, deep eutectic solvents (DES) etc. [113,142]. Next, the C_5 sugar (e.g., xylose) and C_6 sugar (e.g., glucose) obtained from hemicellulose and cellulose depolymerization undergoes dehydration in presence of homogeneous or heterogeneous acid catalyst to yield two important platform molecules, furfural and hydroxymethyl furfural (HMF), respectively. It has been found that Lewis and Brønsted sites are responsible for catalyzing the dehydration reaction of xylose to furfural and the selectivity increases with an increase in the Brønsted to Lewis acid site ratio of the catalyst, which has been confirmed with both heterogeneous (zirconium phosphate catalyst, Zr-P) and homogeneous acids (HCl and ytterbium (III) trifluoromethanesulfonate hydrate, Yb(OTf)$_3$) [143]. Xylose is isomerized to xylulose in presence of Lewis acid and the Brønsted acid is responsible for the dehydration of xylulose to furfural [144]. HMF from glucose or fructose can undergo rehydration to levulinic acid (LA) [145]. Upon hydrogenation, LA can be converted to γ-valerolactone (GVL) over noble metal-based catalysts, particularly Ru-based catalysts [113]. These platform molecules possess highly active functional groups, such as hydroxyl group, carboxyl group, carbonyl group, unsaturated C—C bonds, which make them suitable for various transformations, particularly C—C bond formation. Two carbonyl group containing compounds can undergo aldol condensation reaction over base catalyst (homogeneous/heterogeneous) to increase the C—C chain length by coupling, provided that at least one carbonyl group contains an α-H atom. Acetone, dihydroxyacetone, and glyceraldehyde are some examples of such α-H containing carbonyl group with which 5-HMF, furfural, or methylfuran undergo aldol condensation reaction (Fig. 3). However, α-H atom can be generated in furan by selective hydrogenation to tetrahydrofufural (THFA) and hydroxymethyl tetrahydro furfural (HMTHFA) and can undergo aldol self-condensation to produce long-chain intermediates [146]. Among the base catalysts, NaOH, KOH, NH_4OH are effective and widely applied as homogeneous catalysts and alkali and alkaline earth oxides, basic zeolites, phosphates, and hydrotalcites are employed as heterogeneous catalysts. Heterogeneous catalyst is preferred due to the problem associated with the separation and corrosivity of the aqueous catalyst. Metal (e.g., Pd, Pt, Ru, Ni) incorporated with solid base is a promising bifunctional catalyst as it can be used for both aldolcondensation and the subsequent HDO reactions [141]. Liang et al. [147] investigated the aqueous phase aldol condensation of LA and furfural over MgO and ZnO which yielded δ- and β-furfurylidenelevulinic acids (δ- and β-FDLA), respectively. MgO followed the base-catalyzed mechanism where an enolate ion is firstly formed through the abstraction of a proton from the β- or δ carbon in sodium levulinate molecule by basic sites Mg^{2+}-O^{2-} pairs and surface hydroxyl groups. Then, the enolate anion attacks the carbon atom of the aldehyde group of furfural to form C—C bond. In the acid-catalyzed mechanism, protonation of the carbonyl group was firstly tautomerized sodium levulinate molecule into a nucleophilic enol that attacked the carbonyl group of furfural and finally gave β-FDLA as the dominant product. With ketones, aldol condensation of HMF progresses via the formation of a C_9 alcohol as an intermediate, followed by dehydration to an α,β-unsaturated ketone [148]. CO_2 can act as a catalyst for the production of bio-jet fuel precursors from 5-HMF and acetone through aldol condensation [149]. However, these intermediate products require hydrogenation and HDO over bifunctional metal-acid catalyst (Pd/β-Zeolite, Pt/Al_2O_3-SiO_2, Pd/Al-MCM-41, Pd/H-ZSM-5, Pd/H-β, Pd/Al_2O_3, Pd/SiO_2-Al_2O_3, Pd/H-ZSM-5, Pt/C, Pt/Al_2O_3, and Pt/ZrP) to be used as aviation fuel blend [150].

C—C coupling of furan-based platform molecules to jet fuel range precursors can also be done by Hydroxyalkylation/alkylation (HAA) method over acidic catalyst which involves the protonation of the carbonyl group in furfural or 5-HMF followed by the addition of 1 to 3 furans or 2-methyl furans, forming jet range carbon chain as shown in Fig. 4 [141]. Both homogeneous acids (e.g., H_2SO_4, HCl) [151] and heterogeneous acids, such as acidic resins (e.g., Nafion-212, Nafion-115, Amberlyst-15, Amberlyst-36) [152], acidic zeolites (e.g., H-Y, H-ZMS-5, H-USY) [153], acidic phosphates (e.g., zirconium phosphate) [153], and activated carbon modified with acidic groups (e.g., AC-SO_3H) [154] are employed for HAA reaction. The condensed furanic jet fuel precursors still contain excess oxygen and unstable groups, e.g., –OH groups, C=O bonds,

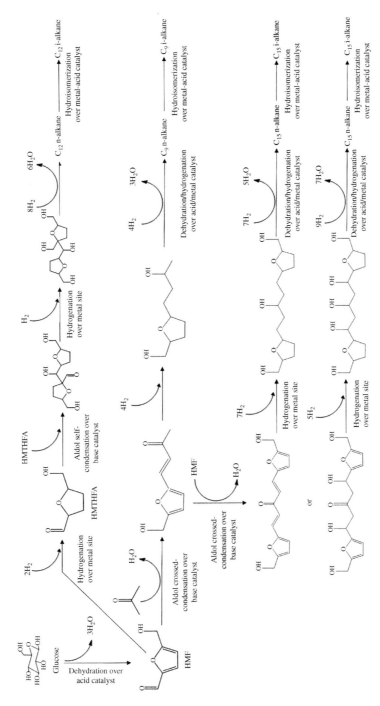

Fig. 3 Effect of catalyst on the production of aviation fuel from lignocellulosic biomass through aldol condensation.

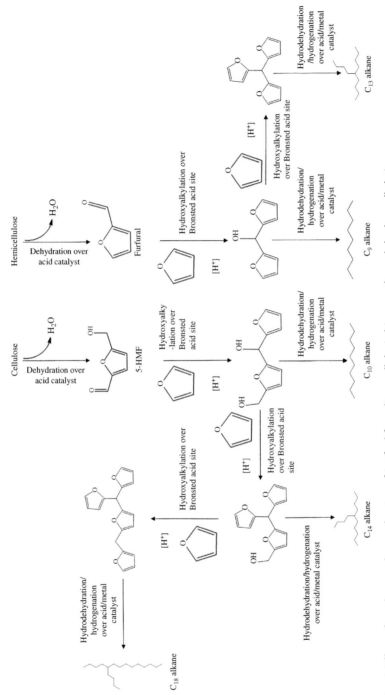

Fig. 4 Effect of catalyst on the production of jet fuel from lignocellulosic biomass through hydroxyalkylation.

and C=C bonds and thus a subsequent HDO is required to convert these oxy compounds to "drop in" fuel. The catalytic HDO of these oxy intermediates involves three steps, such as hydrogenation, ring opening, and deoxygenation. Metal catalysts (e.g., Pd, Pt, Ru, Ni) are responsible for hydrogenation and ring breaking reaction, while acid or base catalysts including HCl, acetic acid, and metaltriflates contribute to only ring opening reaction [141]. Deoxygenation occurs over acid-metal bifunctional catalyst, such as Ni, Co, Mo, Pd, Pt, Ru or W supported on acidic solids, including Al_2O_3, WO_3, $NOPO_4$, $TaOPO_4$, polyoxometalates, and acidic zeolites [150,155].

Having ketone and carboxylic groups, LA shows multiple transformation activities to biojet precursors. Through acid ketonization reaction which involves C—C coupling with the release if CO_2, two LA can be combined to form C_9 ketone which upon HDO can be converted to C_9 hydrocarbons. [121,156]. A variety of catalysts, including metal oxides and zeolites, have been used for the promotion of ketonization reaction [157]. However, Glinski et al. concluded that amphoteric oxides (e.g., CeO_2, MnO_2, La_2O_3) are better than pure acidic or basic oxides after screening 20 different metal oxides supported on silica, including basic, acidic, and amphoteric oxides, for the ketonization of acetic acid [158]. With low lattice energy oxides, such as alkali and alkali earth oxides (e.g., CaO, MgO, BaO), acids form bulk carboxylate salts that can subsequently decompose to ketone, H_2O and CO_2 by thermal treatment [121]. On the other hand, with high lattice energy oxides (e.g., CeO_2, ZrO_2, TiO_2), the reaction proceeds via surface ketonization which involves dehydration of one acid to a highly reactive intermediate ketene that reacts with the other acid to form a six-membered transition state followed by the generation of an enol with the loss of a CO_2. Finally, the enol transforms to ketone through keto-enol tautomerization [156,159]. Apart from ketonization, LA can undergo self-condensation to produce jet fuel precursors (lactone compounds) over Brønsted (trichloroacetic acid) and Lewis acids ($ZnCl_2$) that promote the selective C—C bond formation between two LA [160]. Over acid catalyst, LA can undergo intermolecular dehydration reaction to form an unsaturated lactone called α-angelica lactone (AL) which is more reactive than LA because of having C=C bond and a carbonyl group [161]. The AL isomers can be self-condensed through the conjugated addition reaction between the double bond promoted by alkali catalysts, such as hydroxide or alkoxide salts, active metals, and carbonates such as K_2CO_3, to yield C_{10} or C_{15} oxy-intermediates which can subsequently be converted to C_8–C_{15} hydrocarbons over metal-acid HDO bifunctional catalysts [141,162–164]. AL can also react with furfural through aldol condensation to form aviation fuel precursors. Xu et al. performed the aldol condensation of AL and furan over basic Mn_2O_3 catalyst to C_9 and C_{10} oxygenates which then went through HDO over Pd/C and Pd-FeO$_x$/SiO$_2$ catalysts to C_9 and C_{10} alkanes [165].

The LA produced from C_5 or C_6 sugars can be selectively hydrogenated to GVL which can be used as a building block for the production of aviation fuel precursors and also serves as a good fuel additive due to its high energy density and low vapor pressure. Over solid acid catalyst, SiO_2/Al_2O_3, decarboxylation of GVL proceeds via acid-catalyzed protonation to dismantle the ring structure by cleaving the cyclic ester linkage, followed by proton transfer from the carbenium ion leading to C—C bond scission between the α- and carbonyl carbons and deprotonation to yield equimolar butene and CO_2 [166,167]. The butene can be converted to jet range alkane (C_8–C_{16}) by acid-catalyzed oligomerization over HZSM-5 and Amberlyst-70 catalyst [166]. In another route, GVL can be converted to pentanoic acid (PA) intermediate over bifunctional catalysts (e.g., Pd/Nb_2O_5) that contains acid sites for ring opening reaction and metal sites for hydrogenation reactions [168]. Through ketonization reaction over ceria-zirconia, the PA can be converted to 5-nonanone which upon hydrogenation over metallic catalyst Ru/C produces 5-nonanol. The dehydration of 5-nonanol over acidic catalysts (e.g., acidic zeolites or acidic resins) generates C_9 olefins which can be subsequently hydrogenated to nonane or oligomerized to C_{9+} alkanes [169]. Some important catalysts for the production of aviation grade fuel from lignocellulosic biomass have been tabulated in Table 2.

Unlike the monomers and the platform molecules derived from the dehydration of the monomers of cellulose and hemicellulose, lignin monomers and dimers contain 7–18 carbon atoms which can be converted to cycloalkanes and aromatic hydrocarbons, another two constituents of biojet fuel, simply by HDO reaction without requiring C—C coupling [176,182]. Lignin monomers are connected by various C—C and C—O—C linkages and the function of the catalyst is to break the C—O bonds to release oxygen during HDO in the form of H_2O, CO, CO_2, and methanol through hydrogenolysis, dehydration, decarboxylation, or demethoxylation reactions. Depending on the catalyst, two routes are followed: (1) Direct deoxygenation of lignin monomers to aromatics followed by hydrogenation to produce cycloalkanes (DDO) and (2) hydrogenation of aromatic rings to form cyclohexyl oxy-compounds, and subsequent deoxygenation to yield hydrocarbons (HYD) [141,183,184]. HYD is preferred over catalysts with high hydrogenation activity, such as noble metal (Ru, Pt, Pd)-based catalysts. On the other hand, catalysts that shows high activity in C—O bond cleavage, such as NbO_x in Ru/Nb_2O_5 or Ru/Nb_2O_5-SiO_2, preferentially follow the DDO route [185,186]. However, HYD route is favorable over DDO since cleavage of aromatic C—O bonds is more difficult compared to aliphatic C—O bonds [183]. Bimetallic and bifunctional catalyst, such as Ru-Cu/HY zeolite, are suitable for one pot HDO of lignin to alkanes. The high catalytic performance of Ru-Cu/HY was attributed to (1) high total and strong acid sites, (2) good dispersion of metals, and (3) high affinity for the adsorption of polar fractions, including hydroxyl groups and ether bonds [187]. Small lignin fractions can be transformed to jet range hydrocarbon by controlled coupling reaction that involves C—C bond formation over acid catalyst. Metal catalysts are suitable for hydrogenation reaction and therefore can restrain the coupling reaction to a certain extent [188]. Another important lignin degradation product is phenol which

Table 2 Important catalysts for the production of aviation grade fuel from lignocellulosic biomass.

Catalyst	Reaction condition	Feed	Product	Reference
(i) Aldol condensation over CaO **(ii)** HDO over Pd/HZSM-5	**(i)** $T = 130°C$, $t = 6$ h, $P = 6$ MPa H_2, **(ii)** $T = 260°C$	Furfural and 2-pentanone	**(i)** 98.3% conversion and 86.7% yield **(ii)** 100% conversion and 90% C_9–C_{10} yield	[170]
(i) Aldol condensation over CaO **(iii)** HDO over Co/SiO$_2$, Ni/SiO$_2$, Cu/SiO$_2$	**(i)** $T = 170°C$, $t = 8$ h, $P = 6$ MPa H_2, **(ii)** $T = 350°C$	Furfural and 3-pentanone	**(i)** >80% conversion and 60% yield **(ii)** 100% conversion and 60%–70% C_8–C_{10} yield	[171]
(i) Aldol condensation over NaOH catalyst in THF solvent **(ii)** Diels–Alder reaction (hydrocycloaddition) over Ru/Al$_2$O$_3$ catalyst **(iii)** HDO over Pt/SiO$_2$–Al$_2$O$_3$	**(i)** $T = 25$–$28°C$, $t = 1$ h **(ii)** $P = 5.52$–8.27 MPa H_2, $T = 80$–$140°C$, $t = 6$ h **(iii)** $P = 8.27$ MPa H_2, $T = 300°C$	Furfural and acetone	**(i)** 16% yield of oligomeric compound **(ii)** 5%–20% selectivity in light products and 70%–80% selectivity in heavy products **(iii)** 36.2% selectivity of C_{13}	[172]
Aldol condensation and HDO over MgZr+Pt/Al$_2$O$_3$	**(i)** $P = 1$ MPa N_2, $T = 50°C$, $t = 24$ h **(ii)** $P = 4.5$ MPa H_2, $T = 220°C$, $t = 24$ h	Furfural and acetone	43.2% yield to C_{13}	[173]
(i) Dehydration over ZrP **(ii)** Aldol condensation over CaO and Ca(OH)$_2$	**(i)** $T = 180°C$, $t = 3$ h **(ii)** $T = 180°C$, $t = 5$ h **(iii)** $P = 6$ MPa H_2, $T = 300°C$	Xylose and methyl isobutyl ketone	**(i)** 63.7% yield and 73.9% selectivity of furfural **(ii)** 79.5% yield and 91.8% selectivity of 1-(furan-2-yl)-5-methylhex-1-en3-one over CaO 39.4% yield	[174]

Continued

Table 2 Important catalysts for the production of aviation grade fuel from lignocellulosic biomass—cont'd

Catalyst	Reaction condition	Feed	Product	Reference
(iii) HDO over Ru/HZSM-5			and 83.3% selectivity of 1-(furan-2-yl)-5-methyl-hex-1-en3-one over $Ca(OH)_2$ (iii) 68.4% jet fuel range alkanes ($C_8 - C_{11}$)	
(i) Hydroxyalkylation/alkylation over protonated titanate nanotube (ii) HDO over Ni/HZSM-5	(i) $T = 50°C$, $t = 4$ h (ii) $P = 6$ MPa H_2, $T = 260°C$, WHSV $= 1.3$ h^{-1}	2-Methyl furan and lignocellulosic carbonyl compound (n-butanal)	(i) 76.8% conversion and 73.8% yield of 5,5-(butane-1,1-diyl) bis(2-methylfuran) (ii) About 90% yield of C_9–C_{14}	[154]
(i) Conversion of GVL to butene and CO_2 over SiO_2/Al_2O_3 (ii) Olegomerization of butene and CO_2 over HZSM-5	(i) $P = 3.6$ MPa, $T = 375°C$, WHSV $= 0.9$ h^{-1} (ii) $P = 3.6$ MPa H_2, $T = 225°C$, WHSV $= 0.11$ h^{-1}	GVL	(i) 99% conversion of GVL and 96% yield of butene (ii) 72% yield of C_8–C_{16} alkene	[166]
(i) Aldol condensation over Mn_2O_3 (ii) HDO over Pd/C and Pd-FeO$_x$/SiO$_2$	(i) $T = 80°C$, $t = 4$ h (ii) $P = 6$ MPa, $T = 350°C$	LA and angelica lactone from the hydrolysis/dehydration of the hemicellulose and cellulose	(i) 96% yield of C_{10} oxygenates (ii) 96% yield of C_9 and C_{10} alkanes	[165]
Ru/C co-catalyst in the presence of 63 wt% ZnCl$_2$	$T = 200°C$, $t = 6$ h	Softwood lignin	54 wt% oil containing aromatics and cyclic hydrocarbon	[175]

Catalyst	Conditions	Substrate	Results	Ref.
Metal triflate Hf(OTf)$_4$ and Ru/Al$_2$O$_3$	$P = 4$ MPa N$_2$, $T = 250°$C, $t = 2$ h	Guaiacol (model compound of lignin)	>99 wt% conversion, 82.6% hydrocarbon monomer, >99.9% total hydrocarbon yield, >30 wt% of the hydrocarbons produced cyclohexane and alkylcyclohexanes in the jet fuel range (C$_9$–C$_{18}$)	[176]
MoO$_3$–NiOAl$_2$O$_3$	$P = 2.84$–6.4 MPa H$_2$, $T = 500°$C	Guaiacol (model compound of lignin)	74% yield in aromatics (32% benzene, 30% toluene, 8% xylene, 4% propyl-benzene)	[177]
1% Pt/Hbeta	$P = 0.101325$ MPa H$_2$, $T = 400°$C	Anisole (model compound of lignin)	89.4% yield in aromatics (51.2% benzene, 27.6% toluene, 10.6% xylene)	[178]
HDO over Ru (5 wt%)/SiO$_2$–Al$_2$O$_3$	$P = 4$ MPa H$_2$, $T = 250°$C, $t = 1$ h	Guaiacol (model compound of lignin)	60% yield of cyclohexane	[179]
HDO over Rh (3 wt%)/SiO$_2$–Al$_2$O$_3$	$P = 4$ MPa H$_2$, $T = 250°$C, $t = 1$ h	Guaiacol (model compound of lignin)	57% yield of cyclohexane	[179]
Ru (1 wt%)/HZSM-5	$P = 5$ MPa H$_2$, $T = 200°$C, $t = 4$ h	Anisole (model compound of lignin)	93.4% yield of cyclohexane	[180]
Ni (4.55 wt%)/Al-SBA-15	$P = 6$ MPa H$_2$, $T = 220°$C, $t = 2$ h	Anisole (model compound of lignin)	95% yield of cyclohexane	[181]

demonstrates high alkylation reactivity and can be converted to alkylbenzene over solid acid catalysts. Alkylbenzenes play a crucial role in enhancing the octane number of jet fuel [189].

The byproduct from the lignin HDO can be used as a hydrogen source for the deoxygenation of TG-based feedstock in the absence of external hydrogen supply. Lu et al. investigated an integrated process for simultaneous dehydrogenation of ethylcyclohexane (produced from lignin HDO) and deoxygenation of lipid over bimetallic catalyst Pt-Ir/γ-Al$_2$O$_3$ system, converting two kinds of biomass feedstocks (lignin and lipid) into high-quality bio-aviation fuels in an integrated process [41].

2.4 Roles on sugar-based feedstock

Other than the conversion to biojet fuel through platform molecules (as described in the previous section), sugar-based feedstock can be converted to biojet fuel through alcohol to jet (ATJ) route [136]. In this route (Fig. 5), the biomass sugar extracted from edible feedstock (e.g., sugarcane, corn, beet etc.) or nonedible feedstock (lignocellulosic biomass) are biochemically converted to alcohols which undergo three main steps: (i) dehydration to the corresponding olefin; (ii) oligomerization of the olefins; and (iii) hydrogenation of the oligomerized olefin to the saturated hydrocarbon product [3]. Alcohols produced from the microbial fermentation of biomass sugar include C$_2$ and C$_4$ compounds such as ethanol and butanol (*n*-butanol and isobutanol) [190,191]. Acidic catalysts are the best choice for the dehydration of alcohols. For example, acidic catalysts, such as silica-alumina, silicoaluminophosphates, zeolites, and heteropolyacids have been used to dehydrate the ethanol to ethylene [192–194]. Isobutanol was also converted to isobutylene over acidic alumina catalyst [195]. Water-resistant carbon acidic catalysts can also be employed for the purpose of dehydration in the aqueous environment, eliminating the costly and energy-intensive water removal process [196]. Next, the oligomerization of olefins to jet range carbon chain length olefins are conducted over heterogeneous acidic catalysts such as sulfonic resins, solid phosphoric acid, or zeolites at moderate temperatures and pressures. For instance, oligomerization of isobutene to

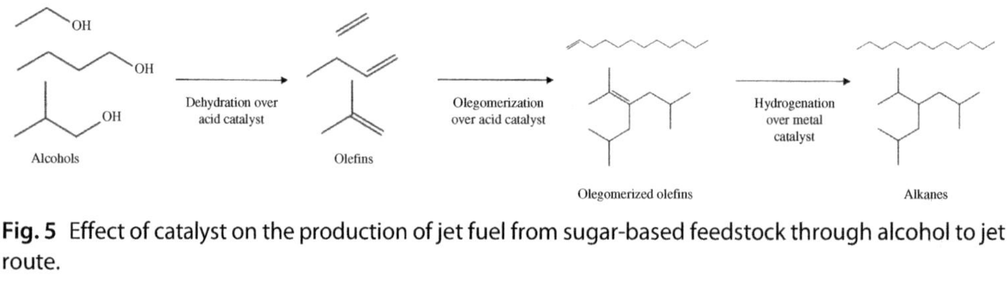

Fig. 5 Effect of catalyst on the production of jet fuel from sugar-based feedstock through alcohol to jet route.

C_8–C_{16} olefins was done over Amberlyst resin [3]. Metal catalysts possess excellent hydrogenation capability and are used to saturate the jet range olefins to jet range alkanes.

3. Roles of catalyst on different stages of biojet fuel production

3.1 Deoxygenation

Deoxygenation is a crucial step to remove the oxygen molecule from oxygenated compounds derived from vegetable oils or biooils to produce *n*-alkanes. *n*-Alkanes are valuable jet fuel components with high energy density, high H/C ratio, and excellent combustion characteristics, although they have poor cold flow properties. The main oxygenated compounds are fatty acids, aldehydes, ketones, platform molecules, alcohols obtained from vegetable oils, pyrolysis bio-oils, lignocellulosic biomass, and sugary feedstocks. The deoxygenation processes involves mainly three types of reactions including HDH, DCO, and DCO_2 as described earlier [67,197,198]. Various transition metals, noble metals, and metal oxides, sulfides and phosphides are employed as deoxygenation catalysts on different catalyst supports, such as Al_2O_3, SiO_2, zeolite, and graphite [43]. Among them, Co or Ni doped MoS_2 has been popular for HDO due to their high catalytic activity and low cost, favorable for large-scale production. Incorporation of Ni modifies the surface of MoS_2 to form Ni-Mo-S phase, which is the active site of $NiMoS_2$. Moreover, sulfur vacancies in metal sulfide catalysts are known for deoxygenation sites. Substitution into S-edges of MoS_2 structure by Ni can create sulfur vacancies and lead to increased deoxygenation activity [44]. Ni or Co also acts as a promoter to enhance the catalytic activity of MoS_2. Metal doping in the sulfide or oxide catalyst promote the catalytic deoxygenation reaction by modifying the electronic and geometric structure of the catalyst active sites by introducing defects such as sulfur vacancy in MoS_2 [199]. Ni or Co doped MoS_2 favors the HDO reaction, whereas Ni or Co on oxide support such as alumina or silica favors the DCO and DCO_2 [200]. Temperature and liquid hourly space velocity (LHSV) also dictate the dominant reaction in the DO process. At a lower temperature and LHSV, HDH was dominant during the DO of palm kernel oil over Ni-MoS_2/c-Al_2O_3, whereas a higher temperature and LHSV promoted DCO and DCO_2 [43]. Metal oxide catalysts also play an important role in HDH reaction in the presence of hydrogen. Oxygen atoms from the oxygenated compounds get captured by the oxygen vacancy sites generated by hydrogen [201].

Hydrotreating process of vegetable oil consumes hydrogen to reduce the unsaturation in TGs, decompose the TGs to fatty acids and glycerol or propane via C—O bond cleavage, and deoxygenate the fatty acids to hydrocarbons. Among the deoxygenation pathways, HDH consumes the highest amount of hydrogen to transform the fatty acids to hydrocarbons for further processing to isomers or cycloalkanes. In general, HDH requires 300–420 m^3 of hydrogen to get 1 m^3 of oil deoxygenated through releasing H_2O [202]. To limit the consumption of hydrogen, researchers came up with nonsulfide catalysts such as Pd, Ru, Pt, N, etc., supported on silica, alumina and activated carbon that performed catalytic deoxygenation reaction through DCO_x which requires less or no hydrogen [203]. Among the studied catalysts, Pd/C showed high activity and selectivity toward the DCO_x pathway in hydrogen free environment [204]. Immer et al. reported that fatty acids can be deoxygenated over Pd/C via decarboxylation in the presence of He, although the reaction was very slow. However, activated carbon-based catalysts come with some limitations as they get rapidly deactivated due to the coke deposition and pore blocking originating form unsaturated and/or long chain hydrocarbons in the absence of hydrogen. Moreover, the regeneration of activated carbon-based catalysts through solvent washing and/or calcination in air is quite difficult [50]. To address this problem, several approaches have been in practice, such as use of mesoporous activated carbon to prevent blocking of pores [205], addition of 10% hydrogen to prevent unsaturated hydrocarbon [53], and recently in-situ hydrogen supply through decomposition of byproduct glycerol [206] and formic acid [50] have been reported. As an alternative to microporous activated carbon support, mesoporous metal oxide (Al_2O_3, MgO, CeO_2, Y_2O_3, and ZrO_2) supported noble catalysts were employed to conduct the DCO_x of fatty acids [66]. Though a high conversion of fatty acids was observed with these catalysts, the selectivity to jet range hydrocarbon was quite low. In their study, Roh et al. observed a 100% conversion of a model fatty acid oleic acid (OA) over MgO-Al_2O_3 supported Ni catalyst, but the selectivity to main products of the deoxygenation-heptadecane or heptadecene—was extremely low [207]. A similar finding was reported during the hydrothermal DCO_x of OA over NiWC/Al-SBA-15 in absence of hydrogen [208]. Fu et al. reported Pt/C catalyst which is capable of converting algal lipid directly to jet fuel through DCO_2 in absence of H_2. To obtain a reasonable activity and selectivity in DCO_x process, Choi et al. [66] reported a novel transition metal oxide supported catalyst WO_x/Pt/TiO_2 (WPT) which is able to produce oxy-free hydrocarbons that can be transformed into biojet fuel. In the absence of hydrogen, the DCO_x occurs via C—C scission and produces CO or CO_2. W impregnation over the Pt/TiO_2 increased the degree of deoxygenation from 36% to 86% under the same reaction condition with trace amount of aromatics and a small amount of isomer hydrocarbons, isoC_{12} and isoC_{17}. Biomass-derived terpenes were used as a hydrogen donor to deoxygenate TGs, fatty esters, and fatty acids to jet fuel-ranged hydrocarbons containing arenes over bimetallic PdNi/HZSM-5 catalyst in the absence of hydrogen [209]. Table 3 reports various catalysts that are efficient in DO reaction.

Apart from coke formation, absence of hydrogen can also decrease the linear alkane portion in the jet fuel which is responsible for the energy content of the fuel. Scaldaferri and Pasa performed the conversion of soybean oil over $NbOPO_4$ in the presence [26] and absence [220] of hydrogen, and observed a higher alkane (46%) and biojet selectivity (62%) in the hydrogen environment compared to that (38% and 58%) of in nitrogen environment. The type and amount of catalyst have considerable effect on the deoxygenation

Table 3 Efficient catalysts for the DO reaction.

Catalyst	Reaction condition	Feed	Product	DO preference	Reference
Pt/γ-Al$_2$O$_3$	$P = 2$ MPa H$_2$, $T = 215$–250°C, WHSV $= 2$ h^{-1}	Palm oil	C$_{15}$ (32.8 wt% yield) and C$_{17}$ (36.3 wt% yield)	Predominance of DCO$_x$ over HDH	[80]
PtRe/USY	$P = 5$ MPa H$_2$, $T = 220$–310°C, WHSV $= 2$ h^{-1}	Palm oil	C$_{15}$–C$_{18}$ (70% yield)	Predominance of HDH and DCO	[40]
NiMo/γ-Al$_2$O$_3$	$P = 4.8$–6.2 MPa H$_2$, $T = 300$–400°C, LHSV $= 2$ h^{-1}	Oil containing C$_{15}$–C$_{18}$ fatty acids	C$_{16}$ and C$_{18}$ (60 wt% concentration) C$_{15}$ and C$_{17}$ (40 wt% concentration)	Predominance of HDH over DCO$_x$	[210]
Pd/C	$P = 4.8$–6.2 MPa H$_2$, $T = 300$–400°C, LHSV $= 2$ h^{-1}	Oil containing C$_{15}$–C$_{18}$ fatty acids	C$_{15}$ and C$_{17}$ (62 wt% concentration) C$_{16}$ and C$_{18}$ (38 wt% concentration)	Predominance of DCO$_x$ over HDH	[210]
Ru/Al$_2$O$_3$, Ni/ SiO$_2$–Al$_2$O$_3$, Pd/γ-Al$_2$O$_3$	$P = 9.5$ MPa H$_2$, $T = 400$°C, $t = 2$ h	Soybean oil	C$_{17}$/C$_{18}$ (39.6, 29.3, 11.9)	Predominance of DCO$_x$ over HDH	[64]
NiMo/γ-Al$_2$O$_3$, CoMo/γ-Al$_2$O$_3$, Pt/γ-Al$_2$O$_3$	$P = 9.5$ MPa H$_2$, $T = 400$°C, $t = 2$ h	Soybean oil	C$_{17}$/C$_{18}$ (2.49, 2.16, 0.92)	Predominance of HDH over DCO$_x$	[64]
Pt/Al$_2$O$_3$/SAPO-11	$P = 3$ MPa H$_2$ $T = 370$–385°C LHSV $= 1$ h^{-1}	Soybean oil	42–48 wt% biojet yield	Predominance of HDH over DCO$_x$	[211]
Pt/C	$T = 360$°C Solvent $=$ water $t = 0.75$ h	Microalgae lipid	Heptadecane selectivity 90%	Predominance of DCO$_2$ in absence of H$_2$	[212]
Pd/C	$P = 1$ MPa H$_2$ $T = 300$°C $t = 5$ h	Macauba oil	85% biojet yield	Predominance of DCO$_x$ over HDH	[213]
Pd/C	$P = 1.86$ MPa H$_2$ $T = 300$°C $t = 4$ h	Licuri oil	80.7% selectivity of n-alkane (C$_9$–C$_{17}$)	Predominance of DCO$_2$ over HDH	[214]

Continued

Table 3 Efficient catalysts for the DO reaction—cont'd

Catalyst	Reaction condition	Feed	Product	DO preference	Reference
Ni-MoS$_2$/γ-Al$_2$O$_3$	$P = 5$ MPa H$_2$ $T = 333°C$ LHSV $= 1$ h^{-1}	Palm kernel oil	92% biojet yield	Predominance of HDH over DCO$_x$	[43]
WO$_x$/Pt/TiO$_2$	$P = 4$ MPa N$_2$ $T = 200°C$ LHSV $= 1.33$ h^{-1}	Jatropha oil	86% efficiency in deoxygenation and conversion to C$_{17}$ alkane	Predominance of DCO$_x$	[66]
Ni/SBA-15	$P = 13$ MPa H$_2$ $T = 300°C$ $t = 1$ h	Pyrolysis biooil	54.9% degree of deoxygenation and 76% biojet yield	Predominance of HDH over DCO$_x$	[215]
NiW/SAPO-11	$P = 6$–8 MPa H$_2$ $T = 375$–$450°C$ LHSV $= 1$ h^{-1}	Jatropha oil	84% hydrocarbon yield (40% aviation kerosene with C$_{15+17}$/C$_{16+18}$ value of 2–4)	Predominance of DCO$_x$ over HDH	[45]
ZSM-5	$T = 500°C$ LHSV $= 3$ h^{-1}	Sunflower oil	77.31% C$_7$–C$_{12}$	DCO$_x$ in absence of H$_2$	[216]
Ni/nano-sized SAPO-11	$P = 4$ MPa H$_2$, $T = 280°C$, $t = 6$ h, LHSV $= 2$ h^{-1}	Palm oil	C$_{15}$/C$_{16}$ is 1.93 and C$_{17}$/C$_{18}$ is 1.2	Predominance of DCO	[74]
Pd/C	$P = 1$ MPa H$_2$, $T = 300°C$	Date seed oil	91.1% conversion to liquid alkane and 30.4% yield of jet fuel	Predominance of DCO$_x$ over HDH	[217]
MoNi/γ-Al$_2$O$_3$	$P = 0.4$–0.8 MPa H$_2$, $T = 350°C$, LHSV $= 1$–20 h^{-1}	Coconut oil	C$_{even}$/C$_{odd}$ < 0.6	Predominance of DCO$_2$	[77]
Pt/ZSM-5	$P = 3$ MPa H$_2$, $T = 380°C$, $t = 2$ h	Soybean oil	100% conversion and 36.1% selectivity to C$_9$–C$_{15}$	Predominance of DCO$_x$ over HDH	[218]
NiMo/SAPO-11	$P = 3$ MPa H$_2$, $T = 380°C$	Jatropha oil	87.8% yield for liquid alkane and 62.3% selectivity for DCO$_x$	Predominance of DCO$_2$	[219]
NiMo/Al$_2$O$_3$	$P = 3$ MPa H$_2$, $T = 380°C$	Jatropha oil	90.1% yield for liquid alkane and 62.3% selectivity for HDH	Predominance of HDH	[219]

pathway. Zhao et al. [75] observed that the use of zinc loaded ZSM-5 inhibited the dehydration during the camelina oil cracking process, whereas promoted the DCO_2 process. Low dehydration results in lower amount of water as well as higher HHV of the final product which is desired for jet fuel. The final gas composition analysis demonstrated that the amount of CO and CO_2 increased with increasing Zn loading, suggesting that increased Lewis site incorporated by metal loading favored DCO and DCO_2 in the deoxygenation process. Catalyst amount and acid site density also play an important role on the deoxygenation process. Scaldaferri and Pasa [26] observed that with increasing catalyst mass and reaction time and temperature, deoxygenation of soybean oil over niobium phosphate catalyst increased. High acid site of Ni-Co/MWCNT favored the DCO and DCO_2 during the DO of *Jatropha curcas* oil [221].

Bimetallic catalyst demonstrates synergistic effect in the deoxygenation of feedstocks. Murata et al. studied the combined effect of noble metal Pt with transition metal Re on a support HZSM-5 zeolite in processing of jatropha oil to biojet fuel [222]. Though the metals existed separately on the support, but they showed a synergistic effect as evidenced by the lowest cracking species compared to that with either Pt/H-ZSM-5 or Re/HZSM-5. Moreover, oil to catalyst ratio had a significant effect on the product yield. With the increasing ratio from 1 to 10, the yield of *n*-alkane (C_{15}–C_{18}) reached from 80% to 2.3%. The C=C bonds were hydrogenated followed by concurrent HDO/DCO/DCO_2 reactions. A similar pathway was followed during hydrotreating of soybean oil over Pd, Pt, Ru supported on Al_2O_3 [64]. The main products were normal paraffinic hydrocarbon ranging from C_8 to C_{18}, with DCO_2/DCO producing odd number of carbon chains (n-C_{15}, n-C_{17}, etc.) and HDO producing even number of carbon chains (n-C_{16}, n-C_{18}, etc.). In terms of conversion and product selectivity, the activity of the catalyst followed the order Pd > Pt > Ru. However, no cyclic or isoparaffin and aromatic compound were not observed, suggesting that the reaction pathway followed up to deoxygenation process.

3.2 Hydrocracking

Catalytic hydrocracking is one of the promising routes to convert vegetable oil and bio-oil to biojet fuel as it produces jet range alkane with high cetane number [223,224]. Hydrocracking of the deoxygenated hydrocarbon products is done to convert the long chain hydrocarbons into desired carbon chain (C_8–C_{16}) and boiling point range of jet fuel. The cracking and isomerization can be carried out parallelly or sequentially where the isomerization is followed by cracking reaction [10]. The conventional catalyst used for the hydrocracking of vegetable oil are mainly sulfide catalysts that include sulfided forms of alumnia- and silica-supported metal and bimetallic catalysts. Among the catalysts, sulfided zeolite showed more catalytic activity in hydrocracking of vegetable oil. However, use of these catalysts require a continuous supply of sulfur containing compounds such as H_2S or dimethyl disulfide to maintain the activity of the catalyst at the desired level [225]. But, these sulfiding agents causes the emission of H_2S and other sulfur containing residues in the environment that is responsible for greenhouse effect and also for creating corrosive environment. To combat the drawbacks of sulfide based catalysts, various nonsulfide zeolites, such as noble metal supported beta-zeolite, SBA-15, ZSM-5, HZSM-5, SAPO-11, and mesoporous-Y zeolite have been found to demonstrate great potentiality in hydrocracking [225]. Other potential catalyst supports include γ-Al_2O_3 and SiO_2. But the unavailability and high cost of noble metals and their high sensitivity to poison and impurities such as sulfur, heavy metals, and oxygenated compounds leading to deactivation have limited their applicability [226]. Therefore, transition metal-based catalysts are experiencing a growing demand for hydrocracking. The hydrocracking of deoxygenated hydrocarbons over metal–acid bifunctional catalysts generally follows the following steps [40]: (i) conversion of paraffin into olefins through dehydrogenation on the metallic sites; (ii) transfer of olefins from the metallic sites to the Brønsted acid sites; (iii) carbenium ions formation on the Brønsted acid sites; (iv) isomerization and cracking (β-scission) of carbenium ions on the Brønsted acid sites; (v) diffusion of the produced olefins to the metal sites; and (vi) transformation of olefins to paraffin through re-hydrogenation on the metal sites. Sometimes, hydrocracking is done in another step after the deoxygenation of the feedstock to take the advantage of catalyst with higher porosity. Rabaev et al. [211] obtained 12 wt% yield of jet fuel and 65.8 wt% yield of heavy fraction from the conversion of soybean oil over $Pt/Al_2O_3/SAPO$-11 catalyst. The heavy fraction was then mild hydrocracked over Ni_2P/HY at 315°C to produce additional 37 wt% yield of jet fraction. Therefore, suitable catalyst for catalytic cracking of vegetable oil are shape selective zeolite-based catalysts and metal oxides. Zeolites have high surface area and adsorption capacity and their Brønsted acid sites are responsible for cracking reactions of the vegetable oil [63,91,227].

Metal loading over a certain amount can suppress the cracking of large hydrocarbon molecules. During catalytic cracking of camelina oil over Zn loaded ZSM-5 catalyst, it was observed that Zn loading at 30% (ZSM-5-Zn-30) blocked the flow of the C_{18} fatty acids from the decomposition of camelina oil TGs [75]. Excess metal loading can decrease the surface area and pore volume and increase the metal oxide aggregation that can lead to deactivation of catalyst [228]. Acidic site on catalyst plays an important role in catalytic cracking. Lewis acid site on Al_2O_3 can result in slight cracking of the carbon chains into jet-fuel range hydrocarbons, facilitating the C—C bond cleavage followed by the C^+ ion generation [41]. However, the acid site must be controlled since high acidity can cause excessive cracking can lead to the formation of unwanted light gases and carbon deposit responsible for deactivating the catalyst. Therefore, an optimum balance between the metal (hydrogenation/dehydrogenation site) and acid (isomerization/cracking site) functions is crucial for achieving ideal catalytic characteristics. Thus the olefinic species on the acid sites get readily re-hydrogenated on the nearby metal sites, suppressing coke, and subsequent catalyst deactivation. [40].

3.3 Hydroisomerization

Upon completion of deoxygenation, the straight chain hydrocarbons are subjected to hydroisomerization and hydrocracking. Hydroisomerization is aimed at improving the cold flow properties and freezing point of biojet fuel. During isomerization, carbocation is created at the double bond site of the hydrocarbon in the presence of solid acid catalyst. The carbocation then rearranges to form methyl or ethyl branch on the hydrocarbon chain, changing the alkane to its isomeric form. Hydroisomerization follows the same reaction mechanism except a hydrogenation step required to saturate the hydrocarbon beforehand [229]. Catalyst with high hydrogenation capability and moderate acidity are the best candidate for optimum hydroisomerization and hydrocracking. The high selectivity for isomerization originates form the accurate balance of acid and hydrogenation activity and the ability of the catalyst to react preferentially with normal paraffins that is dictated by the pore size and topology of the molecular sieves [230]. Conventionally, two step catalytic route is followed to produce aviation fuel from vegetable oil. The first step is hydrogenolysis and deoxygenation TGs in presence of hydrogen to hydrocarbons in the range of diesel fuel (C_{15}–C_{18}) using sulfide form of bifunctional catalysts such as $NiMo/Al_2O_3$, $CoMo/Al_2O_3$, and NiW/Al_2O_3. The hydrocarbons (carbenium ion) then follow isomerization and C—C rupture (cracking) through β-cession process on the Brønsted acid sites of noble metal-based catalyst such as Pt/silica-alumina to produce jet fuel. However, the isomerization depends on the pore size of the acidic catalyst support. Therefore, noble metal supported on nano-beta zeolite catalyst is favorable for branching hydrocarbon [80]. Table 4 lists some important catalysts suitable for isomerization reaction.

Verma et al. [45] reported a single step process for the first time to convert jatropha oil to jet fuel (Jet A-1) over SAPO-11 supported sulfide NiMo and NiW. High degree of unsaturated fatty acid containing vegetable oils such as camelina, sunflower, soybean, jatropha, and algae-derived oils are particularly suitable for the production of jet fuels since these unsaturations facilitate the formation of branched hydrocarbons during hydrotreating, rendering good cold flow properties [232]. Acidic property of catalyst or catalyst support rendered by Lewis and Brønsted acid sites are responsible for catalytic hydroisomerization. The isomerization reaction produces branched alkanes and cyclic compounds that come from the cyclization of the olefins or Diels-Alder reactions. The subsequent dehydrogenation/oxidation of cyclic compounds can produce aromatic compounds [26]. SAPO-11 is known to have high selectivity for hydroisomerization because of its tubular 10-MR zeolite structure and its appropriate acidity [219]. Metal or metal oxide loading on the catalyst support contributes to the formation of Lewis acid

Table 4 List of catalysts suitable for isomerization reaction.

Catalyst	Feed	Product	Reference
Pt/Nano-Beta	Palm oil	i/n-paraffin 7.54; C_8–C_{16} yield of 71.5 wt%	[80]
PtRe/USY	Palm oil	i/n-paraffin 5.1; C_8–C_{14} yield 20%	[40]
NiAg/SAPO-11	Palm oil	i/n-paraffin 1.6; 56.4% liquid product yield	[231]
NiMo/SAPO-11	Jatropha oil	i/n-paraffin 4.2; C_9–C_{14} yield 22%–37.5%	[45]
Ni_2P/HY	Soybean oil	54% isoparaffins in hydrocarbons fraction C_8–C_{18}	[68]
Ni/SAPO-11	Palm oil	50% isomerization selectivity	[69]
NiP/SAPO-11	FAME	14% isomerization rate	[70]
7 wt% Ni/nano-sized SAPO-11	Palm oil	i/n-paraffin 4.32	[74]
Pt/SAPO-11	Coconut oil	>70% isoparaffin in the product	[76]
NiMo/SAPO-11	Jatropha oil	i/n-paraffin 2.1 in C_5–C_{18}	[219]
Sulfided NiW/HZSM-5	Jatropha oil	i/n-paraffin 5.2 in C_9–C_{15}	[78]
Sulfided NiMo/HZSM-5	Algal oil	i/n-paraffin 2.5 in C_9–C_{15}	[78]
Sulfided NiMo/HZSM-5	Jatropha oil	i/n-paraffin 13.5 in C_9–C_{15}	[78]
Ni/MCM-41-APTES (7.5%)-USY	Castor oil	i/n-paraffin 4.4 in C_8–C_{15}	[79]
Ni/HZSM-5	Hydroxyalkylated product of 2-methyl furan and lignocellulosic carbonyl compound (n-butanal)	80% selectivity of C_{14} with 20% isomerization	[154]

sites which are responsible for dehydrogenation reaction leading to isomerization, cyclization, and aromatization [233].

Increasing temperature has an effect on the yield of cyclic alkane and arenes. During the conversion of microalgae biodiesel to biojet fuel over Ni/meso-Y zeolite, Cheng et al. observed 13.7% cycloalkanes and 2.8% arenes in the produced jet fuel at 275°C at the expense of alkanes, whereas at 255°C and 265°C their amount was insignificant to the overall product [109]. However, with further increasing of temperature, the iso-alkanes went through cracking to light hydrocarbons. Therefore, raising reaction temperature in a moderate range could promote the isomerization reaction due to the endothermic nature of the reaction [111].

3.4 Aromatization

Aromatics in jet fuel are crucial as they help the elastomeric seal to be in swelled form, preventing them from drying out and shrinking that could lead to unexpected fuel leakage and subsequent economic and environmental loss [234]. Unlike alkane and branched paraffin in biojet fuel, aromatics are weak hydrogen donor and can form hydrogen bond with weakly polar thioether bond in polythioether (component of fuel tank sealant) and thus lead to swelling of the sealant [235]. It has been reported that jet fuel with aromatic content less than 8% may experience seal contraction leading to leakage and drop in system pressure [236]. Verma et al. observed a significant production (8%) of aromatics during the conversion of Jatropha oil into HRJ over silicoaluminophosphate (SAPO) supported NiMo and NiW and the reason was attributed to the acidity of the SAPO support. The similar observation was reported by Rabaev et al. [211] during the conversion of soybean oil over novel $Pt/Al_2O_3/SAPO$-11 where 15% aromatics were found. They also asserted that polyunsaturation of the fatty acids is the dominant factor in the production of aromatic hydrocarbons. They demonstrated that vegetable oils with high amount of poly-unsaturated fatty acids, such as soybean, sunflower, and camelina oils produced jet fuels containing about 15 wt% aromatics, whereas the content of aromatics in the hydro-treated product of palm or castor oils (containing low amount of unsaturated fatty acids) was only 2 wt%. Only linoleic and linolenic acids produced aromatics while no aromatics were detected in the product of stearic and oleic acids. Production of aromatics reduce the hydrogen consumption as evidenced by the 9.8% reduction in hydrogen consumption during the conversion of jatropha oil over sulfide NiMo/SAPO-11 with 8% aromatic production compared to the process where no aromatics were produced [45]. Moreover, SAPO-11-based catalysts are experiencing growing interest due to their ability to perform HDO, hydroisomerization, and aromatization in a single step which saves the cost of using multiple catalyst and multiple reactor [237]. In another study, it was observed that replacement of Al_2O_3 support with zeolite MFI in conversion of rape oil over Pd–Zn catalyst produced alkane-aromatic hydrocarbon [238]. It is believed that deoxygenated intermediate products undergo cyclization followed by dehydrogenation to produce aromatic compounds. On the moderately acidic support, the unsaturated components in TGs may experience aromatization reaction. As the LHSV increases, less time is left for secondary cracking and hydrogenation which is favorable for the enhancement of the aromatic yield. Over bifunctional catalysts comprising metal site and acid support such as noble metal Pt supported on acidic Al_2O_3, SiO_2-Al_2O_3, SAPO, H-form zeolites, aromatization of alkane involves dehydrogenation of the paraffin on metal site followed by cyclization of the dehydrogenated molecule on the acid sites [239]. However, alumina catalyst showed low selectivity to aromatic production and low stability and hence zeolite with good acidic property was introduced as catalyst support for aromatic production. Acidity of zeolite also plays a parallel role in isomerization and cracking reaction. The smaller is the Pt particle size, the higher is the aromatic yield as small Pt cluster shows low hydrogenolysis property. Unsaturated cyclic alkanes can be converted to aromatics over metal oxide catalyst. Chen et al. transformed lignocellulose derived isophorone to C_8–C_9 aromatic hydrocarbons over MoO_x/SiO_2 catalyst of which the acid sites promoted the rearrangement or cracking reaction beneficial for aromatic hydrocarbons production [201]. The authors also found MoO_x to be more efficient than other oxide catalysts, such as VO_x, WO_x, FeO_x, and CuO supported on SiO_2. Table 5 presents the catalysts that can contribute to the formation of aromatic content in jet fuel.

Aromatization of vegetable oil, mainly TGs with diene and triene hydrocarbon chain, is a multistep process. However, there is no information about the exact sequence of steps that is followed during aromatization. The sequence may be estimated based on the composition of final product, structure of the produced aromatic hydrocarbon, and the knowledge of transformation reactions that the TGs can undergo. According to Rabaev et al. [211], nine steps can occur during the hydrotreatment of TGs aimed at producing aromatic compound. Among them, the most convincing route is as follows: (1) hydrogenolysis of TGs to form fatty acids and propane; (2) dehydrogenation of diene chains to produce fatty acids with triene hydrocarbon chain; (3) conjugation of the double bonds in the triene chain; (4) migration of the double bonds (conjugated system) toward the terminal point of hydrocarbon chain (carboxylic or methyl group); (5) HDO of fatty acids to form hydrocarbon; (6) intermolecular Diels-Alder reaction of conjugated triene system to form cyclohexene ring inside or at the end of hydrocarbon chain; (7) dehydrogenation of cyclohexene to aromatic ring; (8) hydrocracking of alkyl chains; (9) isomerization of alkyl chains.

4. Future look

The deoxygenation of the feedstocks through HDO route requires the supply of external hydrogen which also serves to increase the H/C ratio of alkane to enhance the energy density and to reduce the coke deposition and subsequent catalyst deactivation.

Table 5 Catalysts that are efficient in increasing the aromaticity of jet range fuel.

Catalyst	Reaction condition	Feed	Product (% of aromatics)	Reference
NiAg/SAPO-11	$P = 6$ MPa H_2 $T = 400°C$, WHSV $= 4\ h^{-1}$	Palm oil	9.2%	[231]
Pt/Al$_2$O$_3$/ SAPO-11	$P = 3$ MPa H_2 $T = 370–385°C$ LHSV $= 1\ h^{-1}$	Sunflower and camelina oil	15 wt%	[211]
NiMo/SAPO-11	$P = 6–8$ MPa H_2 $T = 375–450°C$ LHSV $= 1\ h^{-1}$	*Jatropha* oil	8%	[45]
NiW/SAPO-11	$P = 6–8$ MPa H_2 $T = 375–450°C$ LHSV $= 1\ h^{-1}$	*Jatropha* oil	8%	[45]
ZSM-5	$T = 500°C$ LHSV $= 3\ h^{-1}$	Sunflower oil	12%	[216]
Ni/ mesoporous-Y	$P = 3$ MPa H_2, $T = 333°C$	Palm oil	13% yield	[61]
Pt/ZSM-5	$P = 3$ MPa H_2, $T = 380°C$, $t = 2$ h	Soybean oil	100% conversion and 29.2% selectivity to aromatics	[218]
Ru-Cu/HY	$P = 4$ MPa H_2, $T = 250°C$, $t = 2–4$ h	Guaiacol (lignin model compound)	87.5 wt% conversion to mixture of aromatics	[187]
CoMo/Al$_2$O$_3$ and Mo$_2$C/ CNF	$P = 5$ MPa H_2, $T = 300°C$	Lignin oil	25% oxygen-free aromatics of the produced monomers	[240]
Pd/ SiO$_2$-Al$_2$O$_3$-70	$P = 6$ MPa H_2, $T = 160°C$, $t = 3$ h	Benzaldehyde	93.3% aromatic yield (ethyl-benzene)	[241]
Ni$_2$P/SiO$_2$	$P = 0.101325$ MPa H_2, $T = 300°C$	Guaiacol (lignin model compound)	71.54% yield of benzene	[242]

However, external supply of hydrogen comes with associated high cost and safety issues during storage and transportation. Therefore, research in biojet fuel conversion catalyst can be directed to the synthesis of novel multifunctional catalyst that will dehydrogenate a reaction byproduct to produce hydrogen for simultaneous consumption during the hydrotreatment. A novel catalyst with electrocatalytic properties can be designed to produce in-situ hydrogen from the treatment medium water. The HDO pathway generates water vapor that may reduce the acidity of the catalyst that in turn can compromise the hydrocracking of vegetable oil. Zarchin and his coworkers studied the hydroprocessing of soybean oil over Ni_2P/HY catalyst and experienced reduced hydrocracking due to the expelled water from HDO process [68]. Development of a water tolerant hydroprocessing catalyst can overcome the limitation. Searching for inexpensive metal catalysts as an alternative to costly noble metal catalysts in HDO conversion of biomass at low temperatures should get more attention. A variant of HDO catalyst with tolerability against deactivation by feedstock impurities, such as S, N, alkali, P can also be developed.

During aldol condensation reaction for transforming low carbon length carbonyl compounds to higher carbon chain products, the presence of water could significantly affect the catalytic activity of acid and base solid catalysts due to competitive adsorption between the reactants and water molecule [148]. Ngo et al. applied octadecyltrichlorosilane (OTS) coating on MgO catalyst to make it hydrophobic and found an increased catalyst stability, but it decreased the adsorbate-adsorbate interaction and hence the C—C coupling [128]. Catalyst with hydrophobic surface could be a potential solution, provided that the active sites for aldol reaction are not compromised. Moreover, the poor hydrothermal stability of solid base catalyst leads to the leaching of the catalyst and prevents them from using in aqueous phase. Thus, the development of hydrothermal stable solid acids and bases demands future research. With regard to production of biojet fuel form lignocellulosic biomass, a catalyst system should be derived that can convert the platform molecules to biojet precursors in one pot reaction system. Moreover, for direct conversion of sugar derived alcohol to jet over single catalyst, metal-acid support bifunctional catalyst can be used where the dehydration and olegomerization will be done over acid support and the hydrogenation will be occurred on metal site. Since, no literature has been found to conduct this study, ATJ over bifunctional catalyst demands a comprehensive research.

The selective conversion of lignin derived intermediates to high yields of desired products is quite difficult due to their high reactivity rendered by methoxyl groups, phenolic hydroxyl groups, and terminal aldehyde groups and are prone to several side reactions (e.g., C—C coupling to generate chars). Thus, along with the development of suitable pretreatment technologies, searching for highly active catalysts should be paid considerable emphasis to fully unlock the potentiality of lignin.

5. Conclusion

Due to their low cost, easy separation, and efficient recyclability, heterogeneous catalysts are favorable over homogeneous one. Noble metal catalysts play an important role in hydrogenation/dehydrogenation and deoxygenation activity. However, nonprecious transition metal catalysts are widely used for this purpose since the noble metals are costly, very sensitive to reaction products and impurities, and susceptible to catalytic deactivation by poisoning. Bimetallic catalysts are also employed and show higher activity and selectivity than the single metal-based catalyst. Incorporation of a second metal reduces the cracking and catalyst deactivation. Acidic catalysts are suitable for dehydration and isomerization reaction. However, catalyst with moderate acidity is desired as high acidity leads to excessive cracking, coke deposition, and low selectivity to jet range hydrocarbon. Bifunctional metal supported on solid acid catalyst are suitable for deoxygenation and isomerization reaction, completing the conversion of biomass feedstock to biojet fuel in one step. Acidic zeolite is the best choice for the acid support and the acidity can be tuned by controlling the Si/Al ratio. The metal particles must be properly dispersed to avoid rapid hydrogenolysis and agglomeration leading to catalyst deactivation. The deoxygenation and isomerization reactions are carried out in the presence of hydrogen to reduce the coke formation as well as the catalyst deactivation. The hydrogen is supplied externally which is costly and hence, development of a catalyst system that promote a route (e.g., DCO_2) which does not require hydrogen supply is necessary. Continuous search for low-cost catalyst should be done to make the catalytic conversion of biomass to biojet economically and commercially viable.

References

[1] Aviation Benefits Beyond Borders, Annual Report, 2018.
[2] International Air Transport Association, Annual Review, 2017.
[3] M.A. Díaz-Pérez, J.C. Serrano-Ruiz, Catalytic production of jet fuels from biomass, Molecules 25 (4) (2020) 802.
[4] E. Nygren, K. Aleklett, M. Höök, Aviation fuel and future oil production scenarios, Energy Policy 37 (10) (2009) 4003–4010.
[5] U.S. Energy Information Administration, Annual Energy Outlook, 2017.
[6] F. Rosillo-Calle, et al., The Potential Role of Biofuels in Commercial Air Transport-Biojetfuel (No. Task 40: Sustainable International Bioenergy Trade), IEA, Paris, 2012.
[7] J.T. Crawford, et al., Hydrocarbon bio-jet fuel from bioconversion of poplar biomass: techno-economic assessment, Biotechnol. Biofuels 9 (1) (2016) 141.
[8] T.K. Hari, Z. Yaakob, N.N. Binitha, Aviation biofuel from renewable resources: routes, opportunities and challenges, Renew. Sustain. Energy Rev. 42 (2015) 1234–1244.
[9] J.I. Hileman, et al., The feasibility and potential environmental benefits of alternative fuels for commercial aviation, in: Proceedings of the 26th International Congress of the Aeronautical Sciences, 2008.
[10] W.-C. Wang, L. Tao, Bio-jet fuel conversion technologies, Renew. Sustain. Energy Rev. 53 (2016) 801–822.
[11] F. Rosillo-Calle, et al., The Potential and Role of Biofuels in Commercial Air Transport-Biojetfuel, IEA Bioenergy, 2012.
[12] International Air Transport Association, Annual Review, 2015.
[13] C. Gutiérrez-Antonio, et al., Simultaneous energy integration and intensification of the hydrotreating process to produce biojet fuel from *Jatropha curcas*, Chem. Eng. Process. Process Intensif. 110 (2016) 134–145.
[14] C. Gutiérrez-Antonio, et al., A review on the production processes of renewable jet fuel, Renew. Sustain. Energy Rev. 79 (2017) 709–729.
[15] E. Gnansounou, A. Pandey, Classification of Biorefineries Taking Into Account Sustainability Potentials and Flexibility, Elsevier, 2017.

[16] M. Domínguez-Barroso, et al., Diesel oil-like hydrocarbon production from vegetable oil in a single process over Pt–Ni/Al$_2$O$_3$ and Pd/C combined catalysts, Fuel Process. Technol. 148 (2016) 110–116.

[17] X. Zhao, et al., Review of heterogeneous catalysts for catalytically upgrading vegetable oils into hydrocarbon biofuels, Catalysts 7 (3) (2017) 83.

[18] M. Al-Sabawi, J. Chen, S. Ng, Fluid catalytic cracking of biomass-derived oils and their blends with petroleum feedstocks: a review, Energy Fuels 26 (9) (2012) 5355–5372.

[19] J.I. Hileman, et al., Near-Term Feasibility of Alternative Jet Fuels, Rand Corporation, 2009.

[20] S. Khan, et al., A review on deoxygenation of triglycerides for jet fuel range hydrocarbons, J. Anal. Appl. Pyrolysis 140 (2019) 1–24.

[21] X. Wu, et al., Production of jet fuel range biofuels by catalytic transformation of triglycerides based oils, Fuel 188 (2017) 205–211.

[22] P. Kallio, et al., Renewable jet fuel, Curr. Opin. Biotechnol. 26 (2014) 50–55.

[23] S. Blakey, L. Rye, C.W. Wilson, Aviation gas turbine alternative fuels: a review, Proc. Combust. Inst. 33 (2) (2011) 2863–2885.

[24] A.M. Ashraful, et al., Production and comparison of fuel properties, engine performance, and emission characteristics of biodiesel from various non-edible vegetable oils: a review, Energy Convers. Manag. 80 (2014) 202–228.

[25] S.K. Maity, Opportunities, recent trends and challenges of integrated biorefinery: part I, Renew. Sustain. Energy Rev. 43 (2015) 1427–1445.

[26] C.A. Scaldaferri, V.M.D. Pasa, Production of jet fuel and green diesel range biohydrocarbons by hydroprocessing of soybean oil over niobium phosphate catalyst, Fuel 245 (2019) 458–466.

[27] I. Shimada, et al., Deoxygenation of triglycerides by catalytic cracking with enhanced hydrogen transfer activity, Ind. Eng. Chem. Res. 56 (1) (2017) 75–86.

[28] L. Hermida, A.Z. Abdullah, A.R. Mohamed, Deoxygenation of fatty acid to produce diesel-like hydrocarbons: a review of process conditions, reaction kinetics and mechanism, Renew. Sustain. Energy Rev. 42 (2015) 1223–1233.

[29] M.C. Vasquez, E.E. Silva, E.F. Castillo, Hydrotreatment of vegetable oils: a review of the technologies and its developments for jet biofuel production, Biomass Bioenergy 105 (2017) 197–206.

[30] B.P. Pattanaik, R.D. Misra, Effect of reaction pathway and operating parameters on the deoxygenation of vegetable oils to produce diesel range hydrocarbon fuels: a review, Renew. Sustain. Energy Rev. 73 (2017) 545–557.

[31] W.-C. Wang, et al., Hydrocarbon fuels from vegetable oils via hydrolysis and thermo-catalytic decarboxylation, Fuel 95 (2012) 622–629.

[32] W.-C. Wang, et al., Exploration of process parameters for continuous hydrolysis of canola oil, camelina oil and algal oil, Chem. Eng. Process. Process Intensif. 57 (2012) 51–58.

[33] L. Li, et al., Catalytic hydrothermal conversion of triglycerides to non-ester biofuels, Energy Fuels 24 (2) (2010) 1305–1315.

[34] F. Yang, M.A. Hanna, R. Sun, Value-added uses for crude glycerol—a byproduct of biodiesel production, Biotechnol. Biofuels 5 (1) (2012) 1–10.

[35] A. Galadima, O. Muraza, Catalytic upgrading of vegetable oils into jet fuels range hydrocarbons using heterogeneous catalysts: a review, J. Ind. Eng. Chem. 29 (2015) 12–23.

[36] N.H. Tran, et al., Catalytic upgrading of biorefinery oil from micro-algae, Fuel 89 (2) (2010) 265–274.

[37] D. Kubička, L. Kaluža, Deoxygenation of vegetable oils over sulfided Ni, Mo and NiMo catalysts, Appl. Catal. A Gen. 372 (2) (2010) 199–208.

[38] M. Patel, A. Kumar, Production of renewable diesel through the hydroprocessing of lignocellulosic biomass-derived bio-oil: a review, Renew. Sustain. Energy Rev. 58 (2016) 1293–1307.

[39] S. Harnos, G. Onyestyák, D. Kalló, Hydrocarbons from sunflower oil over partly reduced catalysts, React. Kinet. Mech. Catal. 106 (1) (2012) 99–111.

[40] K. Lee, et al., Single-step hydroconversion of triglycerides into biojet fuel using CO-tolerant PtRe catalyst supported on USY, J. Catal. 379 (2019) 180–190.

[41] Y. Lu, B. Ma, C. Zhao, Integrated production of bio-jet fuel containing lignin-derived arenes via lipid deoxygenation, Chem. Commun. 54 (70) (2018) 9829–9832.

[42] H. Zhang, H. Lin, Y. Zheng, The role of cobalt and nickel in deoxygenation of vegetable oils, Appl. Catal. Environ. 160 (2014) 415–422.

[43] V. Itthibenchapong, et al., Deoxygenation of palm kernel oil to jet fuel-like hydrocarbons using Ni-MoS$_2$/γ-Al$_2$O$_3$ catalysts, Energy Convers. Manag. 134 (2017) 188–196.

[44] M.R. De Brimont, et al., Deoxygenation mechanisms on Ni-promoted MoS$_2$ bulk catalysts: a combined experimental and theoretical study, J. Catal. 286 (2012) 153–164.

[45] D. Verma, et al., Diesel and aviation kerosene with desired aromatics from hydroprocessing of jatropha oil over hydrogenation catalysts supported on hierarchical mesoporous SAPO-11, Appl. Catal. A Gen. 490 (2015) 108–116.

[46] R. Kumar, et al., Hydroprocessing of jatropha oil and its mixtures with gas oil, Green Chem. 12 (12) (2010) 2232–2239.

[47] E.-M. Ryymin, et al., Competitive reactions and mechanisms in the simultaneous HDO of phenol and methyl heptanoate over sulphided NiMo/γ-Al$_2$O$_3$, Appl. Catal. A Gen. 389 (1–2) (2010) 114–121.

[48] R. Sotelo-Boyás, Y. Liu, T. Minowa, Renewable diesel production from the hydrotreating of rapeseed oil with Pt/Zeolite and NiMo/Al$_2$O$_3$ catalysts, Ind. Eng. Chem. Res. 50 (5) (2011) 2791–2799.

[49] Q. Tan, Y. Cao, J. Li, Prepared multifunctional catalyst Ni$_2$P/Zr-SBA-15 and catalyzed Jatropha Oil to produce bio-aviation fuel, Renew. Energy 150 (2020) 370–381.

[50] K.-R. Hwang, et al., Bio fuel production from crude Jatropha oil; addition effect of formic acid as an in-situ hydrogen source, Fuel 174 (2016) 107–113.

[51] D.A. Ruddy, et al., Recent advances in heterogeneous catalysts for bio-oil upgrading via "ex situ catalytic fast pyrolysis": catalyst development through the study of model compounds, Green Chem. 16 (2) (2014) 454–490.

[52] Z.N. Eller, Z.N. Varga, J. Hancsók, Renewable jet fuel from kerosene/coconut oil mixtures with catalytic hydrogenation, Energy Fuels 33 (7) (2019) 6444–6453.

[53] J.G. Immer, M.J. Kelly, H.H. Lamb, Catalytic reaction pathways in liquid-phase deoxygenation of C18 free fatty acids, Appl. Catal. A Gen. 375 (1) (2010) 134–139.

[54] Z. Ma, et al., Overview of catalyst application in petroleum refinery for biomass catalytic pyrolysis and bio-oil upgrading, RSC Adv. 5 (107) (2015) 88287–88297.

[55] Y. Liu, et al., Hydrotreatment of jatropha oil to produce green diesel over trifunctional Ni–Mo/SiO$_2$–Al$_2$O$_3$ catalyst, Chem. Lett. 38 (6) (2009) 552–553.

[56] F.P. de Sousa, C.C. Cardoso, V.M. Pasa, Producing hydrocarbons for green diesel and jet fuel formulation from palm kernel fat over Pd/C, Fuel Process. Technol. 143 (2016) 35–42.

[57] T. Morgan, et al., Conversion of triglycerides to hydrocarbons over supported metal catalysts, Top. Catal. 53 (11 − 12) (2010) 820–829.

[58] S. Cheng, et al., Conversion of prairie cordgrass to hydrocarbon biofuel over Co-Mo/HZSM-5 using a two-stage reactor system, Energy Technol. 4 (6) (2016) 706–713.

[59] T.L. Maesen, et al., Alkane hydrocracking: shape selectivity or kinetics? J. Catal. 221 (1) (2004) 241–251.

[60] C. Wang, et al., High quality diesel-range alkanes production via a single-step hydrotreatment of vegetable oil over Ni/zeolite catalyst, Catal. Today 234 (2014) 153–160.

[61] T. Li, et al., Conversion pathways of palm oil into jet biofuel catalyzed by mesoporous zeolites, RSC Adv. 6 (106) (2016) 103965–103972.

[62] J. Zhu, et al., Highly mesoporous single-crystalline zeolite beta synthesized using a nonsurfactant cationic polymer as a dual-function template, J. Am. Chem. Soc. 136 (6) (2014) 2503–2510.

[63] F.A. Twaiq, N.A. Zabidi, S. Bhatia, Catalytic conversion of palm oil to hydrocarbons: performance of various zeolite catalysts, Ind. Eng. Chem. Res. 38 (9) (1999) 3230–3237.

[64] B. Veriansyah, et al., Production of renewable diesel by hydroprocessing of soybean oil: effect of catalysts, Fuel 94 (2012) 578–585.

[65] M. Anand, et al., Optimizing renewable oil hydrocracking conditions for aviation bio-kerosene production, Fuel Process. Technol. 151 (2016) 50–58.

[66] I.-H. Choi, et al., Production of bio-jet fuel range alkanes from catalytic deoxygenation of Jatropha fatty acids on a WOx/Pt/TiO$_2$ catalyst, Fuel 215 (2018) 675–685.

[67] A. Srifa, et al., Production of bio-hydrogenated diesel by catalytic hydrotreating of palm oil over NiMoS$_2$/γ-Al$_2$O$_3$ catalyst, Bioresour. Technol. 158 (2014) 81–90.

[68] R. Zarchin, et al., Hydroprocessing of soybean oil on nickel-phosphide supported catalysts, Fuel 139 (2015) 684–691.

[69] Q. Liu, et al., Hydrodeoxygenation of palm oil to hydrocarbon fuels over Ni/SAPO-11 catalysts, Chin. J. Catal. 35 (5) (2014) 748–756.

[70] C.-Y. Liu, et al., Hydrodeoxygenation of fatty acid methyl esters and isomerization of products over NiP/SAPO-11 catalysts, J. Fuel Chem. Technol. 44 (10) (2016) 1211–1216.

[71] I. Sebos, et al., Catalytic hydroprocessing of cottonseed oil in petroleum diesel mixtures for production of renewable diesel, Fuel 88 (1) (2009) 145–149.

[72] J. Liu, et al., Hydroprocessing of Jatropha oil over NiMoCe/Al$_2$O$_3$ catalyst, Int. J. Hydrogen Energy 37 (23) (2012) 17731–17737.

[73] J. García-Dávila, et al., *Jatropha curcas* L. oil hydroconversion over hydrodesulfurization catalysts for biofuel production, Fuel 135 (2014) 380–386.

[74] Q. Liu, et al., One-step hydrodeoxygenation of palm oil to isomerized hydrocarbon fuels over Ni supported on nano-sized SAPO-11 catalysts, Appl. Catal. A Gen. 468 (2013) 68–74.

[75] X. Zhao, et al., Catalytic cracking of camelina oil for hydrocarbon biofuel over ZSM-5-Zn catalyst, Fuel Process. Technol. 139 (2015) 117–126.

[76] Z. Eller, Z. Varga, J. Hancsók, Advanced production process of jet fuel components from technical grade coconut oil with special hydrocracking, Fuel 182 (2016) 713–720.

[77] T. Kimura, et al., Hydroconversion of triglycerides to hydrocarbons over Mo–Ni/γ-Al$_2$O$_3$ catalyst under low hydrogen pressure, Catal. Lett. 143 (11) (2013) 1175–1181.

[78] D. Verma, et al., Aviation fuel production from lipids by a single-step route using hierarchical mesoporous zeolites, Energy Environ. Sci. 4 (5) (2011) 1667–1671.

[79] S. Liu, et al., Bio-aviation fuel production from hydroprocessing castor oil promoted by the nickel-based bifunctional catalysts, Bioresour. Technol. 183 (2015) 93–100.

[80] M.Y. Kim, et al., Maximizing biojet fuel production from triglyceride: importance of the hydrocracking catalyst and separate deoxygenation/hydrocracking steps, ACS Catal. 7 (9) (2017) 6256–6267.

[81] K.-C. Park, S.-K. Ihm, Comparison of Pt/zeolite catalysts for n-hexadecane hydroisomerization, Appl. Catal. A Gen. 203 (2) (2000) 201–209.

[82] S.I. Sanchez, M.D. Moser, S.A. Bradley, Mechanistic study of Pt–Re/γ-Al$_2$O$_3$ catalyst deactivation by chemical imaging of carbonaceous deposits using advanced X-ray detection in scanning transmission electron microscopy, ACS Catal. 4 (1) (2014) 220–228.

[83] J. Duan, et al., Diesel-like hydrocarbons obtained by direct hydrodeoxygenation of sunflower oil over Pd/Al-SBA-15 catalysts, Catal. Commun. 17 (2012) 76–80.

[84] R.A. Voloshin, et al., Biofuel production from plant and algal biomass, Международный научный журнал Альтернативная энергетика и экология 7–9 (2019) 12–31.

[85] A.F. Miranda, et al., Aquatic plant Azolla as the universal feedstock for biofuel production, Biotechnol. Biofuels 9 (1) (2016) 221.

[86] A. Bahadar, M.B. Khan, Progress in energy from microalgae: a review, Renew. Sustain. Energy Rev. 27 (2013) 128–148.

[87] M.J. Griffiths, S.T. Harrison, Lipid productivity as a key characteristic for choosing algal species for biodiesel production, J. Appl. Phycol. 21 (5) (2009) 493–507.

[88] X. Yang, et al., Carbon distribution of algae-based alternative aviation fuel obtained by different pathways, Renew. Sustain. Energy Rev. 54 (2016) 1129–1147.

[89] P.M. Schenk, et al., Second generation biofuels: high-efficiency microalgae for biodiesel production, Bioenergy Res. 1 (1) (2008) 20–43.

[90] S. Popov, S. Kumar, Renewable fuels via catalytic hydrodeoxygenation of lipid-based feedstocks, Biofuels 4 (2) (2013) 219–239.

[91] M. Saber, B. Nakhshiniev, K. Yoshikawa, A review of production and upgrading of algal bio-oil, Renew. Sustain. Energy Rev. 58 (2016) 918–930.

[92] G.W. Huber, S. Iborra, A. Corma, Synthesis of transportation fuels from biomass: chemistry, catalysts, and engineering, Chem. Rev. 106 (9) (2006) 4044–4098.

[93] H.J. Robota, J.C. Alger, L. Shafer, Converting algal triglycerides to diesel and HEFA jet fuel fractions, Energy Fuels 27 (2) (2013) 985–996.

[94] S.K. Kim, et al., Production of renewable diesel via catalytic deoxygenation of natural triglycerides: comprehensive understanding of reaction intermediates and hydrocarbons, Appl. Energy 116 (2014) 199–205.

[95] Z. Shuping, et al., Production and characterization of bio-oil from hydrothermal liquefaction of microalgae *Dunaliella tertiolecta* cake, Energy 35 (12) (2010) 5406–5411.

[96] A. Demirbas, Liquefaction of biomass using glycerol, Energy Sources A 30 (12) (2008) 1120–1126.

[97] S. Xiu, A. Shahbazi, Bio-oil production and upgrading research: a review, Renew. Sustain. Energy Rev. 16 (7) (2012) 4406–4414.

[98] Y.-P. Xu, et al., Liquid fuel generation from algal biomass via a two-step process: effect of feedstocks, Biotechnol. Biofuels 11 (1) (2018) 83.

[99] Q. Guo, et al., Catalytic hydrodeoxygenation of algae bio-oil over bimetallic Ni–Cu/ZrO$_2$ catalysts, Ind. Eng. Chem. Res. 54 (3) (2015) 890–899.

[100] B. Zhao, et al., Two-stage upgrading of hydrothermal algae biocrude to kerosene-range biofuel, Green Chem. 18 (19) (2016) 5254–5265.

[101] Q. Hu, et al., Microalgal triacylglycerols as feedstocks for biofuel production: perspectives and advances, Plant J. 54 (4) (2008) 621–639.

[102] K. Kaya, et al., Thraustochytrid *Aurantiochytrium* sp. 18W-13a accumulates high amounts of squalene, Biosci. Biotechnol. Biochem. 75 (11) (2011) 2246–2248.

[103] P. Metzger, C. Largeau, *Botryococcus braunii*: a rich source for hydrocarbons and related ether lipids, Appl. Microbiol. Biotechnol. 66 (5) (2005) 486–496.

[104] S. Nagano, et al., Physical properties of hydrocarbon oils produced by *Botryococcus braunii*: density, kinematic viscosity, surface tension, and distillation properties, Procedia Environ. Sci. 15 (2012) 73–79.

[105] Y. Nakaji, et al., Production of gasoline fuel from alga-derived botryococcene by hydrogenolysis over ceria-supported ruthenium catalyst, ChemCatChem 9 (14) (2017) 2701–2708.

[106] K. Zhang, X. Zhang, T. Tan, The production of bio-jet fuel from *Botryococcus braunii* liquid over a Ru/CeO$_2$ catalyst, RSC Adv. 6 (102) (2016) 99842–99850.

[107] S.I. Oya, et al., Catalytic production of branched small alkanes from biohydrocarbons, ChemSusChem 8 (15) (2015) 2472–2475.

[108] K. Murata, et al., Hydrocracking of algae oil into aviation fuel-range hydrocarbons using a Pt–Re catalyst, Energy Fuels 28 (11) (2014) 6999–7006.

[109] J. Cheng, et al., Continuous hydroprocessing of microalgae biodiesel to jet fuel range hydrocarbons promoted by Ni/hierarchical mesoporous Y zeolite catalyst, Int. J. Hydrogen Energy 44 (23) (2019) 11765–11773.

[110] L. Chen, et al., Catalytic hydrotreatment of fatty acid methyl esters to diesel-like alkanes over Hβ zeolite-supported nickel catalysts, ChemCatChem 6 (12) (2014) 3482–3492.

[111] L. Chen, et al., Catalytic hydroprocessing of fatty acid methyl esters to renewable alkane fuels over Ni/HZSM-5 catalyst, Catal. Today 259 (2016) 266–276.

[112] Z. Zhang, J. Song, B. Han, Catalytic transformation of lignocellulose into chemicals and fuel products in ionic liquids, Chem. Rev. 117 (10) (2017) 6834–6880.

[113] F.H. Isikgor, C.R. Becer, Lignocellulosic biomass: a sustainable platform for the production of bio-based chemicals and polymers, Polym. Chem. 6 (25) (2015) 4497–4559.

[114] M. Stöcker, Biofuels and biomass-to-liquid fuels in the biorefinery: catalytic conversion of lignocellulosic biomass using porous materials, Angew. Chem. Int. Ed. 47 (48) (2008) 9200–9211.

[115] J.P. Lange, Lignocellulose conversion: an introduction to chemistry, process and economics, Biofuels Bioprod. Biorefin. 1 (1) (2007) 39–48.

[116] A. Demirbaş, Mechanisms of liquefaction and pyrolysis reactions of biomass, Energy Convers. Manag. 41 (6) (2000) 633–646.

[117] R.M. Santos, et al., Pyrolysis of mangaba seed: production and characterization of bio-oil, Bioresour. Technol. 196 (2015) 43–48.

[118] T. Stedile, et al., Comparison between physical properties and chemical composition of bio-oils derived from lignocellulose and triglyceride sources, Renew. Sustain. Energy Rev. 50 (2015) 92–108.

[119] X. Li, et al., Hydrodeoxygenation of lignin-derived bio-oil using molecular sieves supported metal catalysts: a critical review, Renew. Sustain. Energy Rev. 71 (2017) 296–308.

[120] N. Doassans-Carrère, et al., Comparative study of biomass fast pyrolysis and direct liquefaction for bio-oils production: products yield and characterizations, Energy Fuels 28 (8) (2014) 5103–5111.

[121] T.N. Pham, et al., Ketonization of carboxylic acids: mechanisms, catalysts, and implications for biomass conversion, ACS Catal. 3 (11) (2013) 2456–2473.

[122] J. Wang, et al., Preparation of jet fuel range hydrocarbons by catalytic transformation of bio-oil derived from fast pyrolysis of straw stalk, Energy 86 (2015) 488–499.

[123] W. Won, C.T. Maravelias, Thermal fractionation and catalytic upgrading of lignocellulosic biomass to biofuels: process synthesis and analysis, Renew. Energy 114 (2017) 357–366.

[124] T.N. Pham, D. Shi, D.E. Resasco, Reaction kinetics and mechanism of ketonization of aliphatic carboxylic acids with different carbon chain lengths over Ru/TiO_2 catalyst, J. Catal. 314 (2014) 149–158.

[125] S.D. Randery, J.S. Warren, K.M. Dooley, Cerium oxide-based catalysts for production of ketones by acid condensation, Appl. Catal. A Gen. 226 (1–2) (2002) 265–280.

[126] S. Van de Vyver, Y. Román-Leshkov, Metalloenzyme-like zeolites as lewis acid catalysts for C–C bond formation, Angew. Chem. Int. Ed. 54 (43) (2015) 12554–12561.

[127] R.M. West, et al., Carbon–carbon bond formation for biomass-derived furfurals and ketones by aldol condensation in a biphasic system, J. Mol. Catal. A Chem. 296 (1–2) (2008) 18–27.

[128] D.T. Ngo, et al., Aldol condensation of cyclopentanone on hydrophobized MgO. Promotional role of water and changes in the rate-limiting step upon organosilane functionalization, ACS Catal. 9 (4) (2019) 2831–2841.

[129] G. Liu, B. Yan, G. Chen, Technical review on jet fuel production, Renew. Sustain. Energy Rev. 25 (2013) 59–70.

[130] Q. Yan, et al., Catalytic conversion wood syngas to synthetic aviation turbine fuels over a multifunctional catalyst, Bioresour. Technol. 127 (2013) 281–290.

[131] K. Kumabe, et al., Production of hydrocarbons in Fischer–Tropsch synthesis with Fe-based catalyst: investigations of primary kerosene yield and carbon mass balance, Fuel 89 (8) (2010) 2088–2095.

[132] S.L. Soled, et al., Control of metal dispersion and structure by changes in the solid-state chemistry of supported cobalt Fischer–Tropsch catalysts, Top. Catal. 26 (1–4) (2003) 101–109.

[133] A. Steynberg, Introduction to Fischer-Tropsch technology, in: Studies in Surface Science and Catalysis, Elsevier, 2004, pp. 1–63.

[134] J. Li, et al., Jet fuel synthesis via Fischer–Tropsch synthesis with varied 1-olefins as additives using Co/ZrO_2–SiO_2 bimodal catalyst, Fuel 171 (2016) 159–166.

[135] X. Liu, X. Li, K. Fujimoto, Effective control of carbon number distribution during Fischer–Tropsch synthesis over supported cobalt catalyst, Catal. Commun. 8 (9) (2007) 1329–1335.

[136] H. Wei, et al., Renewable bio-jet fuel production for aviation: a review, Fuel 254 (2019) 115599.

[137] S. Abelló, D. Montané, Exploring iron-based multifunctional catalysts for Fischer–Tropsch synthesis: a review, ChemSusChem 4 (11) (2011) 1538–1556.

[138] M.K. Gnanamani, et al., Fischer–Tropsch synthesis: effect of potassium on activity and selectivity for oxide and carbide Fe catalysts, Catal. Lett. 143 (11) (2013) 1123–1131.

[139] D.G. Miller, M. Moskovits, A study of the effects of potassium addition to supported iron catalysts in the Fischer-Tropsch reaction, J. Phys. Chem. 92 (21) (1988) 6081–6085.

[140] D.M. del Monte, et al., Effect of K, Co and Mo addition in Fe-based catalysts for aviation biofuels production by Fischer-Tropsch synthesis, Fuel Process. Technol. 194 (2019) 106102.

[141] H. Wang, et al., Catalytic routes for the conversion of lignocellulosic biomass to aviation fuel range hydrocarbons, Renew. Sustain. Energy Rev. 120 (2020) 109612.

[142] L.T. Mika, E. Csefalvay, A. Nemeth, Catalytic conversion of carbohydrates to initial platform chemicals: chemistry and sustainability, Chem. Rev. 118 (2) (2018) 505–613.

[143] R. Weingarten, et al., Design of solid acid catalysts for aqueous-phase dehydration of carbohydrates: the role of Lewis and Brønsted acid sites, J. Catal. 279 (1) (2011) 174–182.

[144] V. Choudhary, S.I. Sandler, D.G. Vlachos, Conversion of xylose to furfural using Lewis and Brønsted acid catalysts in aqueous media, ACS Catal. 2 (9) (2012) 2022–2028.

[145] J. Luterbacher, D.M. Alonso, J. Dumesic, Targeted chemical upgrading of lignocellulosic biomass to platform molecules, Green Chem. 16 (12) (2014) 4816–4838.

[146] G.W. Huber, et al., Production of liquid alkanes by aqueous-phase processing of biomass-derived carbohydrates, Science 308 (5727) (2005) 1446–1450.

[147] G. Liang, et al., Selective aldol condensation of biomass-derived levulinic acid and furfural in aqueous-phase over MgO and ZnO, Green Chem. 18 (11) (2016) 3430–3438.

[148] M. Su, et al., Production of liquid fuel intermediates from furfural via aldol condensation over Lewis acid zeolite catalysts, Catal. Sci. Technol. 7 (16) (2017) 3555–3561.

[149] R. Lee, et al., CO_2-catalysed aldol condensation of 5-hydroxymethylfurfural and acetone to a jet fuel precursor, Green Chem. 18 (19) (2016) 5118–5121.

[150] A. Bohre, B. Saha, M.M. Abu-Omar, Catalytic upgrading of 5-hydroxymethylfurfural to drop-in biofuels by solid base and bifunctional metal–acid catalysts, ChemSusChem 8 (23) (2015) 4022–4029.

[151] A. Corma, et al., Production of high-quality diesel from biomass waste products, Angew. Chem. Int. Ed. 50 (10) (2011) 2375–2378.

[152] S. Li, et al., Lignosulfonate-based acidic resin for the synthesis of renewable diesel and jet fuel range alkanes with 2-methylfuran and furfural, Green Chem. 17 (6) (2015) 3644–3652.

[153] G. Li, et al., Synthesis of high-quality diesel with furfural and 2-methylfuran from hemicellulose, ChemSusChem 5 (10) (2012) 1958–1966.

[154] S. Li, et al., Protonated titanate nanotubes as a highly active catalyst for the synthesis of renewable diesel and jet fuel range alkanes, Appl. Catal. Environ. 170 (2015) 124–134.

[155] A. Corma, O. de la Torre, M. Renz, Production of high quality diesel from cellulose and hemicellulose by the Sylvan process: catalysts and process variables, Energy Environ. Sci. 5 (4) (2012) 6328–6344.

[156] E. Karimi, et al., Ketonization and deoxygenation of alkanoic acids and conversion of levulinic acid to hydrocarbons using a Red Mud bauxite mining waste as the catalyst, Catal. Today 190 (1) (2012) 73–88.

[157] S. Ding, Q. Ge, X. Zhu, Research progress in ketonization of biomass-derived carboxylic acids over metal oxides, Huaxue Xuebao 75 (2017) 439.

[158] M. Gliński, J. Kijeński, A. Jakubowski, Ketones from monocarboxylic acids: catalytic ketonization over oxide systems, Appl. Catal. A Gen. 128 (2) (1995) 209–217.

[159] R. Pestman, et al., Reactions of carboxylic acids on oxides: 2. Bimolecular reaction of aliphatic acids to ketones, J. Catal. 168 (2) (1997) 265–272.

[160] Z. Li, et al., Efficient C–C bond formation between two levulinic acid molecules to produce C10 compounds with the cooperation effect of lewis and brønsted acids, ACS Sustain. Chem. Eng. 6 (5) (2018) 5708–5711.

[161] C.G. Lima, et al., Angelica lactones: from biomass-derived platform chemicals to value-added products, ChemSusChem 11 (1) (2018) 25–47.

[162] B. Lu, et al., Production of high value C_{10}–C_{20} products from controllable angelica lactone self-aggregation process, J. Clean. Prod. 162 (2017) 330–335.

[163] F. Chang, S. Dutta, M. Mascal, Hydrogen-economic synthesis of gasoline-like hydrocarbons by catalytic hydrodecarboxylation of the biomass-derived angelica lactone dimer, ChemCatChem 9 (14) (2017) 2622–2626.

[164] O.O. Ayodele, et al., Production of bio-based gasoline by noble-metal-catalyzed hydrodeoxygenation of α-angelica lactone derived Di/Trimers, ChemistrySelect 2 (15) (2017) 4219–4225.

[165] J. Xu, et al., Synthesis of diesel and jet fuel range alkanes with furfural and angelica lactone, ACS Catal. 7 (9) (2017) 5880–5886.

[166] J.Q. Bond, et al., Integrated catalytic conversion of γ-valerolactone to liquid alkenes for transportation fuels, Science 327 (5969) (2010) 1110–1114.

[167] J.Q. Bond, et al., Interconversion between γ-valerolactone and pentenoic acid combined with decarboxylation to form butene over silica/alumina, J. Catal. 281 (2) (2011) 290–299.

[168] J.C. Serrano-Ruiz, D. Wang, J.A. Dumesic, Catalytic upgrading of levulinic acid to 5-nonanone, Green Chem. 12 (4) (2010) 574–577.

[169] J.C. Serrano-Ruiz, et al., Conversion of cellulose to hydrocarbon fuels by progressive removal of oxygen, Appl. Catal. Environ. 100 (1–2) (2010) 184–189.

[170] J. Yang, et al., Synthesis of diesel and jet fuel range alkanes with furfural and ketones from lignocellulose under solvent free conditions, Green Chem. 16 (12) (2014) 4879–4884.

[171] F. Chen, et al., Solvent-free synthesis of C_9 and C_{10} branched alkanes with furfural and 3-pentanone from lignocellulose, Catal. Commun. 59 (2015) 229–232.

[172] H. Olcay, et al., Production of renewable petroleum refinery diesel and jet fuel feedstocks from hemicellulose sugar streams, Energy Environ. Sci. 6 (1) (2013) 205–216.

[173] L. Faba, E. Díaz, S. Ordóñez, One-pot aldol condensation and hydrodeoxygenation of biomass-derived carbonyl compounds for biodiesel synthesis, ChemSusChem 7 (10) (2014) 2816–2820.

[174] B. Pholjaroen, et al., Production of renewable jet fuel range branched alkanes with xylose and methyl isobutyl ketone, Ind. Eng. Chem. Res. 53 (35) (2014) 13618–13625.

[175] H. Wang, et al., $ZnCl_2$ induced catalytic conversion of softwood lignin to aromatics and hydrocarbons, Green Chem. 18 (9) (2016) 2802–2810.

[176] H. Wang, et al., Production of jet fuel-range hydrocarbons from hydrodeoxygenation of lignin over super lewis acid combined with metal catalysts, ChemSusChem 11 (1) (2018) 285–291.

[177] R. Kallury, et al., Hydrodeoxygenation of hydroxy, methoxy and methyl phenols with molybdenum oxide/nickel oxide/alumina catalyst, J. Catal. 96 (2) (1985) 535–543.

[178] X. Zhu, et al., Bifunctional transalkylation and hydrodeoxygenation of anisole over a Pt/HBeta catalyst, J. Catal. 281 (1) (2011) 21–29.

[179] C.R. Lee, et al., Catalytic roles of metals and supports on hydrodeoxygenation of lignin monomer guaiacol, Catal. Commun. 17 (2012) 54–58.

[180] W. Zhang, et al., Hydrodeoxygenation of lignin-derived phenolic monomers and dimers to alkane fuels over bifunctional zeolite-supported metal catalysts, ACS Sustain. Chem. Eng. 2 (4) (2014) 683–691.

[181] T.M. Sankaranarayanan, et al., Hydrodeoxygenation of anisole as bio-oil model compound over supported Ni and Co catalysts: effect of metal and support properties, Catal. Today 243 (2015) 163–172.

[182] H. Wang, et al., Biomass-derived lignin to jet fuel range hydrocarbons via aqueous phase hydrodeoxygenation, Green Chem. 17 (12) (2015) 5131–5135.

[183] Z. Luo, et al., Precise oxygen scission of lignin derived aryl ethers to quantitatively produce aromatic hydrocarbons in water, Green Chem. 18 (2) (2016) 433–441.

[184] Y. Hong, et al., Synergistic catalysis between Pd and Fe in gas phase hydrodeoxygenation of m-cresol, ACS Catal. 4 (10) (2014) 3335–3345.

[185] T. Guo, et al., Direct deoxygenation of lignin model compounds into aromatic hydrocarbons through hydrogen transfer reaction, Appl. Catal. A Gen. 547 (2017) 30–36.

[186] Y. Shao, et al., Selective production of arenes via direct lignin upgrading over a niobium-based catalyst, Nat. Commun. 8 (1) (2017) 1–9.

[187] H. Wang, et al., One-pot process for hydrodeoxygenation of lignin to alkanes using Ru-based bimetallic and bifunctional catalysts supported on zeolite Y, ChemSusChem 10 (8) (2017) 1846–1856.

[188] H. Wang, M. Feng, B. Yang, Catalytic hydrodeoxygenation of anisole: an insight into the role of metals in transalkylation reactions in bio-oil upgrading, Green Chem. 19 (7) (2017) 1668–1673.

[189] F. Cheng, C.E. Brewer, Producing jet fuel from biomass lignin: potential pathways to alkyl-benzenes and cycloalkanes, Renew. Sustain. Energy Rev. 72 (2017) 673–722.

[190] M.A. Rude, A. Schirmer, New microbial fuels: a biotech perspective, Curr. Opin. Microbiol. 12 (3) (2009) 274–281.

[191] S. Atsumi, T. Hanai, J.C. Liao, Non-fermentative pathways for synthesis of branched-chain higher alcohols as biofuels, Nature 451 (7174) (2008) 86–89.

[192] N. Zhan, et al., Lanthanum–phosphorous modified HZSM-5 catalysts in dehydration of ethanol to ethylene: a comparative analysis, Catal. Commun. 11 (7) (2010) 633–637.

[193] Y. Hu, et al., Selective dehydration of bio-ethanol to ethylene catalyzed by lanthanum-phosphorous-modified HZSM-5: influence of the fusel, Biotechnol. J. 5 (11) (2010) 1186–1191.

[194] L. Silvester, et al., Reactivity of ethanol over hydroxyapatite-based Ca-enriched catalysts with various carbonate contents, Catal. Sci. Technol. 5 (5) (2015) 2994–3006.

[195] J.D. Taylor, M.M. Jenni, M.W. Peters, Dehydration of fermented isobutanol for the production of renewable chemicals and fuels, Top. Catal. 53 (15–18) (2010) 1224–1230.

[196] J. Bedia, et al., Ethanol dehydration to ethylene on acid carbon catalysts, Appl. Catal. Environ. 103 (3–4) (2011) 302–310.

[197] J. Han, et al., Carbon-supported molybdenum carbide catalysts for the conversion of vegetable oils, ChemSusChem 5 (4) (2012) 727–733.

[198] S. Phimsen, et al., Oil extracted from spent coffee grounds for bio-hydrotreated diesel production, Energy Convers. Manag. 126 (2016) 1028–1036.

[199] J.V. Lauritsen, et al., Location and coordination of promoter atoms in Co- and Ni-promoted MoS_2-based hydrotreating catalysts, J. Catal. 249 (2) (2007) 220–233.

[200] A. Srifa, et al., Catalytic behaviors of $Ni/\gamma\text{-}Al_2O_3$ and $Co/\gamma\text{-}Al_2O_3$ during the hydrodeoxygenation of palm oil, Catal. Sci. Technol. 5 (7) (2015) 3693–3705.

[201] F. Chen, et al., Catalytic conversion of isophorone to jet-fuel range aromatic hydrocarbons over a $MoOx/SiO_2$ catalyst, Chem. Commun. 51 (59) (2015) 11876–11879.

[202] D. Kubička, Future refining catalysis-introduction of biomass feedstocks, Collect. Czechoslov. Chem. Commun. 73 (8) (2008) 1015–1044.

[203] M. Snåre, et al., Heterogeneous catalytic deoxygenation of stearic acid for production of biodiesel, Ind. Eng. Chem. Res. 45 (16) (2006) 5708–5715.

[204] M. Zhang, et al., Catalytic performance of biomass carbon-based solid acid catalyst for esterification of free fatty acids in waste cooking oil, Catal. Surv. Asia 19 (2) (2015) 61–67.

[205] S. Lestari, et al., Catalytic deoxygenation of stearic acid in a continuous reactor over a mesoporous carbon-supported Pd catalyst, Energy Fuels 23 (8) (2009) 3842–3845.

[206] S.A. Hollak, et al., Hydrothermal deoxygenation of triglycerides over Pd/C aided by in situ hydrogen production from glycerol reforming, ChemSusChem 7 (4) (2014) 1057–1062.

[207] H.-S. Roh, et al., The effect of calcination temperature on the performance of $Ni/MgO–Al_2O_3$ catalysts for decarboxylation of oleic acid, Catal. Today 164 (1) (2011) 457–460.

[208] B. Al Alwan, S.O. Salley, K.S. Ng, Biofuels production from hydrothermal decarboxylation of oleic acid and soybean oil over Ni-based transition metal carbides supported on Al-SBA-15, Appl. Catal. A Gen. 498 (2015) 32–40.

[209] J. Zhang, C. Zhao, A new approach for bio-jet fuel generation from palm oil and limonene in the absence of hydrogen, Chem. Commun. 51 (97) (2015) 17249–17252.

[210] R.-X. Chen, W.-C. Wang, The production of renewable aviation fuel from waste cooking oil. Part I: bio-alkane conversion through hydro-processing of oil, Renew. Energy 135 (2019) 819–835.

[211] M. Rabaev, et al., Conversion of vegetable oils on $Pt/Al_2O_3/SAPO\text{-}11$ to diesel and jet fuels containing aromatics, Fuel 161 (2015) 287–294.

[212] J. Fu, et al., Direct production of aviation fuels from microalgae lipids in water, Fuel 139 (2015) 678–683.

[213] L.N. Silva, et al., Biokerosene and green diesel from macauba oils via catalytic deoxygenation over Pd/C, Fuel 164 (2016) 329–338.

[214] P.H. Araújo, et al., Catalytic deoxygenation of the oil and biodiesel of Licuri (*Syagrus coronata*) to obtain n-alkanes with chains in the range of biojet fuels, ACS Omega 4 (14) (2019) 15849–15855.

[215] S. Oh, et al., Evaluation of hydrodeoxygenation reactivity of pyrolysis bio-oil with various Ni-based catalysts for improvement of fuel properties, RSC Adv. 7 (25) (2017) 15116–15126.

[216] X. Zhao, et al., Catalytic cracking of non-edible sunflower oil over ZSM-5 for hydrocarbon bio-jet fuel, New Biotechnol. 32 (2) (2015) 300–312.

[217] F. Jamil, et al., *Phoenix dactylifera* kernel oil used as potential source for synthesizing jet fuel and green diesel, Energy Procedia 118 (2017) 35–39.

[218] Z. Chunfei, et al., Tuning hierarchical ZSM-5 for green jet fuel production from soybean oil via control of Pt location and grafted TPABr content, Catal. Commun. 155 (2021) 106288.

[219] H. Chen, et al., Effect of support on the NiMo phase and its catalytic hydrodeoxygenation of triglycerides, Fuel 159 (2015) 430–435.

[220] C.A. Scaldaferri, V.M.D. Pasa, Hydrogen-free process to convert lipids into bio-jet fuel and green diesel over niobium phosphate catalyst in one-step, Chem. Eng. J. 370 (2019) 98–109.

[221] N. Asikin-Mijan, et al., Production of green diesel via cleaner catalytic deoxygenation of *Jatropha curcas* oil, J. Clean. Prod. 167 (2017) 1048–1059.

[222] K. Murata, et al., Production of synthetic diesel by hydrotreatment of jatropha oils using Pt − Re/H-ZSM-5 catalyst, Energy Fuels 24 (4) (2010) 2404–2409.

[223] D.R. Shonnard, L. Williams, T.N. Kalnes, Camelina-derived jet fuel and diesel: sustainable advanced biofuels, Environ. Prog. Sustain. Energy 29 (3) (2010) 382–392.

[224] S. Bezergianni, A. Kalogianni, I.A. Vasalos, Hydrocracking of vacuum gas oil-vegetable oil mixtures for biofuels production, Bioresour. Technol. 100 (12) (2009) 3036–3042.

[225] M. Shahinuzzaman, Z. Yaakob, Y. Ahmed, Non-sulphide zeolite catalyst for bio-jet-fuel conversion, Renew. Sustain. Energy Rev. 77 (2017) 1375–1384.

[226] H. Wang, Biofuels Production From Hydrotreating of Vegetable Oil Using Supported Noble Metals, and Transition Metal Carbide and Nitride, 2012.

[227] M. Asadieraghi, W.M.A.W. Daud, H.F. Abbas, Heterogeneous catalysts for advanced bio-fuel production through catalytic biomass pyrolysis vapor upgrading: a review, RSC Adv. 5 (28) (2015) 22234–22255.

[228] Y. Ren, et al., ZnO supported on high silica HZSM-5 as new catalysts for dehydrogenation of propane to propene in the presence of CO_2, Catal. Today 148 (3–4) (2009) 316–322.

[229] S.J. Reaume, N. Ellis, Use of isomerization and hydroisomerization reactions to improve the cold flow properties of vegetable oil based biodiesel, Energies 6 (2) (2013) 619–633.

[230] H. Deldari, Suitable catalysts for hydroisomerization of long-chain normal paraffins, Appl. Catal. A Gen. 293 (2005) 1–10.

[231] C.-H. Lin, Y.-K. Chen, W.-C. Wang, The production of bio-jet fuel from palm oil derived alkanes, Fuel 260 (2020) 116345.

[232] G. Knothe, Biodiesel and renewable diesel: a comparison, Prog. Energy Combust. Sci. 36 (3) (2010) 364–373.

[233] V. Itthibenchapong, A. Srifa, K. Faungnawakij, Heterogeneous catalysts for advanced biofuel production, in: Nanotechnology for Bioenergy and Biofuel Production, Springer, 2017, pp. 231–254.

[234] J.L. Graham, et al., Swelling of nitrile rubber by selected aromatics blended in a synthetic jet fuel, Energy Fuels 20 (2) (2006) 759–765.

[235] K. Chen, H. Liu, The impacts of aromatic contents in aviation jet fuel on the volume swell of the aircraft fuel tank sealants, SAE Int. J. Aerosp. 6 (2013) 350–354 (2013-01-9001).

[236] P.A. Muzzell, et al., The effect of switch-loading fuels on fuel-wetted elastomers, SAE Trans. (2007) 592–606.

[237] N. Chen, et al., Effects of Si/Al ratio and Pt loading on Pt/SAPO-11 catalysts in hydroconversion of Jatropha oil, Appl. Catal. A Gen. 466 (2013) 105–115.

[238] M. Tsodikov, et al., Catalytic conversion of rape oil into alkane-aromatic fraction in the presence of Pd-Zn/MFI, Pet. Chem. 53 (1) (2013) 46–53.

[239] P. Mériaudeau, C. Naccache, Dehydrocyclization of alkanes over zeolite-supported metal catalysts: monofunctional or bifunctional route, Catal. Rev. 39 (1–2) (1997) 5–48.

[240] A.L. Jongerius, P.C. Bruijnincx, B.M. Weckhuysen, Liquid-phase reforming and hydrodeoxygenation as a two-step route to aromatics from lignin, Green Chem. 15 (11) (2013) 3049–3056.

[241] M. Chen, Hydrodeoxygenation of Bio-Oil Model Compounds on Supported Noble Metal Catalysts, University of Sydney, 2013.

[242] S.-K. Wu, et al., Atmospheric hydrodeoxygenation of guaiacol over alumina-, zirconia-, and silica-supported nickel phosphide catalysts, ACS Sustain. Chem. Eng. 1 (3) (2013) 349–358.

CHAPTER 7

Biojet fuels and emissions

Reyes García-Contreras, José A. Soriano, Arántzazu Gómez, and Pablo Fernández-Yáñez

School of Industrial and Aerospace Engineering of Toledo, University of Castilla-La Mancha, Ciudad Real, Spain

1. Biojet fuels properties

1.1 Incorporation of biofuels in aviation transport

The aviation sector represents great demand of fossil fuels [1]. In fact, the energy consumption of jet transport is increasing with a velocity of, approximately, 5% per year during last decades and it is expected to reach more than 23 EJ in 2040 [2–4]. In terms of greenhouse emissions, this means a contribution between 4% and 6% of total transport emissions [3] and a 2% of global CO_2 could be attributed to air transportation [5]. Regarding this, the Directive 2008/101/EC of the European Parliament included the aviation activities in the scheme for greenhouse gas emission allowance trading [6].

If the upward trend in oil prices is also taken into account, closely associated with the unstable socio-political situation of most of crude oil-producing countries, it is necessary to resort to alternative fuels that, in addition to being environmentally friendly, could reduce this climate of energy insecurity in a way.

Biomass has been proved to be a very effective raw material for the production of fossil fuels surrogates [2]. In this sense, fuels derived from biomass are called biofuels and those used specifically for jets are known as biojet fuels or biojets. During last years, it has been possible to verify how biodiesel (defined as esters obtained by means a transesterification process), different bioalcohols or even new paraffinic fuels from Fischer-Tropsch (FT) or hydrotreatment processes have helped to fulfill the increasingly stringent emissions regulations for ground transportation. Moreover, the sooting indexes related to the smoke point, parameters initially used for jet fuels, have been proved to decrease when these biofuels are blended with conventional diesel fuel [7].

Although the production of biojet fuels may involve high costs [5], they present benefits in terms of sustainability (lower CO_2 emissions in life-cycle) and pollutant emissions. For these reasons, they present great potential for using in the aviation sector.

1.2 Biojet fuels standards

The range or limit values allowed for the different characteristics of fossil fuels used in civil aviation in USA, mainly Jet A and Jet A-1 turbine fuels, are specified in the American

Sustainable Alternatives for Aviation Fuels
https://doi.org/10.1016/B978-0-323-85715-4.00009-4

Copyright © 2022 Elsevier Inc.
All rights reserved.

Fig. 1 Evolution of ASTM D7566 standard including new biojets.

Society for Testing and Materials ASTM D1655 standard [8]. In Europe, the corresponding standard for the Jet A-1 is the United Kingdom Ministry of Defense MOD DEF STAN 91-91 [9]. Requirements of jets used in military aviation appear in NATO standards MIL-DTL-5624 (JP-4 and JP-5 fuels) and MIL-DTL-83133 (JP-8 and JP-8 + 100 fuels) but, except for few properties, limits are similar to those specified for civil jets [1,2].

However, since 2009 until now, the ASTM D7566 standard [10] regulates the properties of alternative fuels (Synthetic Paraffinic Kerosene—SPK), used as surrogates of the aforementioned fossil jet fuels, which may be blended (most of them) up to 50% with fossil jet fuels. The first paraffinic fuels included in this standard were those obtained via FT process and those obtained by the Hydrotreatment and subsequent isomerization of Esters and Fatty Acids (HEFA or HRJ). New processes to produce biojets have been included in this standard: fermentation of sugars with later catalytic conversion (fuels called Sugar To Jet, STJ), hydrolysis of biomass or glucose to produce intermediate alcohols followed by dehydration and oligomerization, process known as Alcohol To Jet (ATJ, mainly ethanol and butanol), Catalytic Hydro-Thermolysis (CH) or the new hydroprocessing of bioderived hydrocarbons from oil found in algae *Botryococcus braunii* (HC-HEFA) [11,12]. Fig. 1 shows the inclusion of these new biojets in the ASTM D7566 standard.

In this way, any blend that complies with these regulations is comparable to D1655 fuels, being able to be used, therefore, in aviation turbines without any modification [13–16]. Other biojets produced by recently developed systems, like those obtained by the aqueous phase reforming or the hydrotreated depolymerized cellulosic jets, are not yet included in the ASTM regulation [2]. In Europe, Annex D of MOD DEF STAN 91-91 specifies the requirements for paraffinic surrogates which is equivalent to ASTM D7566. The conversion process of these alternative fuels and their denomination for using in aviation are detailed in the Table 1.

1.3 Properties of biojet fuels

Physicochemical properties of jet fuels must fulfill very stringent requirements if manufacturing companies want to commercialize them. Between them, a high energy content (which implies low fuel consumption and, consequently, larger flight distance) together to good cold flow properties (mainly low viscosity and freezing point to ensure the fuel fluidity at high altitude) are critical. In this sense, ASTM and DEF STAN

Table 1 Main denominations of biojet fuels and conversion processes.

Synthetic Jets included in the ASTM D7566				
Conversion pathway	**Feedstock**	**Production process**	**Denomination**	**Maximum blending**
Gas-To-Jet (Biomass-To-Jet)	Gasified lignocellulosic biomass	Fischer-Tropsch processes and Fischer-Tropsch with aromatics	FT and FT—SKA	50%
Oil-To-Jet	Vegetable oil, Animal Fats, and Used Cooking oils	Hydro-processing technologies: Hydro-Treated Esters and Fatty Acids or Hydro-processed Renewable Jets	HEFA or HRJ	50%
Oil-To-Jet	Algae	Hydro-processing technologies: Hydroprocessing of bioderived hydrocarbons	HH or HC-HEFA	10%
Sugar-To-Jet	Sugary and starchy biomass	Sugar Fermentation or Synthetized isoparaffins	STJ or SIP	10%
Alcohol-To-Jet	Bio-alcohols: Ethanol or butanol	Dehydration of Intermediate alcohols produced by hydrolysis	ATJ	50%
Gas-To-Jet		Catalytic Hydro-Thermolysis	CH	50%
Synthetic Jets not included in the ASTM D7566				
Conversion pathway	**Feedstock**	**Production mechanisms**	**Denomination**	**Maximum blending**
Lignin-To-Jet	Lignocellulosic biomass	Hydro-Treated Depolymerized Cellulosic Jet (Pyrolysis)	HDCJ	–
Sugar/Lignin-To-Jet	Sugary, starchy, or lignocellulosic biomass	Catalytic aqueous phase reforming	APR	–

regulations agree in specifying a minimum value of 42.8 MJ/kg for the lower heating value while a maximum limit of 8 cSt (at $-20°C$) and $-47°C$ for viscosity and freezing point, respectively [7,9,10]. Low densities can also affect the payload capacity of aircraft, especially important when flight distance is long, and the spray characteristics [17]. Because of this, this property is limited to 775–840 kg/m^3 measured at 15°C.

Table 2 Limit values stablished by ASTM and DEF STAN regulations.

Property	Limit	
	Minimum	**Maximum**
Heating value (MJ/kg)	42.8	–
Freezing point (°C)	–	−47°C
Flash point (°C)	38	–
Density measured at 15°C (kg/m^3)	775	840
Viscosity measured at −20°C (cSt)	–	8
Distillation curve (°C)		
T10	—	205°C
Final Boiling Point	—	300°C
Aromatic content (% v/v)	—	25 for fossil jets
	8 for biojets	—
Sulfur content (% w/w)	–	0.3
Smoke point (mm)	25	–

Additionally to these properties, in the fuels quality standards different aspects related to the combustion process (fast evaporation and high combustion efficiency), low soot emissions, and security of fuel storage (minimum fire risk) are considered. Thus, other important characteristics that regulations agree in limiting would be: (i) the sooting tendency measured by means the smoke point technique (minimum 25 mm) and (ii) the fuel volatility related to flash point (minimum 38°C) and the distillation curve [1,16]. A nonvolatile jet fuel, i.e., with high T10, could present problems to ignite whereas if volatility is excessively high, evaporative losses at high altitude could appear. Both effects can finally reduce the combustion efficiency and increase emissions [18].

Moreover, to avoid possible O-ring shrinkage, a certain percentage of aromatic compounds must be present in the fuel, but this percentage is limited up to 25% for smoke purposes [14]. Concerning to Sulfur content, a maximum mass percentage of 0.3 is allowed, much less restrictive than the limit required for fuels used in land vehicles. Table 2 summarizes the limits imposed by the ASTM and DEF STAN regulation for the main jet properties.

These properties are closely related to fuel composition. A conventional fossil jet is mainly composed by n-, iso-, and cycloparaffins (naphthenes) together to different aromatic compounds with a carbon number between 8 and 16 [2]. As each type of paraffin presents different influence (positive or negative) on properties (see Table 3), the final composition will determine if values are within the specifications of regulation [17]. In this way, the higher the amount of n-paraffins and isoparaffins, the higher heating value [12]. However, while the sooting tendency of n-paraffins is very low and, consequently, their smoke point quite high, cycloparaffins present an opposite trend [7]. Concerning to cold flow properties, the worse freezing point is attributed to n-paraffins

Table 3 Effect of type of paraffins on jets properties.

Property	N-paraffins	Iso-paraffins	Cyclo-paraffins
Heating value	Positive (Very High)	Positive (High)	Negative (Medium)
Freezing point	Negative	Positive	Positive
Flash point	Positive	Negative	Positive
Smoke point	Positive	Positive	Negative

[19,20], although hydrocarbon isomerization can improve this property with only a slight penalization in the energy content and smoke point [14].

Moreover, as said previously, the presence of aromatic compounds in order to avoid O-ring failures can disrupt the final properties. In this sense, while the increase of aromatics increases the density and freezing point, a decrease will be observed for the viscosity and smoke point. Concerning to the energy content, similar heating values to those presented for cycloparaffins can be found [17].

Regarding to viscosity and density, although certain influence can be observed with the increase of aromatics, they are more related to the chain length and molecular weight than the compound type, increasing as these two properties become higher [12]. Carbon number (chain length) also plays an important role on the energy content and sooting tendency of the fuel, increasing the heating value and decreasing the smoke point as the carbon length growths [7,12].

Attending to standard specifications for alternative jets (limits similar to those commented for fossil jet fuels), SPK fuels would be appropriate as surrogate fossil jets although their lack in aromatic compounds made necessary to impose a minimum aromatic content of 8% (see Table 2). As it was commented previously, the ASTM D7566 standard establishes that up to 50% of most SPK biojets can be added with conventional jet fuel.

The relative impact of biojet fuels on main fuel properties is presented in Fig. 2. The percentage of variation for blends (50% or 20% v/v for STJ) and neat biojets are depicted using Jet A-1 as baseline. Results for each group of biojet fuels correspond to the average of the values consulted in bibliography [12,21–28]. No data was available for the smoke point of pure FT, STJ, and ATJ fuels.

Cleaner combustion (presented in the increase of smoke point) and higher heating values (Fig. 2A and B, respectively) are important advantages associated to the use of biojet blends when are compared to fossil kerosene. Other benefits are the positive variation of freezing point of SPKs, which means the better cold properties of these fuels (Fig. 2C) together to higher flash points which diminishes the fire risk during handle and storage (Fig. 2D). No notable penalization is obtained in other properties. Only density (with a slight decrease, Fig. 2F) and viscosity (with an increase, especially for STJ, showed in Fig. 2E) can be considered worsened by the use of biojets blends instead of pure Jet A-1.

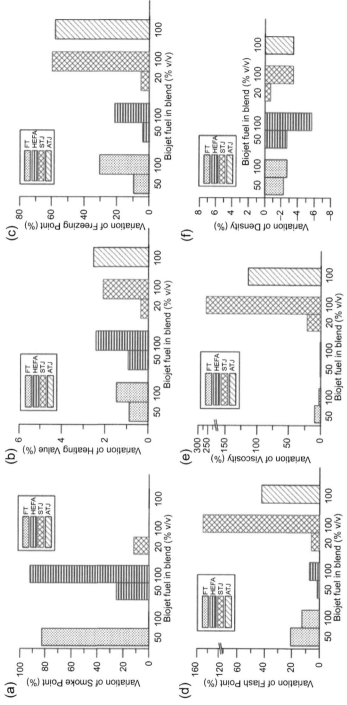

Fig. 2 Relative impact of alternative fuels on main fuel properties, (A) Smoke point, (B) Heating value, (C) Freezing point, (D) Flash point, (E) Viscosity, and (F) Density.

Moreover, apart from the lack in aromatics, the use of biofuel jets for aviation would entail the unfulfillment of the standard limit for these two properties.

In this section, only paraffinic biofuels have been studied because are the only biojets with associated regulation. However, oxygenated fuels widely used in the automotive sector such as biodiesel, due to their bad cold properties (mainly viscosity and freezing point) and slightly lower heating value, have been excluded from ASTM D7566 and MOD DEF STAN 91-91 (Annex D) standards. Although bioalcohols could solve the problem of fluidity at low temperatures, their extremely low energy content also lacks their use as biojets. Nevertheless, different studies can be found in literature where both oxygenated and nonoxygenated fuels (different from SPKs) are blended with conventional jet fuels in order to corroborate if they can be used in aviation purposes [22,29,30].

2. Biojet fuels emissions in jet engines

2.1 Pollutant emissions in jet engines

The purpose of the combustion system of a turbojet engine is to increase the thermal energy of the intake air by a combustion process. To achieve this, the combustion chamber burns the fuel supplied by the spray nozzles, with large volumes of air supplied by the compressor, giving as a result a stream of heated gas. This process must be accomplished in a short length and small cross-section for a wide operating range of air mass flow rates with the maximum heat release, the minimum loss of pressure and a proper temperature distribution [31].

Jet engines work with an overall lean mixture of fuel. However, only part of the air entering the chamber takes part in the combustion. The overall air-fuel ratio is higher than the ratio at which the mixture will burn efficiently so only part of the air entering the chamber is burned with the fuel in what is named the primary combustion zone. In this zone, air passing through the entry section and the swirl vanes interacts with another part of the intake air coming from outside the flame tube body through a selected number of secondary holes in the wall. This interaction leads to a low velocity recirculation region, in the form of a toroidal vortex, in which the flame is stabilized and anchored. An electric plug is needed to start the combustion but after that the flame is self-sustained [31,32].

From this combustion process, the main pollutant emissions generated are carbon monoxide (CO), unburned hydrocarbon (UHC), nitrogen oxides (NO_x) and soot. CO and UHC increase at low power conditions, such as idle and taxi, whereas NO_x and soot are higher at high power conditions such as climb or take-off, where the overall air-fuel ratio is lower. In this type of combustion machines, the emissions of SO_x use to be also important since the quantity of sulfur in the fuel composition is high (three times higher than the limit allowed in the standard of diesel fuel).

In the primary zone, which is fuel rich, CO and UHC are generated but these can be further combusted in the dilution zone with the fresh air intake. Unluckily, the NO_x are formed under high temperatures and it is desirable to cool the flame as quicky as possible and reduce the time in which the combustion takes place. This leads to a trade-off with the formation of the other pollutants [32].

Soot is usually formed also in the primary zone, where they may exist in the fuel-rich regions close to the fuel spray, created due to the recirculation of burned products near the fuel injector. When moving to the high-temperature regions downstream of the primary zone, most of this soot is consumed. Soot formation is influenced more by the fuel properties and atomization process than by the chemical kinetics. Fuel properties such as volatility and viscosity affect the spray penetration, drop size and evaporation rate, leading to an increase of soot [27].

The regulations of aviation emissions started in 1983 with the Committee on Aviation Environmental Protection (CAEP) which is a technical committee of the International Civil Aviation Organization (ICAO) Council. CAEP assists the Council in formulating new policies and adopting new Standards and Recommended Practices (SARPs) related to aircraft noise and emissions, and more generally to aviation environmental impact [33]. At the 2016, in the 10th meeting of CAEP, this institution agreed on the first aircraft CO_2 emission standards applied to new type design aircraft on or after January 2020. Environmental Protection Agency (EPA) proposes to update (from 2020) the existing AIR5715 Aircraft Emissions Calculation Procedure, incorporating the procedure established by ICAO for UHC, CO, NO_x and smoke [33]. The use of biojet fuels presents benefits in terms of sustainability and pollutant emissions [34] compared to fossil jet fuels, making them a potential alternative.

2.2 Biojet fuels in jet engines. A review

Emissions with different biofuels in aviation turbines have been tested in the last years. Most studies are experimental tests carried out in facilities equipped with small turbojet engines, but also theoretical studies taking into account fuel properties with calibrated models reproducing their behavior in a certain engine.

Gaspar and Sousa [24] modeled a small turbofan engine at several flight conditions with a zero-dimensional approach. Several biofuels blends, such as 50% of HEFAs and FT fuels, 30% of ATJ and 10% of SIP, were compared with baseline emissions from Jet A-1. In terms of soot, an almost generalized and significant reduction was found for all flight conditions simulated, i.e., idle, take-off, climb and cruise. This trend is fundamentally associated with the lower aromatics content of these fuels, reflected by a higher hydrogen-to-carbon ratio. Comparing types of SPKs, the highest benefits were obtained with FT kerosenes (between 55% and 70%) followed by HEFA (up to 60% at idle), ATJ and SIP (around 40%). CO and UHC emissions were found to be strongly dependent on the atomization and evaporation properties of the fuel and on the fuel-air ratio, which depends on the operating conditions and on the type of engine. FT biojets slightly reduced CO emissions, results with HEFAs were similar to Jet A, while slightly increase (around 5%) were associated to ATJ and SIP. Nevertheless, simulations for UHC indicated that the use all SPKs generally lead to a slight increase (up to 4%) of emissions for all engine conditions tested, being these differences slightly higher at idle condition (up to 8%). With respect to NO_x emissions, an almost general and significant decrease was found for all flight stages simulated and with all biojets tested.

Gawron and Białecki [25] assessed the impact of a Jet A-1—HEFA camelina blend (48% in volume of biofuel) on the performance and emission characteristics of a miniature turbojet engine (maximum thrust 140 N). With regard to CO emissions, the biggest relative change between the fuels was obtained for a rotational speed of 70,000 rpm, with nearly a 4% reduction compared to Jet A-1. The differences found in NO_x emissions with both fuels did not allow drawing any conclusion, as they did not fall within the scope of statistical data analysis. Badami et al. [22] also tested, in a small-scale turbojet engine, a blend of HEFA from jatropha oil (30% in volume) with Jet A additionally to a Gas To Liquid kerosene (FT) and a baseline Jet-A. CO emissions were found to be slightly higher for the paraffinic fuel and the blend. The results for NO_x were also higher for the alternative fuels except at high engine speeds (near 80,000 rpm). Regarding UHC, the jatropha biodiesel blend presented lower and the GTL higher emissions when compared with Jet A-1.

Different biofuels based on jatropha and camelina oils (HEFA or HRJ) were tested in the work of Sundararaj et al. [27]. The objective of this research was to study the effect of using different biojet fuels blends on gas turbine emissions and performance. These biofuels derived from oils were blended in several concentrations with conventional Jet A-1 and tested in a rig equipped with a can type combustor at two different operating conditions representing idle or low power operation. Inlet temperature and inlet pressure were representative of compressor exit conditions in an actual gas turbine. With increasing camelina concentration, the net heat of combustion increases for resulting in an increased combustion temperature, therefore increasing NO_x emissions. This increase in temperature also reduced CO and UHC emissions, consistent with the trade-off between these emissions. Soot mass concentration was also reduced with the increase in camelina concentration, mainly due to the reduced aromatics content.

For Jatropha-based fuel blends, no clear trend was found since the behavior depended on the test condition. For the lower power tested, CO and UHC emissions increased with higher jatropha concentration. This was reversed for the test point with a higher power operation (but still representative of low power operation), where CO and UHC emissions decreased with the increase in jatropha concentration. Jatropha has a higher net heat of combustion compared to Jet A-1 and camelina, but it also has a much lower viscosity and contains larger hydrocarbons. At the minimum power condition, this alters the spray characteristics, leading to improper mixing, incomplete combustion and reduced primary zone temperatures. Regarding NO_x emissions, an increase with the concentration of both biofuels was found. However, for the jatropha blends, values of NO_x increase as the jatropha concentration is higher, but these results were lower than those obtained with camelina blends. With a 40% increase in camelina concentration, NO_x increased by over 50%, but with a 50% increase in Jatropha concentration, the increase in NO_x is roughly 6%. When 20% Camelina is blended with Jet A-1, NO_x increases three times, and when 20% Jatropha is blended with Jet A-1, NO_x increases four times as compared to 100% Jet A-1. The presence of jatropha, even in relatively small quantities seems to have a detrimental effect on combustion performance, as seen by the resulting primary zone temperature, lower than for camelina blends.

Although biodiesel fuels are not included in the ASTM D7566, different works have evaluated the effect of this biofuel on the emissions generated in aviation turbines. Rehman et al. [35] studied the influence on emissions of jatropha biodiesel and diesel fuel blends in a small single shaft turbine, such as the employed for Auxiliary Power Unit (APU). The blends lowered CO and UHC emissions, which may be due to a better combustion due to the oxygen composition of the biofuel. NO_x emissions obtained were higher than diesel, due to a higher exhaust temperature as consequence of a better combustion.

Saifuddin [36] analyzed the emissions of a 25 kW micro gas turbine blending biodiesel from palm oil with bioethanol and with diesel at different proportions. The CO emissions decreased with increasing biodiesel content, with around a 90% decrease compared to conventional diesel at low engine loads, which is due by the increasing oxygen content of the blends. As expected, the NO_x followed an opposite trend, leading up to four times more emissions for the higher biofuel content blend, the 80:20 blend (16% biodiesel, 4% bioethanol). Blends with 10% of biodiesel fuels with Jet A-1 were tested by Ali et al. [29]. Two biodiesel fuels were selected for this work, one produced from cotton oil and the other form corn oil. Tests were carried out in an Olympus E-start HP small turbojet engine (230 N of maximum torque). The emissions of CO and UHC decreased by 5% and 37%, respectively, compared with conventional Jet A-1. Also, blends of biodiesel notably reduced (up to 75%) SO_2 emissions. In exchange, NO_x and CO_2 emissions were increased by a 27% and up to 11%, respectively, when compared to Jet A-1 emissions.

Only some tests have been published by engine manufacturers. Studies with 25/75 and 50/50 jatropha-algae/Jet A blends in a CFM56-7B engine [37] were done at several engine ratings. Results showed an increase in CO and UHC, which was more pronounced at the highest engine rating, and a decrease in NO_x and soot, which was more pronounced at the lowest engine rating tested. Trends commented for gaseous pollutant emissions are explained by the anticipated reduction in the peak flame temperature

Table 4 Effect of SPKs and biodiesel on pollutant emissions.

| Emissions | SPKs | | Biodiesel |
	HEFA	FT	
CO	Decrease at high load. Similar or increase at low load	Similar or increase at low load	Usually decrease
UHC	Decrease at high load. Similar or increase at low load	Not clear tendency	Usually decrease
NO_x	Not clear tendency	Not clear tendency	Usually increase
Soot	Always decrease	Always decrease	Always decrease

because of the change in H/C ratio in the fuel compared to the conventional kerosene tested, Jet A. Only clear trends are seen for CO, which increased with a higher rating and with a higher biofuel concentration, being the maximum difference (with respect to Jet A) a 9% increase with the 50/50 blend for the highest engine rated tested. For UHC, the maximum increase compared to Jet A was for the 25/75 blend, reaching a 45% at the higher engine rating. However, the increase with the 50/50 blend at the same engine rating was under 30%. The NO_x reduction seemed to be similar at the lowest and the highest engine rating, around 1% and 5% for the 25/75 and 50/50 blend, respectively. The maximum percentual soot reduction, about 30%, was found at the lowest engine rating for the 50/50 blend. At the highest engine rating the maximum reduction was for the 25/75 blend with around a 15% reduction.

Pratt and Whitney [37] tested HEFA and a 50/50 blend with Jet A-1. No significant changes were observed in UHC and CO while soot was significantly reduced with increasing concentration of HEFA. For the pure biofuel, reduction ranged from 100% at minimum power to 80% for maximum thrust. NO_x values were reduced at very low and very high thrust, around 10% for the pure biojet fuel, but no changes were observed at middle-low thrust levels. The 50/50 blend followed the same trend but with a positive offset, obtaining lower NO_x reductions at very low and very high thrusts and increased NO_x levels at middle-low thrust levels.

From the few studies found in the literature reviewed, the effect of different biojet fuels tested in aviation turbines on pollutant emissions are compared to fossil fuels and are commented in Table 4. From the SPKs included in ASTM D7566, HEFA and FT fuels most tested in literature. Biodiesel, although not regulated as biojet, has been also evaluated in aviation turbines.

3. Biojet fuels emissions in compression ignition engines

3.1 Reciprocating engines in aircraft transport

Civil and military aircrafts or airplanes are usually powered by turbines as combustion engines while small planes, civil and military helicopters, or unmanned aerial vehicles use reciprocation combustion engines, also named Aviation Piston Engines [38,39]. These engines can work in two operation modes: spark ignition (gasoline) or compression ignition (diesel). For decades, gasoline was the fuel normally used due to the difficulty to find compression ignition engines with low weight which meet the design characteristics associated to unmanned aerial vehicles or small planes [38]. However, different aspects like the need to use more safety fuels, the lower costs and simplicity of logistics, favored the replacement of gasoline engines with diesel engines. Additionally to these considerations, the use of jet fuels in compression ignitions engines (CIE) increased since the North Atlantic Treaty Organization decided to promote the use of *"Single Fuel Concept (SFC)"* in land based military aircraft, vehicles and equipment when employed on the European battlefield [30]. This idea became in official legislation, being the fuel selected the JP-8 military kerosene jet, but Jet A-1 and JP-5 have been also used in both military vehicles and aircrafts powered by diesel engines.

In land vehicles powered with ignition compression engines, fuels must meet the limits stablished in the fuel quality standard EN 590 [40]. The use of biofuels, which show similar performance and reduce pollutant emissions, has increased in last decades in part due to the appearance of standards, e.g., European Directive 2009/28/EC [41], which establishes that a 10% of energy used in the transport sector must come from renewable sources. Among possible alternative biofuels, biodiesel is the most widely used in these engines which must meet the limits of the EN 14214 [42] while the oxygenated fuels used in turbines has not regulation associated, as it was commented in Section 1.3. Also, different paraffinic fuels have been tested in compression ignition engines. Some of these paraffinic fuels are: (i) Gas-To Liquid (GTL) produced by FT process, which is known as FT-SPK when it is used in aviation turbines and (ii) Hydrogenated Vegetable Oils (HVO) also called HEFA in aviation applications. The paraffinic fuels produced by these two conversion processes have also a specific quality standard EN 15940 [43]. Some properties or this standard share limitation with ASTM D7566, e.g., heating value. However, range limited for density, cold flow properties or viscosity are different and the temperature for measuring the last one is also different ($-20°C$).

The work of Korres et al. [30] was one of the first studies where the emissions produced by biofuels, in a compression ignition, were compared to traditional jet fuels. In this study, a typical Greek diesel fuel, a JP-8 and a biodiesel were tested in a single cylinder compression ignition engine. Additionally to neat fuels, binary blends (biodiesel-diesel, JP-8-biodiesel, JP-5-diesel) and ternary blends were tested. Compared to diesel fuel, kerosene-reduced NO_x emissions and produced higher Particle Matter (PM) emissions (sulfur content of kerosene has a negative effect on soot) while biodiesel showed the opposite trend. Similar study was done by the same authors using the same diesel fuel and biodiesel but replacing the military kerosene JP-8 by the naval application kerosene JP-5 [44]. In this case, PM emissions of JP-5 were lower than diesel fuel but, in binary blends from 60% of JP-5 concentration, slightly increase was observed which may be justified by its lower cetane number. Blends of these fuels (biodiesel, JP-8 and diesel) were tested by Labeckas and Slavinskas [45], obtaining higher opacity values with biodiesel at full loads. The viscosity of the biofuel that reduces the mixing ratio with air and the fuel evaporation, delaying the combustion process and increasing soot, was the main reason provided for explaining this trend.

Performance and emissions obtained with four different fuels (diesel, JP-8, naphtha and biodiesel) were evaluated in a compression ignition engine, at full engine load, by Yontar [46]. The highest values of fuel consumption corresponded to biodiesel while the lowest values were obtained with naphtha, consistent their C/H ratio and heating value. Biodiesel showed potential as alternative fuel because decreased PM emissions while values of NO_x emissions were slightly higher than those of JP-8 and lower than those obtained with diesel fuel. Ashour and Elwardany [47] carried out a study with blends of similar neat fuels (jet fuel, paraffinic solvent, biodiesel and diesel). As biodiesel concentration increased, CO emissions were lower with no penalty in NO_x emissions. Binary blends of biodiesel-diesel and biodiesel-kerosene were evaluated by Ekaab et al. [48]. Blends with 10% and 20% of biodiesel with kerosene emitted the lowest values of PM, CO and UHC while NO_x emissions slightly increased. These results indicated the potential of these blends to compensate the usual trade-off NO_x-PM associated to diesel fuel.

Apart from biodiesel, other alternative fuels have been studied as possible alternative to jet fuel in diesel engines. Millo et al. [49] tested a blend of diesel with a new renewable paraffinic fuel, 2,6,10-trimetyldodecane called Farnesane, produced by means fermentation of biomass-derived sugar blend. The effect of a F30 blend (30% of biofuel) on performance and emissions was evaluated under different engine load operating points. CO and UHC emissions were notably reduced with biofuel blends (except at high loads), values of NO_x were comparable with those of reference fuel and smoke was reduced at medium and high load conditions. In the work of Gowdagiri et al. [50], ignition and emissions characteristics of different diesel and jet surrogates were evaluated. Fuels selected for this work were: a military naval diesel fuel (F-76), a paraffinic biofuel generated from sugar-derived (Farnesane), an HVO produced from alga oil, a GTL diesel fuel produced by algae, two fossil kerosene fuels (Jet A-1 and JP-5), a GTL jet fuel and HRJ fuel produced from cameline. CO emissions decreased as the cetane number increased due to the start of ignition occurs earlier while NO_x emissions increased with fuels with larger fraction of saturations (higher H/C ratio and lower density).

3.2 Pollutant emissions of biojet fuels in compression ignition engines

Soriano et al. [51] evaluated the effect of different alternative fuels on the regulated pollutant emissions UHC, CO, NO_x and PM generated in a diesel engine. These emissions are limited in the homologation standard of land vehicles equipped with both, spark ignition and compression ignition engines. Different fuels were tested in a turbocharged, intercooled, fourcylinder, four-stroke compression ignition engine, which was connected to an eddy current dynamometer brake. Five steady state modes were selected, which are included in the torque-engine speed map representative of the homologation cycle for light-duty vehicles. Characteristics of these engine modes are detailed in Table 5. UHC and NO_x emissions were measured with a flame ionization detector and a chemiluminescence analyzer, respectively. A Scanning Mobility Particle Sizer (SMPS), coupled to a rotating disk diluter, was used to measure Particle Number Concentration (PNC). From these measurements, and considering the particle density proposed in the work of Gómez et al. [52], particle mass emissions were estimated. In Fig. 3, a general sketch of the experimental installation used is shown.

Four neat fuels tested were tested in this work: (i) an ultra-low-sulfur diesel fuel (base fuel), (ii) a biodiesel produced from soybean and palm oils, and (iii) two paraffinic fuels, one a Gas To Liquid (GTL, known as FT for aviation application) obtained from natural gas by means low temperature FT process and provided by Sasol and, the other, a renewable isoparaffinic hydrocarbon called Farnesane. This biofuel, usually considered as jet fuel (defined as STJ or SIJ) is obtained from sugar or its biomass by genetically modified microorganisms, produced and provided by Amyris Biotechnology Inc. Main physicochemical properties of all fuels tested are detailed in Table 6.

The variation of pollutant emissions with respect to diesel fuel is presented in Fig. 4, where values of low load corresponding to the mean values of modes A and C, modes E and G are considered as medium load modes while mode I could represent high load engine mode. The three alternative fuels showed lower UHC emissions than base fuel (Fig. 4A), the more notable reduction being obtained with biodiesel ($\approx 60\%$ except at high load). When both paraffinic fuels are compared, the potential of reduction with Farnesane is higher justified by its faster evaporation (lower volatility). The absence of aromatic compounds in the alternative fuels, and the additional presence of oxygen in

Table 5 Characteristics of engine modes tested.

Mode	Torque (N m)	Engine speed (min^{-1})	Ne (kW)	Engine load
A	10	1000	1.0	Low
C		2400	2.5	
E	60	1700	10.7	Medium
G	110	1000	11.5	
I		2400	27.6	High

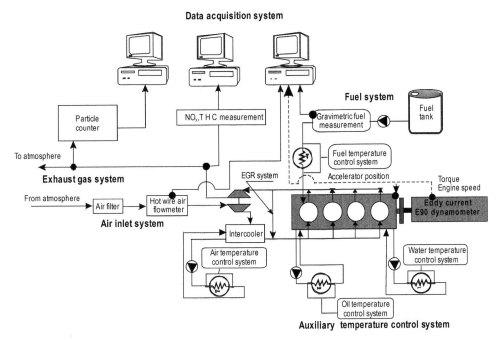

Fig. 3 Schematic of engine test bench.

Table 6 Main properties of fuels tested.

Property	Diesel	GTL	Biodiesel	Farnesane
Molecular formula	$C_{15.18}H_{29.13}$[a]	$C_{16.89}H_{35.77}$[a]	$C_{18.52}H_{34.52}O_2$[b]	$C_{15}H_{32}$
Molecular weight (g/mol)	211.4[c]	238.6[c]	289.25	212.41
H/C Ratio	1.92	2.12	1.86	2.13
Stoichiometric fuel/air ratio	1/14.64	1/14.95	1/12.46	1/14.92
Density at 15°C (kg/m³)	843	771	883	770
Viscosity at 40°C (cSt)	2.97	2.57	4.2	2.32
Lower Heating Value (MJ/kg)	41.37	42.56	37.14	43.39
Cetane number	54.2	>73	53.3	56.7 [49]
Cold Filter Plugging Point (°C)	−17	−7	0	−40
Distillation (vol.)				
10%	207.6	213.9	279.5	243.5
50%	278.2	269.3	282.7	243.8
90%	345.0	340.7	302.2	244.0

[a]Calculated by value of molecular weight and speciation of fuels (in the case of GTL, paraffinic structure is considered).
[b]Calculated from ester composition.
[c]Calculated by AspenTech HYSYS software from CHNS analysis and density value.

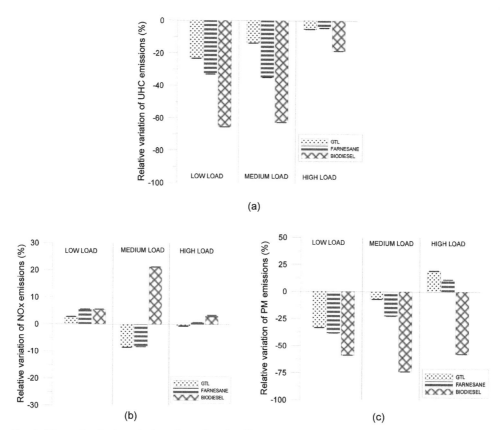

Fig. 4 Alternative fuels variation of regulated pollutant emissions, (A) UHC, (B) NO$_x$, and (C) PM.

the case of biodiesel, are the main justification for the results obtained. It is remarkable that UHC emissions produced at low load are one order of magnitude higher than in the rest of modes due to the low temperature reached in the combustion process and, consequently, in the Diesel Oxidation Catalyst (DOC). This low temperature difficulties the optimum work of DOC to reach its higher efficiency.

Regarding NO$_x$ emissions, Fig. 4B, biodiesel showed the highest values under all engine loads, trend consistent with the most of literature [53–55] and explained by the effect of the oxygen in its composition. The variations associated to paraffinic fuels are lower than 10%. The higher H/C ratio and cetane number of these fuels, compared to the reference diesel, use to imply lower adiabatic flame temperatures [56] but the combustion characteristics (injection, heat release, etc.) have also notable influence in the formation of these compounds. In the case of PM emissions (Fig. 4C), values of biodiesel are the lowest reaching a mean reduction around 60% (between the three engine loads). Again, the absence of aromatic hydrocarbons and the presence of oxygen in its composition reduce the formation of soot nuclei. Lower PM emissions were also observed with paraffinic fuels, especially at low load, reaching reductions of 38% with Farnesane and 33% with GTL. Only at high load, a slightly increase of PM were obtained with both fuels.

As conclusion, among the three alternative fuels tested in this work, Farnesane has potential as biofuel for using in reciprocating engine (although it was usually considered as jet fuel) since its use implies reductions in PM emissions (except a high load) without considerable increments in NO_x emissions, which is beneficial compared to the traditional trade-off NO_x-PM associated to diesel fuel.

3.3 Reactivity and genotoxicity of soot

Apart from the regulated pollutant emissions established in the vehicle standard homologation, other aspects such morphology, nanostructure and oxidation or biological reactivity of soot, are important in order to evaluate parameters related to the toxicity or dangerousness of soot. Most of the literature about this topic studied the effect of biodiesel as alternative fuel. Arias et al. [57] evaluated the effect of biodiesel (neat and blended in 10% and 20% of volume with diesel) on 17 polycyclic aromatic hydrocarbon (PAH) compounds and on the ecotoxicity through a *Daphnia pulex* mortality test. Results indicated that the use of biodiesel decreased PAH compounds but the ecotoxicity was higher. The effect of biodiesel with *n*-butanol were tested by Yilmaz and Davis [58] to evaluate regulated pollutant emissions and PAHs. Authors concluded that the toxicity of PAHs was higher in blends with more than 20% of alcohol.

The soot generated by the four neat fuels studied in the work detailed in the previous section [51], was characterized by means different techniques in the work of Soriano et al. [59]. The aim of this work was to evaluate the oxidation reactivity and morphology of soot generated in mode G (higher soot emission), information which may be necessary to: (i) explain the genotoxicity ad mutagenicity of PM, (ii) explain the formation and composition of aerosols in the environment air and their possible impact on the climate change, and (iii) the design of Diesel Particle Filter (DPF).

The techniques used to study nanostructure and morphology of soot were: (i) a SMPS was used to measure the particle number size distributions, (ii) X-Ray diffraction (XRD) which provided the lattice parameter and the size of graphene lamellae (fringe), (iii) the Raman spectroscopy (RS) to measure the defects of the edge and basal plane on the graphene layer, and (iv) the tortuosity, the layers and the sizes of primary particles and agglomerates were obtained by means a high resolution transmission electron microscopy (HRTEM). Related to proximal analysis, soot oxidation reactivity and active surface area (ASA), two techniques were selected: a Thermogravimetric Analysis (TGA) where the application of a temperature ramp allowed to determine the volatile organic fraction (VOF) and by means the Fourier Transform Infrared spectroscopy (FTIR) the chemical

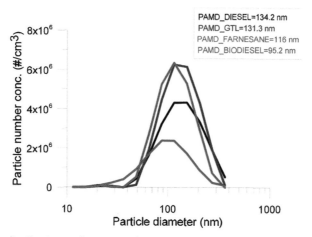

Fig. 5 Particle size distribution and mean agglomerate particle diameter.

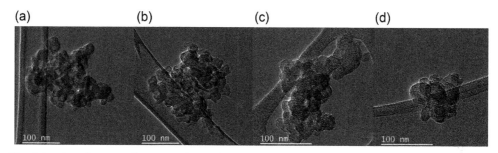

Fig. 6 Images of soot agglomerate obtained with (A) diesel, (B) GTL, (C) Farnesane, and (D) biodiesel.

compounds located in the soot surface (which have influence in its reactivity) were identified.

Particle size distributions for all fuels tested are showed in Fig. 5, where the values of Particles Agglomerate Mean Diameter (PAMD) for each fuel are also included. Particle number concentration of biodiesel was the lowest, two times lower that that corresponding to diesel fuel (reference), being its particle mean diameter also the lowest. Both paraffinic fuels, Farnesane and GTL, showed similar results in number concentration but the particle mean diameter was slightly lower with the biofuel.

Images of the particle agglomerate were taken for all fuels with HRTEM using a magnification of $40,000 \times$. As can be observed in the Fig. 6, the size of particle agglomerates of biodiesel was the lowest while the soot obtained with diesel and GTL fuels showed the highest sizes. This trend corroborates the results derived from the particle size distributions.

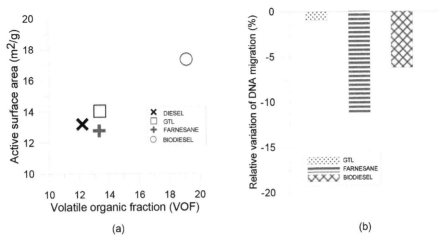

Fig. 7 (A) ASA and VOF and (B) relative variation in DNA migration.

From the thermogravimetric analysis, values of ASA, VOF and ash content were calculated. ASA parameter provides information about the active sites which are available for the possible oxygen chemisorption and it is calculated following the method proposed by Arenillas et al. [60]. Comparing results of ASA for the different fuels tested (Fig. 7A), it is clear that soot of biodiesel presents the highest oxidation reactivity, followed by GTL, diesel and Farnesane. Values of VOF, also shown in Fig. 7A, are consistent with the trend commented for ASA and it is also indicative of the reactivity of soot, as it is commented in literature [61–63]. Some authors indicate that small traces of metal ashes may favor the soot oxidation [64]. In this sense, the high percentage of ash in the biodiesel soot (16%) is consistent with the higher reactivity commented from the results of ASA and VOF compared to the other fuels (around 2%), although the presence of ash has also negative effect in after-treatment devices [65]. The differences in the reactivity between biodiesel and diesel are quite clear but paraffinic fuels also show slightly higher reactivity than reference fuel, being the literature scarcer for these types of fuels [66].

In addition to the evaluation of oxidation reactivity and morphology of soot generated by these fuels, the genotoxicity and mutagenicity of PM was evaluated by these authors [67]. The genotoxicity was considered as the Deoxyribonucleic Acid (DNA) breaks or genetic damage, which is evaluated with the alkaline comet assay, initially proposed by Singh et al. [68] and modified by McNamee et al. [69]. 3 PM concentrations of each fuel were selected to carry out the alkaline comet assay and values of DNA migration obtained showed significant differences ($P < .05$ with respect to negative control) in DNA migration for each. Mean values of the three concentrations tested with each fuel were compared to diesel, results showed in Fig. 7B. Reductions of 6% and 11% in DNA migration were obtained with biodiesel and Farnesane, respectively, while results of GTL were similar to diesel fuel. Trends corresponding to mutagenicity index were consistent with this, as is concluded in [67].

Comparing values of Fig. 7A and B, interesting trends can be observed. The higher reactivity of biodiesel PM, which is beneficial for a faster oxidation (for example, when it is cumulated in DPF), could indicate a higher biological reactivity which would have a negative effect in the human and animal health. However, lower DNA migration is associated to biodiesel, consistent with other works [57,70]. Regarding Farnesane PM, its oxidation reactivity is slightly higher than reference fuel and its biological reactive is lower than the other fuels tested.

4. Strengths and weaknesses of the use of biojet fuels

Synthetic Paraffin Kerosene fuels (SPKs) produced from renewable raw material are considered as surrogates of fossil kerosene and they must fulfill the ASTM 7566 standard for their use in aviation engines. Apart from these hydrocarbons, other biofuels such as biodiesel have been also tested as alternative fuel (always blended with a reference kerosene).

According to the results presented in this chapter, the most important strengths of the use of biojet fuels are the following:
- SPKs show higher heating value, higher smoke point and lower freezing point compared to fossil kerosene. These trends imply lower fuel consumption (lower CO_2 emissions), lower sooting tendency, and better cold flow properties.
- Regarding emissions generated in both, aviation turbines and compression ignition engines, the use of these biofuels implies reductions in the soot emitted in comparison with fossil kerosene, mainly justified by their lower aromatic content. Not clear trends are associated to UHC and CO emissions, although reductions used to be observed at high load. These advantages are additional to the lower CO_2 emissions in life-cycle associated to fuels produced by renewable raw material.
- Farnesane fuel, known as SPJ or SIP when it is used in aviation applications, not only decreases soot emissions but also presents a higher oxidation reactivity (faster elimination in DPFs) and lower biological one (lower risk for human health).

By the contrary, the use of biojets shows some disadvantages:
- The production process of biojet fuels may involve higher costs than those corresponding to fossil kerosene.
- NO_x emissions usually increase with the use of biodiesel, while not clear trend is obtained with SPKs (being HEFA and FT fuels the usually tested in aviation turbines).

References

[1] N. Yilmaz, A. Atmanli, Sustainable alternative fuels in aviation, Energy 140 (2017) 1378–1386.
[2] H. Wei, W. Liu, X. Chen, Q. Yang, J. Li, H. Chen, Renewable bio-jet fuel production for aviation: a review, Fuel 254 (2019) 115599.

[3] M. Wise, M. Muratori, P. Kyle, Biojet fuels and emissions mitigation in aviation: an integrated assessment modeling analysis, Transp. Res. Part D Transp. Environ. 52 (2017) 244–253.

[4] B.H.H. Goh, C.T. Chong, Y. Ge, H.C. Ong, Progress in utilisation of waste cooking oil for sustainable biodiesel and biojet fuel production, Energy Convers. Manag. 223 (2020) 113296.

[5] G. Louise, A.F.M.C.J. Lagerkvist, Preferences for bio jet fuel in Sweden: the case of business travel from a city airport, Sustain. Energy Technol. Assess. 29 (2018) 60–69.

[6] Directive 2008/101/EC of the European Parliament and of the Council of 19 November 2008 Amending Directive 2003/87/EC so as to Include Aviation Activities in the Scheme for Greenhouse Gas Emission Allowance Trading Within the Community, Off. J. Eur. Union L8 (2009).

[7] A. Gómez, J.A. Soriano, O. Armas, Evaluation of sooting tendency of different oxygenated and paraffinic fuels blended with diesel fuel, Fuel 184 (2016) 536–543.

[8] ASTM D1655—20c. Standard Specification for Aviation Turbine Fuels.

[9] Defence Standard 91-91. Issue 7. Incorporating Amendment 3Turbine Fuel, Kerosine Type, Jet A-1. NATO Code: F-35 Joint Service Designation: AVTUR.

[10] ASTM D7566—20b. Standard Specification for Aviation Turbine Fuel Containing Synthesized Hydrocarbons.

[11] Sustainable Aviation Fuel: Technical Certification—IATA. https://www.iata.org/contentassets/d13875e9ed784f75bac90f000760e998/saf-technical-certifications.pdf.

[12] G.B. Han, J.H. Jang, M.H. Ahn, B.H. Jung, Recent application of bio-alcohol: bio-jet fuel, in: Alcohol Fuels—Current Technologies and Future Prospect, IntechOpen, 2019, p. 4.

[13] C. Zhang, X. Hui, Y. Lin, C.J. Sung, Recent development in studies of alternative jet fuel combustion: Progress, challenges, and opportunities, Renew. Sustain. Energy Rev. 54 (2016) 120–138.

[14] D. Kang, D. Kim, V. Kalaskar, A. Violi, A.L. Boehman, Experimental characterization of jet fuels under engine relevant conditions–part 1: effect of chemical composition on autoignition of conventional and alternative jet fuels, Fuel 239 (2019) 1388–1404.

[15] W.C. Wang, L. Tao, Bio-jet fuel conversion technologies, Renew. Sustain. Energy Rev. 53 (2016) 801–822.

[16] C.R. Ranucci, H.J. Alves, M.R. Monteiro, C.L. Kugelmeier, R.A. Bariccatti, C.R. de Oliveira, E.A. da Silva, Potential alternative aviation fuel from jatropha (*Jatropha curcas* L.), babassu (*Orbignya phalerata*) and palm kernel (*Elaeis guineensis*) as blends with Jet-A1 kerosene, J. Clean. Prod. 185 (2018) 860–869.

[17] C.J. Chuck, J. Donnelly, The compatibility of potential bioderived fuels with Jet A-1 aviation kerosene, Appl. Energy 118 (2014) 83–91.

[18] E.E. Elmalik, B. Raza, S. Warrag, H. Ramadhan, E. Alborzi, N.O. Elbashir, Role of hydrocarbon building blocks on gas-to-liquid derived synthetic jet fuel characteristics, Ind. Eng. Chem. Res. 53 (5) (2014) 1856–1865.

[19] A.R. Glasgow, E.T. Murphy, C.B. Willingham, F.D. Rossini, Purification, purity, and freezing points of 31 hydrocarbons of the API-NBS series, J. Res. Natl. Bur. Stand. 37 (2) (1946) 141–145, https://doi.org/10.6028/jres.037.003.

[20] A.J. Streiff, J.C. Zimmerman, L.F. Soule, M.T. Butt, V. Sedlak, C. Willingham, F. Rossini, Purification, purity, and freezing points of 30 hydro-carbons of the API-standard and API-NBS series, J. Res. Natl. Bur. Stand. 41 (4) (1948) 323–357.

[21] H.S. Han, C.J. Kim, C.H. Cho, C.H. Sohn, J. Han, Ignition delay time and sooting propensity of a kerosene aviation jet fuel and its derivative blended with a bio-jet fuel, Fuel 232 (2018) 724–728.

[22] M. Badami, P. Nuccio, D. Pastrone, A. Signoretto, Performance of a small-scale turbojet engine fed with traditional and alternative fuels, Energy Convers. Manag. 82 (2014) 219–228.

[23] Evaluation of Amyris Direct Sugar to Hydrocarbon (DSHC) Fuel, Pratt & Whitney, 2014.

[24] R.M.P. Gaspar, J.M.M. Sousa, Impact of alternative fuels on the operational and environmental performance of a small turbofan engine, Energy Convers. Manag. 130 (2016) 81–90.

[25] B. Gawron, T. Białecki, A. Janicka, T. Suchocki, Combustion and emissions characteristics of the turbine engine fueled with HEFA blends from different feedstocks, Energies 13 (5) (2020) 1277, https://doi.org/10.3390/en13051277.

[26] N.M. Mazlan, M. Savill, T. Kipouros, Evaluating NOx and CO emissions of bio-SPK fuel using a simplified engine combustion model: a preliminary study towards sustainable environment, Proc. Inst. Mech. Eng. G J. Aerosp. Eng. 231 (5) (2017) 859–865.

[27] R.H. Sundararaj, R.D. Kumar, A.K. Raut, T.C. Sekar, V. Pandey, A. Kushari, S.K. Puri, Combustion and emission characteristics from biojet fuel blends in a gas turbine combustor, Energy 182 (2019) 689–705.

[28] P. Lobo, D.E. Hagen, P.D. Whitefield, Comparison of PM emissions from a commercial jet engine burning conventional, biomass, and Fischer–Tropsch fuels, Environ. Sci. Technol. 45 (24) (2011) 10744–10749.

[29] A.H.H. Ali, M.N. Ibrahim, Performance and environmental impact of a turbojet engine fueled by blends of biodiesels, Int. J. Environ. Sci. Technol. 14 (6) (2017) 1253–1266.

[30] D.M. Korres, E. Lois, D. Karonis, Use of J P-8 Aviation Fuel and Biodiesel on a Diesel, SAE 2004-01-3033, 2004.

[31] D. Mattingly, W.H. Heiser, D.T. Pratt, K.M. Boyer, B.A. Haven, Aircraft Engine Design, third ed., The American Institute of Aeronautics and Astronautics, Inc., 2018. ISBN (print): 978-1-62410-517-3.

[32] The Jet Engine, ISBN 9780902121232, ED. Rolls-Royce. Technical Publications Department Rolls-Royce, 1996.

[33] Implementation of the Latest CAEP Amendments to ICAO Annex 16 Volumes I, II and III.

[34] R.C. Neves, B.C. Klein, R.J. da Silva, M.C.A.F. Rezende, A. Funke, E. Olivarez-Gómez, A. Bonomi, R. Maciel-Filho, A vision on biomass-to-liquids (BTL) thermochemical routes in integrated sugarcane biorefineries for biojet fuel production, Renew. Sustain. Energy Rev. 119 (2020) 109607.

[35] A. Rehman, D.R. Phalke, R. Pandey, Alternative fuel for gas turbine: esterified jatropha oil–diesel blend, Renew. Energy 36 (10) (2011) 2635–2640.

[36] N. Saifuddin, H. Refal, P. Kumaran, Performance and emission characteristics of micro gas turbine engine fuelled with bioethanol-diesel-biodiesel blends, Int. J. Automot. Mech. Eng. 14 (2017) 4030–4049.

[37] International Coordinating Council of Aerospace Industries Associations (Ed.), Impact of alternative fuels on aircraft engine emissions, ICAO Conference on Aviation and Alternative Fuels. Rio de Janeiro, Brazil, 16–18 November, 2009.

[38] P. Hooper, Initial development of a multi-fuel stepped piston engine for unmanned aircraft application, Aircr. Eng. Aerosp. Technol. 73 (5) (2001) 459–465.

[39] L. Chen, S. Ding, H. Liu, Y. Lu, Y. Li, A.P. Roskilly, Comparative study of combustion and emissions of kerosene (RP-3), kerosene-pentanol blends and diesel in a compression ignition engine, Appl. Energy 203 (2017) 91–100.

[40] EN 590:2014 + A1:2017. Automotive fuels—Diesel—Requirements and Test Methods.

[41] Directive 2009/28/EC of the European Parliament and of the Council of 23 April 2009 on the Promotion of the Use of Energy from Renewable Sources and Amending and Subsequently Repealing Directives 2001/77/EC and 2003/30/EC.

[42] EN 14214:2013 V2 + A1:2018. Liquid Petroleum Products—Fatty Acid Methyl Esters (FAME) for Use in Diesel Engines and Heating Applications—Requirements and Test Methods.

[43] EN 15940:2016. Paraffinic Diesel Fuel from Synthesis or Hydrotreatment—Requirements and Test Methods.

[44] D.M. Korres, D. Karonis, E. Lois, M.B. Linck, A.K. Gupta, Aviation fuel JP-5 and biodiesel on a diesel engine, Fuel 87 (2008) 70–78.

[45] G. Labeckas, S. Slavinskas, Combustion phenomenon, performance and emissions of a diesel engine with aviation turbine JP-8 fuel and rapeseed biodiesel blends, Energy Convers. Manag. 105 (2015) 216–229.

[46] A.A. Yontar, Injection parameters and lambda effects on diesel jet engine characteristics for JP-8, FAME and naphtha fuels, Fuel 271 (2020) 117647.

[47] M.K. Ashour, A.E.M. Elwardany, Addition of two kerosene-based fuels to diesel–biodiesel fuel: effect on combustion, performance and emissions characteristics of CI engine, Fuel 269 (2020) 117473.

[48] N.S. Ekaab, N.H. Hamza, A.T. Chaichan, Performance and emitted pollutants assessment of diesel engine fuelled with biokerosene, Case Stud. Therm. Eng. 13 (2019) 100381.

[49] F. Millo, S. Bensaid, D. Fino, S.J. Castillo Marcano, T. Vlachos, B.K. Debnath, Influence on the performance and emissions of an automotive Euro 5 diesel engine fueled with F30 from Farnesane, Fuel 138 (2014) 134–142.

[50] S. Gowdagiri, X. Cesari, M. Huang, M.A. Oehlschlaeger, A diesel engine study of conventional and alternative diesel and jet fuels: ignition and emissions characteristics, Fuel 136 (2014) 253–260.

[51] J.A. Soriano, R. García-Contreras, D. Leiva-Candia, F. Soto, Influence on performance and emissions of an automotive diesel engine fueled with biodiesel and paraffinic fuels: GTL and biojet fuel farnesane, Energy Fuels 32 (2018) 5125–5133.

[52] A. Gómez, O. Armas, G.K. Lilik, A. Boehman, Estimation of volatile organic emission based on diesel particle size distributions, Meas. Sci. Technol. 23 (2012) 105305.

[53] S. Rajkumar, J. Thangaraja, Effect of biodiesel, biodiesel binary blends, hydrogenated biodiesel and injection parameters on NOx and soot emissions in a turbocharged diesel engine, Fuel 240 (2019) 101–118.

[54] U. Rajak, T.N. Verma, Effect of emission from ethylic biodiesel of edible and non-edible vegetable oil, animal fats, waste oil and alcohol in CI engine, Energy Convers. Manag. 166 (2018) 704–718.

[55] J.G. Ge, H.Y. Kim, S.K. Yoon, N.J. Choi, Reducing volatile organic compound emissions from diesel engines using canola oil biodiesel fuel and blends, Fuel 218 (2018) 266–274.

[56] K. Narayanaswamya, H. Pitsch, P.A. Pepiot, Component library framework for deriving kinetic mechanisms for multi-component fuel surrogates: application for jet fuel surrogates, Combust. Flame 165 (2016) 288–309.

[57] S. Arias, F. Molina, J.R. Agudelo, Palm oil biodiesel: an assessment of PAH emissions, oxidative potential and ecotoxicity of particulate matter, J. Environ. Sci. 101 (2021) 326–338.

[58] M. Yilmaz, S.M. Davis, Polycyclic aromatic hydrocarbon (PAH) formation in a diesel engine fueled with diesel, biodiesel and biodiesel/n-butanol blends, Fuel 181 (2016) 729–740.

[59] J.A. Soriano, J.R. Agudelo, A.F. López, O. Armas, Oxidation reactivity and nanostructural characterization of the soot coming from farnesane—a novel diesel fuel derived from sugar cane, Carbon 125 (2017) 516–529.

[60] A. Arenillas, F. Rubiera, C. Pevida, C. Ania, J. Pis, Relationship between structure and reactivity of carbonaceous materials, J. Therm. Anal. Calorim. 76 (2004) 593–602.

[61] A.L. Boehman, J. Song, M. Alam, Impact of biodiesel blending on diesel soot and the regeneration of particulate filters, Energy Fuels 19 (2005) 1857–1864.

[62] H. Zhang, O. Pereira, G. Legros, E.E. Iojoiu, M.E. Galvez, Y. Chen, P. Da Costa, Structure-reactivity study of model and biodiesel soot in model DPF regeneration conditions, Fuel 239 (2019) 373–386.

[63] J. Rodríguez-Fernández, M. Lapuerta, J. Sánchez-Valdepeñas, Regeneration of diesel particulate filters: effect of renewable fuels, Renew. Energy 104 (2017) 30–39.

[64] M. Lapuerta, F. Oliva, J.R. Agudelo, A.L. Boehman, Effect of fuel on the soot nanostructure and consequences on loading and regeneration of diesel particulate filters, Combust. Flame 159 (2) (2012) 844–853.

[65] P. Oungpakornkaew, P. Karin, R. Tongsri, K. Hanamura, Characterization of biodiesel and soot contamination on four-ball wear mechanisms using electron microscopy and confocal laser scanning microscopy, Wear 458–459 (2020) 203407.

[66] A. Tapia, S. Salgado, P. Martin, F. Villanueva, R. García-Contreras, B. Cabañas, Chemical composition and heterogeneous reactivity of soot generated in the combustion of diesel and GTL (Gas-to-Liquid) fuels and amorphous carbon Printex U with NO2 and CF3COOH gases, Atmos. Environ. 177 (2018) 214–221.

[67] J.A. Soriano, R. García-Contreras, J. de la Fuente, O. Armas, L.Y. Orozco-Jiménez, J.R. Agudelo, Genotoxicity and mutagenicity of particulate matter emitted from diesel, gas to liquid, biodiesel, and farnesane fuels: a toxicological risk assessment, Fuel 282 (2020) 118763.

[68] N.P. Singh, M.T. McCoy, R.R. Tice, E.L. Schneider, A simple technique for quantitation of low levels of DNA damage in individual cells, Exp. Cell Res. 175 (1) (1988) 184–191.

[69] J.P. McNamee, J.R. McLean, C.L. Ferrarotto, P.V. Bellier, Comet assay: rapid processing of multiple samples, Mutat. Res. 466 (1) (2000) 63–69.

[70] B. Novotna, J. Sikorova, A. Milcova, M. Pechout, L. Dittrich, et al., The genotoxicity of organic extracts from particulate truck emissions produced at various engine operating modes using diesel or biodiesel (B100) fuel: a pilot study, Mutat. Res. Genet. Toxicol. Environ. Mutagen. 845 (2019) 403034.

CHAPTER 8

Governance and policy developments for sustainable aviation fuels

Marina Efthymiou[a] and Tim Ryley[b]
[a]Dublin City University Business School, Dublin, Ireland
[b]Griffith Aviation, School of Engineering & Built Environment, Griffith University, Brisbane, QLD, Australia

1. Introduction

Sustainable Aviation Fuels (SAFs) are identified as the only low-carbon fuels available for aviation in the short to mid-term, with other solutions like cryogenic hydrogen fuel and electric-power aircraft unlikely to be commercially ready before 2030. The first commercial flights using SAF were performed in 2011, and since 2016 regular SAF supply has been introduced at various airports (e.g., Oslo, Los Angeles, and Stockholm). More than 250,000 flights have used various amounts of blend-in SAF and more than 40 commercial airlines have used certified SAF. By 2050, improved operations, aircraft technology, and use of SAF will allow carbon dioxide (CO_2) emissions reductions of 9%, 25%, and 41%, respectively, according to ICAO.

The focus of this chapter is on the governance and policy rather than technical aspects of developing SAF. It starts with an introductory argument for the environmental need for SAF governance and policy. The subsequent sections outline the Governance surrounding SAF and provide a more specific policy framework, followed by some ongoing policy lessons from these developments. Finally, conclusions are provided.

This topic needs to be set in the broader context of **sustainable development**. Implicit in the sustainable development concept, is the requirement to meet "the needs of the present without compromising the ability of future generations to meet their own needs" [1, p. 41]. This future dimension approach aligns with the long-term planning and development of alternative aviation fuels. Sustainable development does have a core tension between economic and environmental goals, although the original concept also incorporates other elements such as population growth, poverty, technology, and social organization. Sustainability solutions are also meant to incorporate three "pillars," according to economic development, social development, and environmental protection aspects [2, p. 11]. Using this **sustainability pillar framework**, aviation is arguably economically and socially sustainable, but it is the environmental sustainability challenge which is greatest and hence the focus of this chapter.

Sustainable Alternatives for Aviation Fuels
https://doi.org/10.1016/B978-0-323-85715-4.00010-0
Copyright © 2022 Elsevier Inc.
All rights reserved.
201

Table 1 General environmental impacts from aviation.

Environmental impact	Description
1. Climate change and air quality	Global (e.g., carbon dioxide) and local (e.g., carbon monoxide, nitrogen oxide) pollutants
2. Energy and mineral resources	Energy resources (mainly oil-based), infrastructure construction materials
3. Land resources	Infrastructure land
4. Water resources	Pollution (surface run-off, oil), change to water systems from infrastructure construction
5. Solid waste	Scrapped aircraft, waste oil/tires
6. Biodiversity	Impact on wildlife habitats from infrastructure construction
7. Noise and vibration	Proximity airports, as well as access by connecting roads and railway lines
8. Built environment	Structural damage to infrastructure, property damage, building corrosion from local pollutants
9. Health	Death and injuries from accidents as well as local pollutants and noise disturbance

The **general environmental impacts from aviation** are listed in Table 1. Most of the focus on environmental sustainability concerns relates to the first impact in the Table of climate change. Yet when considering sustainable aviation fuels, the limited resource of oil is also a major concern. The other impacts listed in the Table show the wide range of overlapping environmental impacts of aviation across the full range of spatial scales from local right up to global implications.

There has always been a globally changing climate, but in recent times it has been possible to statistically link changes associated with the increased level of key greenhouse gas emissions to human activity. This is causing an increase in the earth's average temperature. Most scientists and the wider population would agree that there is the **climate change** phenomenon (see Ref. [3]).

Aviation is a major contributor to greenhouse gas emissions. Even though aviation-related emissions make up a small proportion of the global contribution to climate change, there is a major concern relating to the growth in air travel [4]. The aviation sector will take an increasingly significant proportion of any carbon budget and is one of the hardest to reduce emissions at a time when a major reduction is required [5].

Much debate focuses on the scale of the problem and how to respond. This includes technological solutions and the implementation of aviation fuels, one of a suite of **mitigation measures** to reduce environmental impacts. The focus on fuels is partly due to the lack of development possibilities with aircraft design over the timeframe required for environmental improvements. Aircrafts have a very long lifespan and existing aviation infrastructures are very difficult to adapt and update.

2. Governance of sustainable aviation fuels

With the primary environmental concerns relating to the global issues of climate change and limited oil resources, the Governance of sustainable aviation fuels principally concerns international bodies. The main relevant bodies, the International Civil Aviation Organization, the European Commission, and the International Air Transport Association, are reviewed in this section.

The 1999 report "Aviation and the Global Atmosphere" published by the Intergovernmental Panel on Climate Change (IPCC) is an important milestone for governance and policy formation of climate change mitigation. The report found that the impact from aviation will be far greater in 2050 unless new technologies and operational modes are developed. In 1983, the **International Civil Aviation Organization (ICAO)** established the Committee on Aviation Environmental Protection (CAEP). CAEP assists ICAO in formulating new policies and adopting new international Standards and Recommended Practices (SARPs) relating to aircraft noise and emissions. It includes an Alternative Fuels Task Force (AFTF) that provides technical input. ICAO resolution A38-18 on climate change recognizes the role and importance of alternative fuels and the needs for coordinated policies and sustainability criteria. In 2016, the ICAO Assembly adopted Resolution A39-2: Consolidated statement of continuing ICAO policies and practices related to environmental protection—Climate change. ICAO set a 2% annual fuel efficiency improvement through 2050 and carbon-neutral growth from 2020 onwards goals (Fig. 1). A basket of measures that includes SAF can assist aviation to reach these two global goals.

The most significant in terms of effect and scale of an environment-related project is ICAO's Carbon Offsetting and Reduction Scheme for International Aviation (CORSIA), a market-based measure. With CORSIA, aircraft operators should comply

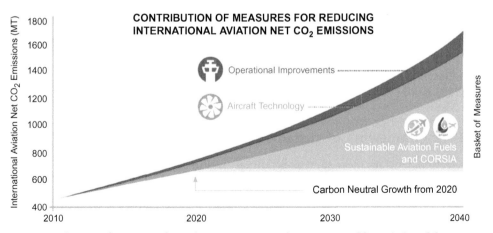

Fig. 1 Contribution of measures for reducing international aviation net CO_2 emissions [6].

with emission commitments by offsetting emissions through "offset credits" from crediting mechanisms and "allowances" from emissions trading schemes. Aircraft operators can claim emission reductions from the use of biofuels only when the fuel complies with the ICAO sustainability criteria and is categorized as SAF [6,7].

The **European Commission** has a long-established trading scheme, the primary Governance at a subinternational level. The European Union Emissions Trading Scheme (EU ETS) is one of the main instruments used by the EU to reach the statutory reduction of GHG targets. In 2008, the European Parliament and the Council adopted a new law, Directive 2008/101/EC, amending the EU ETS (Directive 2003/87/EC) to include aviation activities [8]. In a similar manner to CORSIA, the EU ETS provides an economic incentive to aircraft operators to use SAF by attributing them zero emissions under the scheme and thereby reducing the number of ETS allowances it has to purchase. The EU recognizes that while SAF use is still limited, the contribution of SAF to decarbonizing aviation is important. From 2013 to 2020, a total budget of €464 million has been provided to R&D to study advanced biofuels and other renewable sources, of which €25 million has been specifically allocated to SAF [9]. EU has a target of 40% SAF by 2050.

A significant issue with the governance of biofuels is that while the scientific understanding of the negative effects of biofuels is increasing, stricter sustainability criteria have been imposed. This leads to high levels of uncertainty and investment risk for biofuel producers. Moreover, there is no standardization for the biofuel's sustainability criteria. In the case of SAF, where the majority of aircraft operators need to comply with CORSIA, ICAO sustainability requirements provide some guidance to SAF users and producers.

In 2009, the **International Air Transport Association (IATA)**, as the trade association for the world's airlines, has set three environmental targets: (a) an average improvement in fuel efficiency of 1.5% per year from 2009 to 2020, (b) a cap on net aviation CO_2 emissions from 2020 (carbon-neutral growth), and (c) a reduction in net aviation CO_2 emissions of 50% by 2050, relative to 2005 levels. In order to reach the 2050 target, SAF are required. The main role of IATA in relation to SAF is recommending and influencing policy, as well as providing expert input through relevant working groups.

3. Policy framework of sustainable aviation fuels (SAF)

This section develops more specific SAF interventions within a policy framework which emerges from the Governance structures previously discussed.

Fossil fuels are an exhaustible source of energy for aircraft and fuel is a significant cost for commercial aviation that represents 30% of their total operating cost. Airlines respond to the increasing public and political pressure to reduce their emissions by investing in fuel-efficient technologies and operational practices [10,11]. Frequently, they follow fuel-hedging strategies or, in some rare cases, invest in vertical integration by buying oil refineries (e.g., Delta airlines). With biofuels being significantly more expensive than fossil fuels, arguably two to three times more (A typical price for fossil-based aviation fuel would be €600/tonne, the price of aviation biofuel produced from used cooking oil can be in the range of €950–€1015/tonne [9]), airlines have no significant financial incentive to use them and trigger the supply of SAF. Yet, some carriers like Lufthansa and KLM have used biofuels as part of specific projects that aim to trial SAF. Policy frameworks can, therefore, drive the production and use of biofuels due to the lack of economic incentives for the aircraft operators and encourage a more sustainable path for aviation.

Climate change and environmental degradation have been acknowledged as threats by regulators who have implemented several **regulations for airlines** like the Emissions Trading Scheme [8,10] and service providers like the Single European Sky [12]. Governments are committed to sustainability targets. Therefore, they need to encourage initiatives that reduce carbon emissions. The European Green Deal for example aims to boost the efficient use of resources by moving to a circular economy and to restore biodiversity and cut pollution. The ReFuelEU initiative aims to boost the supply and demand for sustainable aviation fuels in the EU. The decarbonization of air transport is a central element in the agenda, with sustainable aviation fuels playing a key role.

Countries have set indicative **targets for the share of renewable fuels** in their total fuel consumption. Many biofuels policies have been developed and implemented to encourage the supply and demand for biofuels. Many of the policies are not specific to aviation or even exclude aviation. Yet these policies influence SAF as they encourage biofuel production and biofuel R&D overall. Across developed and developing counties there have been goals to implement biofuel policies for energy supply diversity (e.g., Malaysia's Fuel Diversification Policy), environmental improvement, creation of new outlets or demand for agricultural products, regional development and enhanced economic activity (e.g., South Korea) reasons [13]. There are a variety of supply-side (e.g., subsidies to SAF producers) and demand-side (e.g., SAF mandates) policy tools that can support the "missing SAF market" issue. Ebadian et al. [14] categorize the policies to technology-push and market-pull types of policies.

Policy intervention in the case of biofuels is justified based on the welfare maximization argument whereby the government intervenes to correct market failure (e.g., an externality) and improve allocative efficiency. Biofuels without regulatory support would remain an underinvested private sector with lacking skills and capacity as well as high uncertainty. From a political economy point of view, public intervention is a rent-seeking behavior of politicians and bureaucrats. In any case, public intervention should aim to correct the market failure in the most effective and efficient way. Considering that environmental externalities are diverse and not easily quantified, policies are often complex and can even fail.

Several initiatives and policies of the federal government promote alternative fuels. The main regulatory framework in the United States is the Energy Policy Act that introduced the Renewable Fuel Standard (RFS) and the Farm Bill. The policy schemes are concentrated in fossil fuel/biofuel blend mandates, tax credits for producers and suppliers, financial help for new and expansion of existing biofuel supply and production facilities, reduction of import duties for some kind of biofuels, incentives to manufactures for production and sale of alternative fuel vehicles and funding for R&D [15]. RFS, for example, requires US transportation fuel to contain a minimum volume of renewable fuel. The RFS is a market-based compliance system in which involved parties have to submit credits (Renewable Identification Numbers known as RINs) to cover their obligations. The statute focuses on four renewable fuel categories—conventional biofuel, advanced biofuel, cellulosic biofuel, and biomass-based diesel—each with its own target volume (Renewable Volume Obligation) legislated by the Energy Independence and Security Act of 2007 (EISA). The Biorefinery Assistance Program, established within the Farm Bill, provides a loan guarantee for companies that turn waste into renewable jet fuel.

The most important legislative measures related to biofuels in Europe are the Renewable Energy Directive (RED), the Energy Taxation Directive and the Directive on the Quality of Petrol and Diesel Fuels (FQD). RED (2009/28/EC) requires that 20% of the EU's energy needs should come from renewable sources by 2020, and includes a target for the transport sector of 10% from biofuels while it considers sustainability criteria. EU member states adopt individual targets following current shares and other indicators (such as GDP) and SAF contribute to achieving these targets. A specific multiplier of 1.2 applies to the quantity of SAF supplied, in calculating its contribution toward renewable energy targets [9]. The revised Renewable Energy Directive 2018/2001/EU, commonly known as RED II has a new target of at least 32% renewable energy in the EU by 2030 and ranges from a low of 10% in Malta to a high of 49% in Sweden. RED II has also enhanced the sustainability and GHG criteria for produced and consumed bioliquids in the EU. To qualify biofuels as renewable energy sources, fuels have to achieve a 65% greater reduction in emissions against a fossil fuel baseline of 94 g CO_2e/MJ. Transport must comply with these criteria for them to count toward the overall 14% target and to be eligible for financial support by the authorities.

There are also transport dedicated policy initiatives and regulations like the Directive 2003/30/EC [16], on the **promotion of the use of biofuels** or other renewable fuels for transport. The European Union has launched the "Initiative Towards Sustainable Kerosene for Aviation" (ITAKA) to produce sustainable bio-jet-fuel at a large enough scale to test its use in normal flight operations. The "Aviation Initiative for Renewable Energy" in Germany (AIREG) and "Bioqueroseno" in Spain, are also pursuing the development of a sustainable biojet fuel industry. The European Advanced Biofuels Flightpath (EABF) initiative, introduced in 2011 and part of the EU strategic energy technology plan, aimed to establish a functioning supply chain for bio-kerosene and achieve 2 million tons of sustainable biofuels to be used in aviation by 2020. EABF among others focuses on the establishment of financing structures, certification frameworks, and acceleration of targeted research and innovation.

Considering the lack of economic viability of biofuels due to high production cost, supply-side policies can encourage the production of biofuels. **Subsidies** to the energy sector are a very common practice. The world's total, direct energy sector subsidies (including those to fossil fuels, renewables and nuclear power) are estimated to have been at least USD 634 billion in 2017 with the cost of unpriced externalities and the direct subsidies to fossil fuels exceeding the subsidies for renewable energy by a factor of 19 [17]. Direct subsidies for biofuels were only 6% (USD 38 billion) of the total amount and mainly came from the European Union (USD 11.4 billion) and the United States (USD 14.1 billion).

Biofuel supply subsidies can be direct subsidies per output of biomass (e.g., a subsidy to a farmer according to the volume of production) or can be general input subsidies that may not be direct subsidies for the production, but have an indirect effect on the production costs of biomass (e.g., subsidy on fertilizers). Therefore, subsidies can take many forms, for example, free access to water or subsidized irrigation water, crop subsidies, labor subsidies, below-market access to public lands, load guarantees, and investment incentives. IATA [18] suggests that economic instruments are beneficial for starting a market, but become expensive in sustaining the market.

Technology-push policies encourage early-stage technology development such as R&D, demonstration, and commercialization of SAF, whereas market-pull policies support low-risk precommercial and commercial biofuel technologies [14]. It is essential to push **funding of research on technology**. A good example of a supply-side policy to biofuels production is the Biomass Crop Assistance Program (BCAP) in the United States. The BCAP provides three types of payments to producers of dedicated energy crops. It provides feedstock establishment payments that lower the planting cost by 50%, annual land rental payments for up to 5 years and matching payments of up to 45 Mg^{-1} of feedstock partially covering the cost of collecting, harvesting, storing, and transporting biomass to a biorefinery for up to 2 years. The BCAP financing lowered the expected cost and investment risk for the biofuel industry program and motivated farmers to switch to lignocellulosic biomass (LCB) crops. Denmark has allocated €14 million as deficit guarantee for new biorefineries [14].

The supply-side policies aimed at the reduction of the supply cost including the cost of infrastructure. This can be in the form of **capital grants** partly financing a renewable fuel installation. A system of guaranteed loans underwritten by authorities or capital allowances is another option. Load guarantee programs have primarily funded biofuels in the later stages of development and demonstration, whereas grants are mainly used for SAF technologies at the pilot and demonstration phases. A tender selection process that grants a license can reduce the risk for the firms while ensuring some robustness in the funding process. Commercialization of the biofuels should rely on technology-neutral financing to encourage competition for funds between different technologies and innovative projects. Public-private partnerships to commercialize the production of biofuels may be a way forward. International collaboration is also common. In 2011, the United States and Brazil signed a Memorandum of Understanding to cooperate on the development of renewable aviation biofuels.

The **distribution of biofuels** is also of importance. The aviation industry wants to use the same distribution systems as fossil fuels, yet distribution remains challenging as it needs technical capacity and core skills for deploying and handling SAF. International agreements are key to delivering biofuels to big international hub airports. The most challenging part of this process is the transportation from the refinery to the airport, but there is also the handling at the airport environment. Existing supply systems have supplied biofuel to airlines (e.g., at Oslo Gardermoen Airport and San Francisco International Airport). SAF distribution can be encouraged by funding schemes.

A **guaranteed price for biofuel** produced is less used. The most well-known example is that of Indonesia. As part of the National Biofuel Policy, the state-owned oil company Pertamnina sold 5% biofuel blends to local markets at the same price as fossil fuel. The key policy instruments increasing the consumption of SAF are **blending mandates** and **tax exemptions**. Brazil has tax incentives for biofuel producers, blenders and users. Noh et al. [15] suggest that tax exemptions are easier to be applied to production and distribution rather than consumption and that blending mandates in aviation can be applied by using tradable certificates like for example the RINs (Renewable Identification Number is a numeric code assigned to a biofuel package, identifying its vintage, volume, and fuel classification.) in the United States. The Malaysian government introduced the Pioneer Status of Investment Tax Allowance (ITA) under the Promotion of Investments Act of 1986, whereby a company with pioneer status is granted tax exemption on at least 70% of the income derived from the biodiesel production for five years.

Low-carbon fuel standards (LCFS), initially implemented in California in 2011 and later to other states in the United States, require a 10% reduction in the carbon intensity of transportation fuels by 2020 and a 20% reduction by 2030. This policy includes both fossil fuels and biofuels. Fuels that can be produced at a lower carbon intensity compared to their petroleum-based counterparts generate higher carbon credits and therefore they reward efficiency and encourage ongoing innovation in biofuels.

Fuel blending mandates are a very common policy implemented at both the national/federal and state/provincial level aiming to encourage both SAF supply and demand. Ebadian et al. [14] determined that increasing blending mandates in Brazil, Canada, Japan, the Netherlands, South Korea, Sweden, and the United States are growing the production and use of biofuels, whereas mandates that remained stable in Austria, Denmark, Germany, and South Korea led to a fairly flat biofuels production and use.

They argue that the lack of blending mandates for biodiesel in Japan and ethanol in South Korea is to blame for the slow market development.

Several countries have now introduced blending mandates for aviation fuels at the supplier level. Canada has a federal mandate of 5% for ethanol and 2% biodiesel, whereas British Columbia, Alberta, Saskatchewan, Manitoba, and Ontario have established a blending requirement of 5%–8.5% for ethanol and 2%–4% for biodiesel. According to Ebadian et al. [14] the use of ethanol and biodiesel has increased at an annual rate of 30% and 10%, respectively, since 2010. Finland has set a 30% blending mandate for SAF by 2035. France has a target of 2% by 2025, 5% by 2030, and 50% by 2050. The Norwegian government set a 0.5% blending mandate that will increase to up to 30% by 2030. While blending mandates are widely used, they are not enough to grow or maintain biofuel markets due to lack of or high cost of feedstock, shortage of infrastructure, and sustainability concerns according to Ebadian et al. [14]. Mandates are largely based on the volume or energy content of biofuel rather than the decarbonization potential. Table 2 summarizes the strengths and limitations of selective biofuels policy instruments.

4. Policy lessons on sustainable alternative fuels

Policies are frequently accused by environmental groups of not being **ambitious** enough and properly implemented. **Political feasibility** is a common concern for regulatory authorities. In policies that address a global concern collaboration is needed and many times the decision-makers of the various countries may have diffident beliefs, targets, and priorities. CORSIA, for example, has been characterized of having a low ambition and being too compromised. Reaching an agreement among 193 member states for CORSIA is an important step toward decarbonizing air transport.

Moreover, the initial policy frameworks and regulations did not consider some shortcomings related to the development of biofuels mainly due to the lack of scientific understanding. Firstly, regulations ignored the effect of biofuels on food production. Biofuels competed with **food production** in terms of land. Farmers switched to biofuels resulting in an increase in food prices. For example, BCAP subsidies may lead more food cropland to be converted for biofuel crops, creating the unintended consequence of land competition for food versus fuel use affecting the food supplies and prices globally. With certain regions facing issues with limited or lack of arable land, biofuel production can have a significant effect on food production capability. Thus, subsidies for biofuels need to comply with strict sustainability principles as they may unintentionally encourage land-use changes.

Indirect environmental effects related to the biofuel production are of concern to the regulators [15]. The use of nitrogen fertilizers to increase the produced quantities of biofuels is connected to N_2O emissions. Indirect Land Use Change (ILUC) is also

Table 2 Strengths and limitations of selective biofuels policies [9,14].

Policy instrument	Strengths	Limitations
Biofuel blending mandates	- Effective for developing a biofuel market at the early stages - Effective in establishing biofuels markets and in shielding biofuels from low oil prices - Greater certainty of increased development - Broadly effective to support relatively mature technologies, as they create a demand for biofuels, which is typically met with commercial conversion technologies such as conventional ethanol or biodiesel	- Need to balance costs of infrastructure while demand is low in the early stages - Need suitable governance to ensure compliance - Not necessarily as useful in expanding/maintaining markets - Not necessarily successful for meeting GHG reduction targets - Limited in their capacity to pull early-stage technologies into the market, since these are often not yet fully commercially viable, or are typically more expensive to be produced commercially—struggling to compete against first-generation conventional biofuels
Excise duty reductions/exemptions	- Increases the competitiveness of biofuels with fossil fuels, especially at early stages of development, if fossil and renewable fuels are taxed differently - Can also be considered for the production of biomass such as dedicated biomass crops (e.g., switchgrass) to ensure sufficient feedstock for the production of conventional and advanced biofuels and ultimately the achievement of mandates for use - Broadly effective to support relatively mature technologies, as they create a demand for biofuels, which is typically met with commercial conversion technologies such as conventional ethanol or biodiesel	- As fuel excise rates vary, this may not be a strong enough driver to foster the biofuels market as a stand-alone policy - Limited in their capacity to pull early-stage technologies into the market, since these are often not commercially viable, or are typically more expensive to be produced commercially—struggling to compete against first-generation conventional biofuels

Table 2 Strengths and limitations of selective biofuels policies [9,14]—cont'd

Policy instrument	Strengths	Limitations
Research, development, and demonstration funding and financial de-risking measures	– Necessary to support early market technology development and initial commercial projects with longer-term market potential but high investment risk – Successful in de-risking technology and catalyzing private investment for subsequent stages, somewhat sparing public budgets as technologies advance into commercial stages	– Financial risks associated with potential project failures

another acknowledged significant issue with negative effects on the environment. While the scientific understanding of biofuels and their effects is achieved, regulators amend the regulation to incorporate sustainability criteria. The European Commission, for example, includes ILUC emission in the assessments of the GHG effect of biofuels. RED II sets limits on high ILUC-risk biofuels with a significant expansion in land with high carbon stock. Member states cannot reach the targets by using these biofuels, but they can still use biofuels certified as low ILUC-risk. CORSIA eligible fuels (CEF) include SAF and Lower Carbon Aviation Fuels (LCF). CORSIA has defined sustainability criteria for CEF based on two themes, GHGs and Carbon stock (see Table 3).

Policies should focus on global measures rather than local measures to avoid **carbon leakage**. Carbon leakage is the increase in CO_2 emissions outside of countries with domestic policy mitigation divided by the reduction of emissions of these countries [19]. For instance, the increase in local prices of fuel resulting from mitigation policies can lead to a shift of production to areas with less stringent mitigation rules (or no rules at all), which increases emissions in these regions and therefore carbon leakage. For example, the decline in global demand for fossil fuels and the consequent reduction in the prices of fossil fuels can lead to an increased consumption of fossil fuels in countries that have not taken steps to mitigate and, therefore, a risk of carbon leakage. However, the investment attitude in many developing countries may be such that they are not ready yet to benefit from such a leakage.

Policies aiming at reducing GHG emissions are well intended, but in many cases can result in unintended and undesirable consequences. A subsidy for biomass, for example,

Table 3 CORSIA sustainability criteria for eligible fuels [7].

Theme	Principle	Criteria
Greenhouse gases (GHG)	Principle: CORSIA eligible fuel should generate lower carbon emissions on a life cycle basis.	Criterion 1: CORSIA eligible fuel shall achieve net greenhouse gas emissions reductions of at least 10% compared to the baseline life cycle emissions values for aviation fuel on a life cycle basis.
Carbon stock	Principle: CORSIA eligible fuel should not be made from biomass obtained from land with high carbon stock.	Criterion 1: CORSIA eligible fuel shall not be made from biomass obtained from land converted after January 1, 2008 that was a primary forest, wetlands, or peatlands and/or contributes to the degradation of the carbon stock in primary forests, wetlands, or peatlands as these lands all have high carbon stocks Criterion 2: In the event of land-use conversion after January 1, 2008, as defined based on IPCC land categories, direct land-use change (DLUC) emissions shall be calculated. If DLUC greenhouse gas emissions exceed the default induced land-use change (ILUC) value, the DLUC value shall replace the default ILUC value.

can lead to a reduced consumer price for biofuel and consequently an increase to biofuel consumption, but not necessarily to fuel substitution. Nonglobal level pressure to limit the demand of fossil fuel without a pressure to reduce the supply of fossil fuel, will result to a reduction of fossil fuel prices to increase the demand from countries that have no GHG reduction targets and therefore can lead to an increase of extraction volume. This can also cause an intertemporal effect where contemporary fossil fuel supplies increase the oil extraction levels in anticipation of the future impact on demand. In summary, the supply-side effect of fossil fuel suppliers may dominate the substitution effect from fossil fuels to biofuels. This does not lead to a GHG emissions reduction and may contribute to accelerating global carbon emissions. This effect on green demand-side policies is commonly known as the "**Green Paradox**," initially introduced by Sinn [20] and later investigated by various researchers. Several parameters influence "green paradox" with the most important being the imperfect competition in the fossil fuels market, political feasibility, and the timing of policy implementation.

5. Conclusions

Initially at an international level, Governance structures have been implemented to assist SAF development, initially at a global level. Over time, they have been accompanied by more specific policy interventions, with appropriate funding mechanisms. There is a geographic spread of SAF policy implementation, more closely aligned to regions and countries with more of a vested interested in alternative fuel production and distribution, plus areas with more environmental awareness such as in Europe.

Aircraft operators in response to environmental market-based measure obligations (i.e., CORSIA and EU-ETS) aim to use SAF in order to reach their targets. Yet, until concerns related to cost and environmental impacts are addressed, SAF will not gain a significant share of the fuel market. Ebadian et al. [14] argue that both technology-push and demand-pull policies are fundamental in increasing the rate of introduction and diffusion of new environmental technologies. A balance between encouraging and forcing change toward greater use of SAF is also required, although a greater level of enforcement will be required moving forward as the environmental imperative increases. Individual environmental policies have been criticized about their efficiency, effectiveness and occasionally their political feasibility and ambition. Considering the advantages and limitations of individual policies, as demonstrated by the three biofuels policy instrument comparison in Table 2, policy combinations can correct the inefficiencies in sustainability behaviors [11]. Therefore, a mixed policy approach is the best way forward for SAF deployment and scaling up.

For the successful implementation of a suite of SAF policies, the four lessons emerging from this study need to be considered: make them politically effective, efficient and feasible; be aware of indirect environmental effects; avoid carbon leakage; and try to avoid "green paradox" situations. There seems to be forward momentum in the implementation of such SAF policies, although the speed of this change depends on external economic and environmental pressures faced at different levels of Governance from global aviation bodies through to national and local policymakers.

References

[1] World Commission on Environment and Development, Our Common Future, 1987. Bruntland Report.
[2] United Nations General Assembly, 2005 World Summit Outcome, 2005. New York.
[3] IPCC, Climate Change 2014. IPCC 5th Assessment Synthesis Report, 2014, Available online at: https://ar5-syr.ipcc.ch/. (Accessed 9 December 2020).
[4] A.W. Schafer, I.A. Waitz, Air transportation and the environment, Transp. Policy 34 (2014) 1–4.
[5] S. Gossling, A. Humpe, The global scale, distribution and growth of aviation: implications for climate change, Glob. Environ. Chang. 65 (2020) 102194.
[6] ICAO, Destination Green: The Next Chapter—2019 Environmental Report Aviation and Environment, ICAO, Montreal, 2019. Available online at: https://www.icao.int/environmental-protection/Documents/ICAO-ENV-Report2019-F1-WEB%20(1).pdf. (Accessed 19 November 2020).

[7] ICAO, CORSIA Sustainability Criteria for CORSIA Eligible Fuels, ICAO, Montreal, 2019. Available online at: https://www.icao.int/environmental-protection/CORSIA/Documents/ICAO%20document%2005%20-%20Sustainability%20Criteria.pdf. (Accessed 19 November 2020).

[8] M. Efthymiou, A. Papatheodorou, EU Emissions Trading scheme in aviation: policy analysis and suggestions, J. Clean. Prod. 237 (2019) 117734.

[9] EASA, European Aviation Environmental Report 2019, EASA, Cologne, 2019.

[10] M. Efthymiou, The fundamentals of environmental regulation of aviation: a focus on EU emissions trading scheme, Aeronaut. Aerosp. Open Access J. 5 (1) (2021) 9–16, https://doi.org/10.15406/aaoaj.2021.05.00122.

[11] M. Efthymiou, A. Papatheodorou, Environmental policies in European aviation: a stakeholder management perspective, in: T. Walker, A.S. Bergantino, N. Sprung-Much, L. Loiacono (Eds.), Sustainable Aviation, Springer Nature/Palgrave Macmillan, Cham, 2020, pp. 101–125.

[12] M. Efthymiou, A. Papatheodorou, Environmental considerations in the single European sky: a Delphi approach, Transp. Res. A Policy Pract. 118 (2018) 556–566.

[13] OECD, Biofuel Support Policies: An Economic Assessment, Organisation for Economic Co-operation and Development, Paris, 2008.

[14] M. Ebadian, S. van Dyk, J.D. McMillan, J. Saddler, Biofuels policies that have encouraged their production and use: an international perspective, Energy Policy 147 (2020) 111906.

[15] H.M. Noh, A. Benito, G. Alonso, Study of the current incentive rules and mechanisms to promote biofuel use in the EU and their possible application to the civil aviation sector, Transp. Res. Part D Transp. Environ. 46 (2016) 298–316.

[16] Directive 2003/30/EC of the European Parliament and of the Council of 8 May 2003 on the promotion of the use of biofuels or other renewable fuels for transport.

[17] M. Taylor, Energy Subsidies: Evolution in the Global Energy Transformation to 2050, International Renewable Energy Agency, Abu Dhabi, 2020.

[18] IATA, IATA Sustainable Aviation Fuel Roadmap, IATA, Montreal, 2015. Available online at: https://www.iata.org/contentassets/d13875e9ed784f75bac90f000760e998/safr-1-2015.pdf. (Accessed 23 November 2020).

[19] M.H. Babiker, Climate change policy, market structure, and carbon leakage, J. Int. Econ. 65 (2) (2005) 421–445.

[20] H.W. Sinn, The Green Paradox: A Supply-Side Approach to Global Warming, MIT press, 2012.

CHAPTER 9

Life cycle assessment of biojet fuels

Qing Yang[a,b,c] and Fuying Chen[a]
[a]State Key Laboratory of Coal Combustion, Huazhong University of Science and Technology, Wuhan, Hubei, PR China
[b]China-EU Institute for Clean and Renewable Energy, Huazhong University of Science and Technology, Wuhan, PR China
[c]John A. Paulson School of Engineering and Applied Sciences, Harvard University, Cambridge, MA, United States

1. Introduction

With the development of economic globalization and the advancement of human science and technology, air transportation is playing an increasingly significant role in economic trade and interpersonal communication. Therefore jet fuel has also become the most important transportation fuel next to motor gasoline and diesel [1]. According to the report by the International Air Transport Association (IATA), in 2019, there were 4.54 billion air passengers in the world, while the freight volume was reduced to 61.2 million tons, the main reason for which was increased trade friction [2]. And it was estimated by a report released in 2016 that passenger traffic would double, compared to the current level in the next two decades [3]. The increasing demand for air transport significantly contributes to the rapid growth of global demand for aviation fuels. It was expected that the consumption of jet fuel would increase by another 10 quadrillion Btu between 2010 and 2040 [4]. As the world's second-largest consumer of jet fuels, China's consumption increased by more than 11% in 2006–17, and jet fuel consumption in 2017 increased nearly three times compared with 2006 [5]. Traditional aviation fuel is mainly derived from the fractionation of petroleum, which is a nonrenewable energy source [6]. Therefore, the sustainable development of the aviation industry will depend on the development of renewable energy alternatives to fuel continued growth and demand in the sector.

In 2019, worldwide, flights generated 915 million tons of CO_2, accounting for 12% of all transport sources and 2% of all anthropogenic CO_2 emissions [7]. IATA had made three major greenhouse gas (GHG) emission reduction commitments: first, the average annual combustion efficiency of aviation fuel would be increased by 1.5% during 2009–20; second, carbon emissions would no longer increase by 2020; third, compared with 2005, GHG emissions would be reduced to 50% in 2050 [8]. However, simply improving the existing fuel efficiency and airline operating efficiency cannot solve the problem or effectively meet emission targets. Therefore, it is imperative we realize the sustainable development of the aviation industry, which requires we find renewable

Copyright © 2022 Elsevier Inc.
All rights reserved.

and low-carbon alternative aviation fuels. The technology of producing aviation fuel using biomass as a raw material has attracted widespread attention.

Biomass is the only carbon-containing renewable energy that can directly absorb CO_2 from the environment and then produce organic matter. Biojet fuel refers to aviation fuel synthesized directly or indirectly from biomass. For aviation, biofuels can not only reduce the dependence on fossil fuels but also decrease whole life cycle emissions because of the carbon neutrality of feedstock [9]. The use of biofuels may reduce CO_2 emissions by up to 80% during its life cycle [10]. In addition, when biofuels, instead of traditional petroleum fuels, are used in aircraft, it is not necessary to change the structural design of aircraft engines, only to replace oil products. Therefore, it is convenient to replace them [11]. According to the actual operation, biofuel is not inferior to traditional petroleum fuel in terms of energy efficiency, driving, and maneuverability [12]. Based on the above advantages, biojet fuels have been widely studied and some of them have been put into use successfully.

Life cycle assessment (LCA), also known as the resource and environmental status analysis, is a kind of overall resource consumption and environmental impact analysis of the whole process of product technology or system from raw material mining, production, processing, packaging and transportation, consumption, recycling and final treatment [13]. Over the past decade, LCA has been widely used to assess the sustainability of alternative aviation fuels. A complete traditional LCA method usually needs to construct the whole product supply chain, which is called cradle-to-grave analysis. When LCA is applied to aviation fuel, there are four typical boundary settings, including Well-to-Wake (WtWa), Well-to-Pump (WtP), Well-to-Gate (WtG), and Gate-to-Gate (GtG). However, in order to evaluate fuel-based products, this type of research is mainly referred to as WtWa [14].

This chapter then summarizes the feedstocks and preparation processes of biofuels, as well as their techno–economic and environmental impacts. Finally, the current situation and prospects of biofuels are discussed.

2. The technology of biofuels production

2.1 The feedstocks for the production of biofuels

The biomass raw materials used in the production of biojet fuels are rich in types and sources, and different production technologies require different feedstocks. The feedstocks for the first generation of biojet fuels production come from edible crops. For instance, corn and wheat, which will cause the problem of competition for food and land with humans [15]. The second generation mainly includes nonedible crops and lignocellulosic biomass, such as linseed oil, jatropha oil, and waste animal fat. At the same time, some waste by-products from industrial production, including crude oil for the paper industry, soap residue from the edible oil refinery, petroleum sediment, and acid oil, can also be applied to produce biojet fuels [16]. Lignocellulosic biomass, which includes forest wastes, agricultural residues, halophytes, short-term growing crops, and municipal solid waste, can also be converted into biofuels indirectly [13]. Because of its abundant annual production, lignocellulose is considered to be the most suitable biomass material for the long-term production of biofuels [17]. The third generation of biofuels is mainly algae, which is considered one of the most prospective feedstocks for biofuel production because of its relatively high oil content and reduced land requirements [18].

2.2 The main technological process of biofuels

At present, there are a variety of processes for generating biojet fuels at home and abroad, most of which are to convert biomass into intermediate products through pretreatment and then synthesize jet fuels by modifying the intermediate products. According to the different types of feedstocks, it is mainly divided into the following methods.

Based on the different oil components, aviation fuel can be converted by hydrogenated esters and fatty acids (HEFA) and catalytic hydrothermolysis (CH). Lignocellulosic biomass conversion can be divided into Hydroprocessed depolymerized cellulosic jet (HDCJ), Fischer-Tropsch (FT), and lignin to jet processes. The conversion process of sugars to alkane fuels can be carried out directly through anaerobic fermentation without being converted to ethanol. The main process is called Direct Sugar to Hydrocarbons (DSHC) or Direct Fermentation of Sugar to Jet (DFSTJ). Besides the biochemistry pathway, Aqueous Phase Reforming (APR) included in thermochemistry way can also realize the conversion from sugar to jet fuel. Alcohol, including methanol, ethanol, butanol, or isobutanol can also be synthesized into aviation fuel through a battery of reactions such as dehydration, polymerization, and hydrogenation, which referred to Alcohol to jet (ATJ) process.

3. Economic evaluation

Different conversion paths have different energy efficiency and energy consumption. Neuling and Kaltschmitt [19] compared four different biokerosene production processes using two different types of feedstocks in northern Germany. The overall efficiency of the HEFA process was 90%, which was higher than that of ATJ (30%–68%), Biogas-to-Liquids (56%), and Biomass-to-Liquids (35%–38%). Li et al. [20] analyzed a 1000 t/a synthetic system of hydrocarbons (C8–C15) for biojet fuel series synthesized by the APR method. The technical and economic analysis of biomass residue utilization in the catalytic conversion process was carried out. The results indicated that when the tail gas and acid hydrolysis residual generate steam at 85% energy conversion efficiency, the thermal efficiency of the whole energy conversion including process or treatment energy input and feedstocks energy input was about 31.8%. However, when biojet fuel was the only energy export product, the value was 13.5%. Diederichs et al. [21] compared three processes to produce jet fuel from different feedstocks. And the energy efficiency of FT, ATJ, and HEFA was 37.2%, 29.6%–37.1%, and 75.3%, respectively. Michailos [22] estimated the

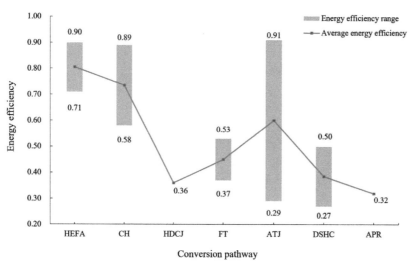

Fig. 1 Energy efficiency of major biojet fuel conversion technologies.

techo-economic and LCA of biojet fuel based on bagasse through DSHC pathway. And the energy efficiency was calculated to be 26.5%. The efficiency of several major conversion technologies has been summarized by Wei et al. [23] which can be seen in Fig. 1.

3.1 Production costs

Junqueira et al. [24] estimated the cost of ethanol production in different periods. In the short term, the cost of second-generation ethanol was higher than that of the first generation. However, in the long run, even considering some uncertainties in technology and the market, second-generation ethanol was more competitive and had lower production costs than first generation ethanol. Neuling and Kaltschmitt [19] adopted a bottom-up techno-economic approach to quantify all aspects of biofuel production, in which four cultivation technologies were set. Then the production costs were estimated to 3.22–12.05 USD/gallon. For the whole supply chain, consumption-related costs dominated the production costs, and these costs mainly came from the feedstock supply costs. In addition, large investment cost differences would also have a certain impact on production costs [19]. Therefore, different production processes or the use of different feedstocks in the same production process will lead to differences in production costs.

3.2 Minimum jet fuel selling price (MJSP)

3.2.1 HEFA process

Wang [25] studied that based on the fact that the daily output of *Jatropha curcas* fruit was 2400 tons, the MJSP of jet fuel derived from jatropha was 5.42 USD/gallon. When jatropha oil was adopted as feedstock, MJSP was 5.74 USD/gallon. Factors, including feedstock costs, refinery capital costs, by-product credits, and energy costs, can explain the differences in MJSP based on fruit and oil scenarios. Additionally, the sensitivity analysis indicated that the cost of feedstock had caused a great degree of uncertainty to the production cost, causing that a variety of MJSP had been produced. By increasing the oil content from 33% to 48%, MJSP could be reduced by 22%. In addition, factors, such as plant capacity, reactor construction, and catalyst price or loading, may also lead to high levels of uncertainty. These measures, including improving the way in which by-products were used or sold, could help reduce the cost of producing biojet fuel [25]. Other research, based on the waste oils and animal fats, the MJSP for yellow grease and tallow derived HEFA fuels were calculated to be 3.33–4.01 and 3.98–4.73 USD/gallon, respectively [26]. Camacho Gonzalez [27] evaluated the MJSP of jet fuel production based on waste vegetables in Mexico as 2.36 USD/gallon. In another study, Diederichs et al. [21] evaluated the hydrotreating of vegetable oil to produce jet fuel as the main product. And the MJSP was 6.89 USD/gallon. Pearlson et al. [28] estimated that the MJSP of soybean oil was 3.82–4.39 USD/gallon based on different factory scale. The research conducted by Tao et al. [29] estimated the MJSP of camelina, pennycress, jatropha, castor bean, and yellow grease was 10.98, 6.43, 3.82, 9.43, and 4.80 USD/gallon, respectively. The difference of oil prices were the main reason for the great change of MJSP. Klein-Marcuschamer et al. [30] reported that the MJSP of microalgae was 31.98 USD/gallon. Base on the sensitivity analysis of technology and market development, the price would be reduced to 9.17 USD/gallon. The species of feedstock have a great influence on the cost of the HEFA process. According to previous studies, vegetable oils, including camellia oil and soybean oil, can realize lower costs than microalgae. Nevertheless, HEFA derived from algae as a promising technology may become a viable method for producing biojet fuel.

3.2.2 FT process

Feedstock also has a significant impact on the cost of the biofuels produced through the FT process. The cost of this process using biomass as a feedstock is higher than that of using coal and natural gas to produce liquid fuels, mainly because biomass has the characteristics of low energy density and long feedstock transportation distances. The FT is the process that accounts for the largest proportion of the fixed capital cost in the aviation fuel production path. Diederichs et al. [21] assessed the technology and economy of jet production from FT synthesis after gasification, and the MJSP was 7.57 USD/gallon. For this process, changes in feedstocks and fixed capital costs would lead to an overall change of more than 10% in MJSP. Fixed capital costs and by-products would cause high uncertainty in the whole process [21]. The MJSP of another plant producing jet fuel from biomass gasification followed by FT process was 6.23 USD/gallon [31]. The plant adopted a circulating fluidized bed gasifier and could process 864 tons of wood chips in 1 day. The reason may be that this situation produced a large number of by-products, involving gasoline and liquefied petroleum gas, which could significantly counteract the cost of biojet fuel [31]. Suresh et al. [32] calculated the Median MJSP of jet fuel from municipal solid waste as 3.75, 6.74, and 4.54 USD/gallon, corresponding to FT, Plasma FT, and ATJ upgrading processes. When Liu [13] considered by-products, the MJSP of aviation kerosene was 4.45 USD/gallon, which was 28.32% lower than that without considering by-products. Reducing the cost of biomass was the most effective way to improve the economy of aviation fuel. The MJSP of the FT pathway is smaller than that of the HEFA process in the previous section. Nevertheless, the FT process requires more capital so that it is easily affected by the cost of feedstock. At the same time, an important strategy to produce by-products, involving other fuels and electricity, can observably reduce the cost of bio-aircraft fuel.

3.2.3 DSHC process

Klein-Marcuschamer et al. [30] studied that when sugarcane was used as feedstock, MJSP was 7.17 USD/gallon. The feedstock cost accounted for 70.89% of the jet fuel cost. Through the analysis of technology and market development sensitivity, it was found that the price was expected to drop to 4.00 USD/gallon. In another study conducted by de Jong et al. [33] some different ways to generate jet fuel adopting biomass as feedstock were compared. The MJSP of the DSHC process derived from forest waste and wheat straw was significantly higher than other methods, ranging from 18.55 to 20.61 USD/gallon and 24.74 to 26.80 USD/gallon, respectively. Nevertheless, the DSHC procedure may produce the high-value intermediate isoprenoids, which can be applied in the production of cosmetics, flavors, fragrances, lubricants, and biopharmaceuticals. Farnesene was one of them, which was 22.53 USD/gallon. The feedstock is the main factor affecting the economy of DSHC, and sugarcane may be the best one. Meanwhile, compared with producing biojet fuel, this process is being considered to be more favorable for high-value intermediates.

3.2.4 ATJ process

Diederichs et al. [21] made a comparison of the technology economics of three approaches, including syngas fermentation, lignocellulose biochemical transformation, and sugarcane biochemical transformation. And the MJSP were 7.73, 10.64, and 7.88 USD/gallon, respectively. The results showed that compared with other processes, the MJSP of the lignocellulose biochemical pathway was much higher due to the high enzyme cost. In the following research, the cost of enzymes needs to be further reduced, meanwhile, the total yield or concentration of ethanol culture medium needs to be further increased. For sugarcane biochemical transformation, the cost of feedstock and fixed capital, stream factor, lowest acceptable internal rate of return, and the price of electric product were the factors to influence MJSP. In another study, Staples et al. [34] evaluated the lowest selling price of progressive fermentation with sugarcane, switchgrass, and corn grain, as feedstocks. The results showed the MJSP were 2.31–9.96, 4.13–23.85, and

3.18–13.82 USD/gallon. The cost of feedstocks is the main factor in the breakeven price fluctuation of sugarcane and corn grain, while the capital cost is the main factor for the breakeven price fluctuation of switchgrass [34]. The MJSP was estimated as 4.32–6.78 USD/gallon in the case of producing ethanol. And capital investment will change from 356 to 1026 million USD due to different facilities capacity [35]. Atsonios et al. [31] assessed the production routes of aviation fuel from *n*-Butanol and isobutanol, MJSP of which were 7.49 and 6.44 USD/gallon, respectively. According to the sensitivity analysis, the equipment scale, biomass feedstock cost, and by-products were the chief factors affecting MJSP. The biochemical pathway, which on account of Acetone-Butanol-Ethanol fermentation, may be the most economical choice in the researched processes [31]. The average breakeven prices of jet fuel for alcohol production with sugarcane, switchgrass, and corn grain were 3.65, 5.21, and 3.84 USD/gallon, respectively [36]. Vela-García et al. [37] simulated the process of producing biojet fuel (triisobutane) from cellulose isobutane by ATJ-SPK thermochemical route. The MJSP was evaluated as 5.03 USD/gallon. Sensitivity analysis showed that lignin selling price had the greatest impact on the MJSP. When selling high-purity lignin, a competitive MJSP would be obtained [37]. The economics of jet fuel derived from ethanol procedure is primarily affected by the pathway of alcohol production, however, the MJSP of biochemical method is relatively small. From the economic prospect, sugarcane and starch are suitable feedstocks.

3.2.5 Other conversion process

de Jong et al. [33] assessed the technologic and economic feasibility of applying forest residues or wheat straw as feedstocks to generate through CH and pyrolysis technology, the MJSP of which ranged from 3.66 to 5.06 and 5.23 to 7.15 USD/gallon, respectively. The fuel costs of the two methods were relatively low due to high yield and moderate equipment costs. And the sensitivity analysis showed the change of yield may have a significant impact on the MJSP of the CH pathway. Nevertheless, feedstocks price and total capital investment can easily affect the price of pyrolysis fuel. Natelson et al. [38] established a refinery model, which produced 76,000 m^3 of hydrocarbons per year by hydrolysis decarboxylation of camellia oil for jet fuel production. In 2014, the total cost was estimated as 283 million dollars. Based on the benchmark assumption, the breakeven price was 2.46 USD/gallon. When decarboxylation catalysts, solvents, and other uncertain replacement parts were applied, the yearly operating supply and laboratory costs would be double. Breakeven prices of jet fuel ranged from 2.48 to 3.23 USD/gallon. The research conducted by Olcay et al. [39] adopted APR process to convert cellulosic biomass, the results of which showed that the MJSP of jet fuels was 1.00–6.31 USD/gallon. Li et al. [20] also assessed the technical economics of producing biojet fuel by APR pathway. Corncob was adopted as feedstock and the MJSP was 4.66 USD/gallon. Based on the sensitivity analysis, catalyst life and corn cob price were the main reasons that affected the fuel cost of the process. Bond et al. [40] researched the method of producing renewable aviation fuel series alkanes with biomass as feedstock through integrated catalytic treatment. The MJSP of distillate fuel was 4.75 USD/gallon. If the suggestions for further cost reduction, which include reducing the catalyst cost of hydrodeoxygenation reactor, developing conversion rate of lignin and humin, increasing the output and recovery rate of levulinic acid and low xylose, and enhancing the circulation structure of water flow, are achieved, MJSP would be 2.88 USD/gallon. Zhou [41] constructed the production process of aviation fuel by direct hydrogenation of lignin. The calculated MJSP was 6.34 USD/gallon. The results indicated that the MJSP decreased with the increase of feedstock scale and the yield of products. In addition, through the sensitivity analysis of the components of production cost, it was feasible to reduce MJSP by reducing labor cost and management cost. The Table 1 gives MJSP of several conversion pathways. The jet fuel meets the American Society of Testing Materials (ASTM) standard with a

Table 1 MJSP of different pathway for biojet fuel production.

Pathway	Feedstock	MJSP (USD/gallon)
HEFA	Jatropha oil	3.82–5.74
	Jatropha curcas fruit	5.42
	Waste oils and animal fats	3.33–4.80
	Waste vegetables oil	2.36–7.06
	Microalgae	31.98
	Camelina oil	3.06–31.01
	Soybean oil	3.82–4.39
	Pennycress oil	6.43
	Castor bean oil	9.43
	Used cooking oil	5.02–5.44
FT	Dry wood chips	6.23
	Municipal solid waste	3.75–6.74
	Corn stalk	4.45
	Lignocellulose	7.16–7.76
	Sugarcane	2.54–2.99
DSHC	Sugarcane	7.17
	Forest residues	18.55–20.61
	Wheat straw	24.74–26.80
ATJ	Municipal solid waste (thermochemical)	4.54
	Lignocellulose (Biochemistry)	10.06–10.91
	Lignocellulose (Thermochemistry)	5.03–7.92
	Sugar cane (Biochemistry)	2.31–9.96
	Corn grain (Biochemistry)	3.18–13.82
	Switchgrass (Biochemistry)	4.13–23.85
	Poplar (Biochemistry)	4.32–6.78
APR	Corncob	4.66
	Red maple wood	1.00–6.31

density of 775–840 kg/m^3, and the average exchange rate (0.753 €/USD) in 2013 euro-dollar is adopted.

3.3 Energy consumption assessment

Cox et al. [42] indicated that the energy consumption of microalgae oils, sugarcane, and pongamia to produce jet fuel process was 1.0, 1.7, and 1.1 MJ/MJ. Poplar is a short-rotation woody crop, which can be applied as feedstock for jet fuel production through the intermediate ethanol process [43]. The fossil fuel usage for hydrogen produced by steam reforming of natural gas, lignin gasification, and petro-jet was 0.78–0.84, 0.71–1.00, and 1.2 MJ/MJ [43]. In another research, the energy consumption for jet fuel production through FT and HEFA pathway was 0.06 and 0.34 MJ/MJ [44]. In other study conducted by Guo et al. [45] chlorella, isochrysis, and nannochloropsis were taken as research objects. The results showed that with the increase of lipid content, fossil fuel consumption ranged from 0.32 to 0.68 MJ/MJ. Shonnard et al. [46] considered several options for adopting camellia as a sustainable feedstock for producing advanced biofuel. Cumulative energy demands for each camelina biofuel product were 1.2–1.4 MJ/MJ.

4. Environmental impact assessment

As people pay more and more attention to environmental issues, the production restrictions and standards of biojet fuel will become more and more significantly stricter over time. There have been some studies focused on evaluating the environmental impact of biojet fuel production, such as GHG emissions, water use, land use, and other environmental impacts. Due to the variety and composition of biomass feedstocks, the technological process and results will be different to some extent. The following section reviews the impact of biofuels on different aspects of the environment.

4.1 Greenhouse gas emissions

The original purpose of producing biojet fuel is to reduce GHG emissions induced by fossil fuels. Biojet fuels are sometimes referred to as carbon neutral products. But the whole production system requires direct and indirect inputs related to carbon emissions in each process. It is essential to conduct a full LCA of the emissions generated during the production of biojet fuel. The LCA can clearly identify changes in GHG emissions caused by the application of alternative aviation fuels, and can also conduct assessment and guidance on changes in GHG emissions to optimize production systems. Due to different feedstocks and reaction processes, different conversion pathways will produce different GHG emissions. Making comparison with different production processes will contribute to determining a preferable jet fuel production platform. At the industrial level, a complete LCA involves the planting, harvesting, and transportation of feedstocks, changes in land use, the production, and transportation of auxiliary chemicals, the biorefinery process, and fuel storage, transportation, distribution and combustion [43]. Variability in LCA is inherent in imprecise LCA procedures and changes in numerical inputs. Stratton et al. took the production of diesel and jet fuel from 14 different feedstocks as an example, the results of which showed that the GHG emission through the FT pathway was -1.6–18.2 g $CO_{2\text{-eq}}$/MJ. And Stratton et al. [47] explained that subjective choices such as the use and distribution of coproduct may be more important sources of variability than fuel production process and energy use, therefore, LCA results should be regarded as a range rather than a point value. The GHG emissions of different pathway to produce biojet fuel are given by Table 2. GHG emission of conventional jet fuel is 90 g $CO_{2\text{-eq}}$/MJ.

4.1.1 Oil

Li and Mupondwa [48] studied the hydrotreating process of camellia oil. The results showed that the global warming potential (GWP) was 3.06–31.01 g $CO_{2\text{-eq}}$/MJ, in which oil extraction was the largest part of CO_2 emission. Camellia oil caused 76%–84% of GHG emissions when the available oil production was the lowest, compared with 28%–46%, when that was higher. In terms of waste oil and animal fat, GWP was 16.8–21.4 g $CO_{2\text{-eq}}$/MJ, equivalent to 76%–81% reduction in GHG emissions [26]. Based on WtWa analysis, Han et al. [49] indicated that compared with oil injection, the GHG emission reduction of oil injection process was 41%–63%, in which the GHG emissions in the production stage were far less, compared with the fertilizer use and feedstocks collection stages. And feedstocks and the scale of the plant were also the main factors in analyzing CO_2 emissions. In another study, compared with jet fuel-based fossil, microalgae oil and pongamia oil product could reduce the GHG emissions by 53% and 43% [42]. Fortier et al. [50] assessed the hydrothermal liquefaction process based on microalgae oil, which could reduce GHG emissions by 76%. Because of the highly sensitive distance from the refinery, the location of the plant was an important factor affecting emissions. The study by Obnamia et al. [51] investigated the life cycle GHG emissions of rapeseed grown in western Canada to produce jet fuel by HEFA. The results showed that GHG emissions were 44–48 g $CO_{2\text{-eq}}$/MJ, but when land use and land management changes were taken into account, the value became 16 and 58 g $CO_{2\text{-eq}}$/MJ. Zemanek et al. [52] reviewed the results and methods for 20 LCAs of oilseed feedstocks from the HEFA pathway. The results reported ranged widely between -18.3 and 564.2 g $CO_{2\text{-eq}}$/MJ, with feedstocks, by-product distribution methods, and land-use changes were considered to be the main causes of the change. It has been demonstrated that feedstock and the plant scale have significant impacts on GHG emissions.

4.1.2 Lignocellulose

Neuling and Kaltschmitt [19] indicated that both ATJ and Biomass-to-Liquids processes based on lignocellulose have good GHG emission reduction benefits, and their GHG emissions are 31.6–38.2 g $CO_{2\text{-eq}}$/MJ. APR process was established to covert red maple

Table 2 GHG emission of different pathway to produce jet fuel.

Pathway	Feedstock	GHG emission (g $CO_{2\text{-eq}}$/MJ)
HEFA	Jatropha oil	40.00–73.50
	Palm oil	33.30–52.00
	Waste oils and animal fat	16.80–21.40
	Camellia oil	−17.03–60.00
	Rapeseed oil	44.00–97.90
	Soybean oil	21.00–564.20
	Microalgae oil	27.00–38.00
	Pongamia oil	47.00
	Carinata oil	34.50
	Sunflower oil	37.90
	Pennycress oil	−18.30–44.90
	Canola oil	−12.00–49.00
	Algae	27.00
ATJ	Wheat straw	31.60
	Wheat grain	71.50
	Municipal solid waste	52.70
	Sugarcane	−27–19.7
	Corn grain	47.5–117.5
	Switchgrass	11.7–89.8
	Lignocellulose	64.54
	Residual woody	19.41
	Poplar	32.00–73.00
	Corn	72.00–78.00
	Corn stover	23.00–28.00
FT	Municipal solid waste	32.90–62.30
	Corn stalk	9.90–54.44
	Woody residues	5.00
	Lignocellulose	13.50
APR	Red maple wood	31.60–104.50
	Corn stover	35.00–61.00
HDCJ	Corn stover	21.60–40.50
DSHC	Sugarcane	22.00–80.00
	Bagasse	46.20
CH	Microalgae oil	21.20–39.30

wood into chemicals and liquid fuels, in which the WTW GHG emissions of jet fuel were estimated as 31.6–104.5 g $CO_{2\text{-eq}}$/MJ [39]. The GHG emissions of jet fuel produced by the US municipal solid waste through three thermochemical conversion pathways, including FT, Plasma FT, and ATJ upgrading, were calculated as 32.9, 62.3, and 52.7 g $CO_{2\text{-eq}}$/MJ [32]. The study conducted by Vela-García et al. [37] assessed the potential of producing triisobutene from cellulosic isobutanol as a blending component of biojet fuel. The GHG emissions were 65 g $CO_{2\text{-eq}}$/MJ, representing 28% less than conventional JetA1. FT–Synthetic paraffinic kerosene from forest residues was estimated to reduce GHG emissions by 94%, while the HEFA pathway from soybean reduced by 52% [44]. The research adopted a Woods-to-Wake LCA approach, assessing the environmental impact of feedstocks recovery, production, and utilization of residual lignocellulosic fuel. The results indicated that the impact of its on global warming was 78% better than that of petroleum-based aviation fuels [53]. And the for jatropha pathway, the GWP was 40 g $CO_{2\text{-eq}}$/MJ, which was 55% lower than that of traditional jet fuel [54]. The GHG emission of cellulose fast pyrolysis process was 55%–85% lower than that of traditional jet fuel, and the hydrogen production pathway had a great impact on the emission [55]. Budsberg et al. [43] adopted jet fuel produced from intermediate ethanol from poplar, a short-term woody crop. The GWP values of different hydrogen production methods were different. The GWP for hydrogen generated by steam reforming of natural gas, lignin gasification, and petro-jet was 60–66, 32–73, and 93 g $CO_{2\text{-eq}}$/MJ. The results showed that the category of biorefinery and the use of jet fuel were the two main sources contributing to GWP [43]. The GHG emissions of the FT synthesis biojet fuel system were evaluated as 54.44 g $CO_{2\text{-eq}}$/MJ, which was about 39.5% less than conventional aviation fuel. The direct GHG emissions from the soil during the growth stage of corn stalks, CH_4 in sewage, and caused by fertilizer use and electricity consumption in the production phase were the main contributors [13]. In a different study, applying corn stover as feedstock to produce jet fuel through the FT and pyrolysis process would reduce GHG by 89% and 68%–76% [49]. In summary, the difference of feedstock and production process will certainly have an impact on the results.

4.1.3 Sugar

The Cradle to grave GHG emissions of biojet fuel from bagasse through DSHC technology can be reduced by approximately 47% compared to fossil jet fuel [22]. Bressanin et al. [56] designed an integrated sugarcane distillery, which adopted biomass gasification and FT synthesis technology and could reduce GHG emissions by 85%–95% compared to fossils. Staples et al. [34] evaluated GHG emissions from ATJ processes using sugarcane and corn grain as feedstocks. The GHG emissions of these feedstocks were −27.0–19.7 and 47.5–117.5 g $CO_{2\text{-eq}}$/MJ, respectively, compared with 90.0 g $CO_{2\text{-eq}}$/MJ of conventional aviation fuel. It had been found that feedstock yield had an important impact on GHG emissions and direct land-use change emissions [34]. Other research has shown the GHG emission reduction of sugarcane to produce jet fuel was estimated as 73% compared with aviation kerosene, which was higher than that of oil feedstock [42].

4.1.4 Ethanol

Han et al. [57] studied the WtWa results of four emerging alternative jet fuels, which included ATJ conversion of corn and corn stalk, and sugar-to-spray conversion of corn stalk from biological and catalytic. For the ATJ approach, integration (processing corn or corn stalk as raw material) and distribution (processing ethanol as raw material) were examined of two plant designs. The results showed that feedstocks were the key factor of GHG emissions by the ATJ process, resulting in the estimated GHG emissions of 16% and 73% less than that of oil injection. As for the sugar-to-spray approach, GHG emissions were 59% lower. Da Silva et al. [58] conducted a LCA of the future first- and second-generation ethanol scenarios (2020–30) compared with current ethanol production, in which environmental impact categories included abiotic depletion, global warming, human toxicity, ecotoxicity, photochemical oxidation, acidification, and eutrophication. The studies showed that compared with the current production of sugarcane and ethanol, the no-tillage of sugarcane cultivation could bring good environmental benefits in the future. Junqueira et al. [24] indicated that compared with gasoline, both first generation and second generation (also known as ethanol production of lignocellulose) ethanol can reduce the impact of climate change by more than 80%.

4.2 Water footprints

Rapid growth in biofuel production could cause other environmental problems, such as water shortages. The research conducted by Xie et al. [59] aimed to assess the life cycle water footprint of several potential nonedible raw materials, including Chinese cassava, sweet sorghum, and *J. curcas*. In this study, different water footprint types were considered, including blue water, green water, and gray water. The results of water footprint showed that the growth of feedstocks was the most intensive process of water footprint, while the conversion and transportation of biofuels made little contribution to the total water footprint. The estimated life cycle water footprints of cassava ethanol, sweet sorghum ethanol, and *J. curcas* seed biodiesel were 73.9–222.2, 115.9–210.4, and 64.7–182.3 m^3/GJ of biofuel, respectively. Similarly, the study by Shi et al. [60] took blue, green, and gray water consumption footprints into account. The estimated water footprint of North Dakota jet fuel derived from rapeseed was 131–143 m^3/GJ, using energy allocation. Cox et al. [42] assessed the environmental impact of Australian pongamia and microalgae producing biojet fuels through the HEFA process and sugarcane through the DSHC process. The results showed that the water consumption of sugarcane was higher, which was 15.6 m^3/GJ and 147.0 m^3/GJ under economic allocation and system expansion. The other two pathways had little impact on water resource utilization. In terms of system expansion, the water consumption of pongamia and microalgae were 11.8 and 13.9 m^3/GJ, higher than 5.5 m^3/GJ and 6.4 m^3/GJ of economic allocation. Carter [61] quantified the freshwater consumption of four plantings and two extraction technology groups. For each planting and extraction type, low, baseline, and high scenarios were adopted to assess the variability of each performance indicator. The results showed that compared with other scenarios, the performance of open raceway pond cultivation with wet extraction was the most favorable. Its freshwater consumption was 0.38 m^3/GJ. Staples et al. [62] researched

Table 3 Water consumption of jet fuel production from different pathway.

Pathway	Feedstock	Water consumption (m³/GJ)
HEFA	Microalgae oil	6.40–13.90
	Pongamia oil	5.50–11.83
	Rapeseed	57.91–143.00
	Jatropha	66.45–75.03
	Soybean	63.53–106.79
DSHC	Sugarcane	15.60–147.0
	Corn	76.46–85.81
	Switchgrass	92.40–104.74
ATJ	Corn	1.60–1.90
	Corn stover	0.83–0.88
APR	Corn stover	0.40–1.23

water consumption in the production of medium distillate transportation fuel in the United States. For producing jet fuel derived from rainfed biomass, the mid water consumption of jatropha, soybean, and rapeseed was 66.45, 63.65, and 57.91 m³/GJ, respectively. As for irrigation biomass, due to a large amount of water used for irrigation, the mid water use raised to 75.03, 106.79, and 102.65 m³/GJ, respectively. This study also indicated that the mid water consumption of aviation fuel produced by the DSHC process for sugarcane, corn, and switchgrass was 71.73–86.57, 76.46–85.81, and 92.40–104.74 m³/GJ, respectively. Water consumption of several biojet fuels production pathways is listed in Table 3.

4.3 Other environmental issues

In terms of land use, Da Silva et al. [58] indicated that the combined production of first- and second-generation ethanol would require less agricultural land, but would not perform better than the predicted first-generation ethanol, although the uncertainty was relatively high. The study conducted by Cox et al. [42] assessed biofuels, in which sugarcane, plankton, and microalgae were used as feedstocks. The land use intensities of the sugarcane, pongamia, and microalgae were 5.1–38.9, 4.5–7.8, and 6.8–7.0 m²/100 MJ, respectively.

Also, Cox et al. [42] analyzed the eutrophication, the results of which showed that sugarcane had a higher effect on eutrophication (0.015–0.044 kg PO_{4eq}/100 MJ), compared with plankton (0.009–0.012 kg PO_{4eq}/100 MJ) and microalgae (0.007–0.011 kg PO_{4eq}/100 MJ). Ganguly et al. [53] assessed the environment impact of jet fuel derived from residual woody, including eutrophication, smog, acidification, and respiratory effects. Table 4 gives other environmental issues for jet fuel production.

Table 4 Other environmental issues for jet fuel production.

Feedstock	Pathway	Land use (m²/GJ)	Eutrophication	Smog (kg O_{3eq}/GJ)	Acidification (kg SO_{2eq}/GJ)	Respiratory effects (kg $PM_{2.5eq}$/GJ)
Sugarcane	DSHC	0.51–3.89	0.0015–0.0044 (kg PO_{4eq}/GJ)	–	–	–
Plankton	HEFA	0.45–0.78	0.0009–0.0012 (kg PO_{4eq}/GJ)	–	–	–
Microalgae	HEFA	0.68–0.70	0.0007–0.0011 (kg PO_{4eq}/GJ)	–	–	–
Residual woody	ATJ	–	1.12 (kg N_{eq}/GJ)	9.17	0.39	–0.34

5. Biojet fuels development

5.1 Current polices and attitudes of countries

With the rapid increase in global aviation fuel demand and the global aviation industry's task of reducing emissions, the large-scale development and utilization of biojet fuel have been included in many countries and regions in their aviation industry sustainable development strategic goals.

Brazil was the first country to use biofuels in aviation. As early as 2004, Embraer's all ethanol-powered agricultural aircraft was certified airworthiness. It can be said that 2011 is a turning point for biofuels. The Global Fuel Standards Agency has approved the use of biofuels in civil aviation. Since then, Airbus, Boeing, and Embraer have participated in a number of aviation flights using biofuels. For the U.S. Department of Defense's alternative fuel program, some goals have been set by the U.S. Air Force to test and certify that all aircraft and systems have a 50:50 alternative fuel mixture, and to ensure that 50% of the domestic aviation fuel comes from the extraction of alternative fuel mixtures by 2025 [63]. China has become the fourth country with independent research and development of the production technology of bio aviation fuel. In 2014, the Civil Aviation Administration of China issued the project approval letter of No.1 biological aviation fuel technical standard to Sinopec, which marked that the domestic No.1 biological aviation kerosene had officially obtained airworthiness approval and could be put into commercial use [64]. The European Union recently declared programs for a revised Renewable Energy Directive (2020–30), including incentives for the use of sustainable aviation fuel. Alternative aviation fuel regulations have been introduced in Indonesia. Starting in 2018, the proportion of alternative aviation fuels will be increased from 2% to 5% by 2025. Groups around the world are working together to develop biojet fuel deployment measures. From 2011 to 2015, 22 airlines used 50% biojet fuel blends on more than 2500 commercial passenger flights [65]. In 2016, approximately 4.5 million liters of sustainable aviation fuel were produced globally, twice the amount of 2015. At the same time, there were more than 3000 flights using hybrid biojet fuels [10]. Renewable jet fuel consumption may increase significantly in the next 10 years, depending on policy incentives, technology development, and biomass supply [66]. Some technologies, such as FT, fatty acids, and hydrogenated esters, are currently in the commercial stage, however, other emerging technologies require further research and development before they can be fully utilized [23]. Fig. 2 illustrates the Commercial Aviation Alternative Fuels Initiative (CAAFI) Fuel Readiness (FRL) scale, which measures technological maturity for certification and commercial use [14]. Key milestones include proof of concept (FRL 3), preliminary technical and evaluation (FRL 4), scaling from laboratory to pilot (FRL 5), full-scale technical evaluation (FRL 6), certification by the ASTM (FRL 7), commercial and business model approval (FRL 8), and commercial factory fully operational (FRL 9) [14].

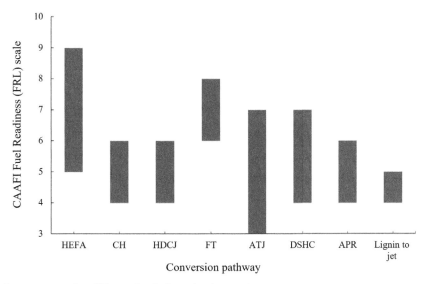

Fig. 2 Current status for different jet fuel production pathways.

5.2 Challenges and potential risks of biojet fuels

The rapid increase in aviation fuel demand has placed more and more attention on bio aviation fuel. However, in order to replace fossil jet fuel completely, many challenges still need to be overcome, including feedstock availability, economy, and sustainability [23].

From the perspective of long-term development, the availability of feedstock makes a major dare to the commercial-scale application and deployment of sustainable jet fuel. The feedstocks for biojet fuel production should have the characteristics, including large output, wide distribution, a wide range of acquisition, no competition with grain production, and no deforestation. Additionally, low-cost feedstocks are required for mass production [67]. It is necessary to incorporate the development of feedstocks production into government support policies. Innovation in feedstocks, including lessening land requirements, and water quality control and nutrition, will be critical to the large-scale adoption of alternative aviation fuels [68].

The economy is the main obstacle to be overcome in the development of alternative aviation fuels [69]. The production of lignocellulosic fuels carries a lower GHG burden, but seems to require more energy input than vegetable oil. Technological advances are likely to improve the situation. Second-generation biofuels have good prospects in terms of reducing GHG emissions, but for some analytical approaches, their overall energy efficiency performance is relatively poor. Therefore, policymakers need to set clear priorities when deciding how to invest, assessing the GHG reduction potential of alternative jet fuels alone is not enough [70]. Currently, the price of biofuels is above the level of traditional fossil fuels. In order for airlines to purchase fuel, it is necessary to provide policy incentives or compensation mechanisms for the environmental benefits inherent to biojet fuels, which can narrow the price gap and create a promising market, thus attracting investors and reducing perceived risks [68]. The factors that affect the economic feasibility of biojet production mainly include feedstocks prices, operating costs, equipment costs, conversion efficiency or product yield, and the selling prices of distillate fuels and by-products [40]. Therefore, efforts to further reduce costs should focus on feedstock productivity, cheap catalysts, equipment selection and distribution, optimal reaction conditions, plant scale, and recycling and utilization of by-products [33,40]. The key to increasing competition for alternative aviation fuels is to reduce market prices so that they can be competitive with conventional aviation fuels. This will require further support and investment in research and development, as well as technology demonstration and expansion. The widespread promotion of an aviation carbon tax would also increase the competitiveness of biojet fuels.

The sustainability of biojet fuel is also an important issue. Commercial-scale development may have additional impacts on the environment, society, and economy. For instance, biomass production may cause major changes to land use, water consumption, biodiversity, soil degradation, and rural societies and local communities [68]. It is necessary to formulate policies for the sustainable use of alternative fuels, including setting sustainability goals and implementing related special measures, especially in terms of the indirect effects of large-scale use of alternative fuels, such as land-use changes and food security [71].

HEFA and FT synthesis have certain advantages in existing feedstocks, technical status, environmental impact, and economic evaluation, which are considered to have the most potential in biojet fuel production technology in the long run. However, before these two technologies could be put into large-scale production, more work needs to be done. As the cost of emerging technologies decreases, there are many possibilities for the future development of biojet fuels.

6. Conclusion

Air transportation plays an important role in interpersonal communication and cargo trade. In recent years, with the continuous increase of world air transportation, the demand for aviation fuel is also growing rapidly. Conventional aviation fuel not only relies heavily on fossil energy but also produces a lot of GHG emissions, which cannot meet the sustainable development principles and emission reduction targets of the aviation industry. Bio-fuel not only comes from renewable energy but also can effectively reduce GHG emissions in the life cycle, which has attracted wide attention. Although the life cycle results are slightly different due to the influence of feedstocks, production processes and so on, bio-fuels still have huge potential for future development compared with traditional fuels. However, favorable GHG reduction benefits do not always coincide with high energy efficiency. In the future, it will be able to develop more rapidly and compete with traditional aviation fuels by reducing its market price, including further policy support and technological progress.

Acknowledgments

This work was supported by National Natural Science Foundation of China (No. 51622604), the Foundation of State Key Laboratory of Coal Combustion (No. FSKLCCA1902), and the Double first-class research funding of China-EU Institute for Clean and Renewable Energy (No. 3011120016).

References

[1] G.X. Yao, Thoughts on accelerating the development of China's bio aviation fuel industry, Sino-Global Energy 16 (4) (2011) 18–26.

[2] International Air Transport Association, Statistical Report of Air Transport Industry in 2019, 2020.

[3] International Air Transport Association, Forecast Report of Air Passenger Transport in the Next 20 Years, 2016.

[4] US Energy, Information Administration, International Energy Outlook, 2016.

[5] China National Aviation Fuel Group Limited, China Aviation Kerosene Consumption Index, 2018.

[6] G. Liu, B. Yan, G. Chen, Technical review on jet fuel production, Renew. Sustain. Energy Rev. 25 (2013) 59–70, https://doi.org/10.1016/j.rser.2013.03.025.

[7] Air Transport Action Group, Facts & Figures, 2020.

[8] C. Zhang, Utilization of renewable energy lignin to increase added value of products, Chem. Ind. Eng. Soc. China 1994 (2015) 292.

[9] D.B. Agusdinata, F. Zhao, K. Ileleji, D. Delaurentis, Life cycle assessment of potential biojet fuel production in the United States, Environ. Sci. Technol. 45 (21) (2011) 9133–9143, https://doi.org/10.1021/es202148g.

[10] International Air Transport Association, Annual Review, 2017.

[11] M. Mofijur, M.A. Hazrat, M.G. Rasul, H.M. Mahmudul, Comparative evaluation of edible and non-edible oil methyl ester performance in a vehicular engine, Energy Procedia 75 (2015) 37–43, https://doi.org/10.1016/j.egypro.2015.07.134.

[12] T. Zhan, Z.B. Deng, Evaluation and countermeasures of sustainable aviation fuel development, Technol. Innov. Appl. 2 (2019) 151–152.

[13] W. Liu, Life Cycle Assessment and Economic Analysis of Jet Fuel Produced by Biomass Gasification Fischer-Tropsch Synthesis, 2018. https://kns.cnki.net/kcms/detail/detail.aspx?dbcode=CMFD&dbname=CMFD201901&filename=1018782728.nh&v=rIilr%25mmd2BWbWaYNJd1F9hWzjvHbExngXO3AkYqd6J3%25mmd2FnQsSGCarlNZaUv8fWY119JKY.

[14] B.W. Kolosz, Y. Luo, B. Xu, M.M. Maroto-Valer, J.M. Andresen, Life cycle environmental analysis of "drop in" alternative aviation fuels: a review, Sustain. Energy Fuels 4 (7) (2020) 3229–3263, https://doi.org/10.1039/c9se00788a.

[15] A. Gasparatos, P. Stromberg, K. Takeuchi, Sustainability impacts of first-generation biofuels, Anim. Front. 3 (2) (2013) 12–26, https://doi.org/10.2527/af.2013-0011.

[16] A. Kumar, S. Sharma, Potential non-edible oil resources as biodiesel feedstock: an Indian perspective, Renew. Sustain. Energy Rev. 15 (4) (2011) 1791–1800, https://doi.org/10.1016/j.rser.2010.11.020.

[17] L. Uorn, Billion-Ton Update: Biomass Supply for a Bioenergy and Bioproducts Industry, Office of Entific & Technical Information Technical Reports, 2011.

[18] L.D. Zhu, Z.B. Xu, L. Qin, Z.M. Wang, E. Hiltunen, Z.H. Li, Oil production from pilot-scale micro-algae cultivation: an economics evaluation, Energy Sources Part B Econ. Plan. Policy 11 (1) (2016) 11–17, https://doi.org/10.1080/15567249.2015.1052594.

[19] U. Neuling, M. Kaltschmitt, Techno-economic and environmental analysis of aviation biofuels, Fuel Process. Technol. 171 (2018) 54–69, https://doi.org/10.1016/j.fuproc.2017.09.022.

[20] Y. Li, L. Chen, X. Zhang, et al., Process and techno-economic analysis of bio-jet fuel-range hydrocarbon production from lignocellulosic biomass via aqueous phase deconstruction and catalytic conversion, Energy Procedia 105 (2017) 675–680, https://doi.org/10.1016/j.egypro.2017.03.374.

[21] G.W. Diederichs, M. Ali Mandegari, S. Farzad, J.F. Görgens, Techno-economic comparison of biojet fuel production from lignocellulose, vegetable oil and sugar cane juice, Bioresour. Technol. 216 (2016) 331–339, https://doi.org/10.1016/j.biortech.2016.05.090.

[22] S. Michailos, Process design, economic evaluation and life cycle assessment of jet fuel production from sugar cane residue, Environ. Prog. Sustain. Energy 37 (3) (2018) 1227–1235, https://doi.org/10.1002/ep.12840.

[23] H. Wei, W. Liu, X. Chen, Q. Yang, J. Li, H. Chen, Renewable bio-jet fuel production for aviation: a review, Fuel 254 (2019), https://doi.org/10.1016/j.fuel.2019.06.007.

[24] T.L. Junqueira, M.F. Chagas, V.L.R. Gouveia, et al., Techno-economic analysis and climate change impacts of sugarcane biorefineries considering different time horizons, Biotechnol. Biofuels 10 (1) (2017), https://doi.org/10.1186/s13068-017-0722-3.

[25] W.C. Wang, Techno-economic analysis of a bio-refinery process for producing Hydro-processed Renewable Jet fuel from Jatropha, Renew. Energy 95 (2016) 63–73, https://doi.org/10.1016/j.renene.2016.03.107.

[26] G. Seber, R. Malina, M.N. Pearlson, H. Olcay, J.I. Hileman, S.R.H. Barrett, Environmental and economic assessment of producing hydroprocessed jet and diesel fuel from waste oils and tallow, Biomass Bioenergy 67 (2014) 108–118, https://doi.org/10.1016/j.biombioe.2014.04.024.

[27] D.C. Gonzalez, Techno-Economic Analysis of Jet Fuel Production From Waste Vegetable Oil in Mexico, 2016.

[28] M. Pearlson, C. Wollersheim, J. Hileman, A techno-economic review of hydroprocessed renewable esters and fatty acids for jet fuel production, Biofuels Bioprod. Biorefin. 7 (1) (2013) 89–96, https://doi.org/10.1002/bbb.1378.

[29] L. Tao, A. Milbrandt, Y. Zhang, W.C. Wang, Techno-economic and resource analysis of hydroprocessed renewable jet fuel, Biotechnol. Biofuels 10 (1) (2017), https://doi.org/10.1186/s13068-017-0945-3.

[30] D. Klein-Marcuschamer, C. Turner, M. Allen, et al., Technoeconomic analysis of renewable aviation fuel from microalgae, *Pongamia pinnata*, and sugarcane, Biofuels Bioprod. Biorefin. 7 (4) (2013) 416–428, https://doi.org/10.1002/bbb.1404.

[31] K. Atsonios, M.A. Kougioumtzis, K.D. Panopoulos, E. Kakaras, Alternative thermochemical routes for aviation biofuels via alcohols synthesis: process modeling, techno-economic assessment and comparison, Appl. Energy 138 (2015) 346–366, https://doi.org/10.1016/j.apenergy.2014.10.056.

[32] P. Suresh, R. Malina, M.D. Staples, et al., Life cycle greenhouse gas emissions and costs of production of diesel and jet fuel from municipal solid waste, Environ. Sci. Technol. 52 (21) (2018) 12055–12065, https://doi.org/10.1021/acs.est.7b04277.

[33] S. de Jong, R. Hoefnagels, A. Faaij, R. Slade, R. Mawhood, M. Junginger, The feasibility of short-term production strategies for renewable jet fuels—a comprehensive techno-economic comparison, Biofuels Bioprod. Biorefin. 9 (6) (2015) 778–800, https://doi.org/10.1002/bbb.1613.

[34] M.D. Staples, R. Malina, H. Olcay, et al., Lifecycle greenhouse gas footprint and minimum selling price of renewable diesel and jet fuel from fermentation and advanced fermentation production technologies, Energy Environ. Sci. 7 (5) (2014) 1545–1554, https://doi.org/10.1039/c3ee43655a.

[35] J.T. Crawford, C.W. Shan, E. Budsberg, H. Morgan, R. Bura, R. Gustafson, Hydrocarbon bio-jet fuel from bioconversion of poplar biomass: techno-economic assessment, Biotechnol. Biofuels 9 (1) (2016), https://doi.org/10.1186/s13068-016-0545-7.

[36] G. Yao, M.D. Staples, R. Malina, W.E. Tyner, Stochastic techno-economic analysis of alcohol-to-jet fuel production, Biotechnol. Biofuels 10 (1) (2017), https://doi.org/10.1186/s13068-017-0702-7.

[37] N. Vela-García, D. Bolonio, A.M. Mosquera, M.F. Ortega, M.J. García-Martínez, L. Canoira, Techno-economic and life cycle assessment of triisobutane production and its suitability as biojet fuel, Appl. Energy 268 (2020), https://doi.org/10.1016/j.apenergy.2020.114897.

[38] R.H. Natelson, W.C. Wang, W.L. Roberts, K.D. Zering, Technoeconomic analysis of jet fuel production from hydrolysis, decarboxylation, and reforming of camelina oil, Biomass Bioenergy 75 (2015) 23–34, https://doi.org/10.1016/j.biombioe.2015.02.001.

[39] H. Olcay, R. Malina, A.A. Upadhye, J.I. Hileman, G.W. Huber, S.R.H. Barrett, Techno-economic and environmental evaluation of producing chemicals and drop-in aviation biofuels via aqueous phase processing, Energy Environ. Sci. 11 (8) (2018) 2085–2101, https://doi.org/10.1039/c7ee03557h.

[40] J.Q. Bond, A.A. Upadhye, H. Olcay, et al., Production of renewable jet fuel range alkanes and commodity chemicals from integrated catalytic processing of biomass, Energy Environ. Sci. 7 (4) (2014) 1500–1523, https://doi.org/10.1039/c3ee43846e.

[41] S.S. Zhou, Process simulation and techno-economic analysis of lignin to jet fuel and its co-production in a cellulosic ethanol, Biorefinery (2017).

[42] K. Cox, M. Renouf, A. Dargan, C. Turner, D. Klein-Marcuschamer, Environmental life cycle assessment (LCA) of aviation biofuel from microalgae, *Pongamia pinnata*, and sugarcane molasses, Biofuels Bioprod. Biorefin. 8 (4) (2014) 579–593, https://doi.org/10.1002/bbb.1488.

[43] E. Budsberg, J.T. Crawford, H. Morgan, W.S. Chin, R. Bura, R. Gustafson, Hydrocarbon bio-jet fuel from bioconversion of poplar biomass: life cycle assessment, Biotechnol. Biofuels 9 (1) (2016), https://doi.org/10.1186/s13068-016-0582-2.

[44] F. Carvalho, F.T.F. da Silva, A. Szklo, J. Portugal-Pereira, Potential for biojet production from different biomass feedstocks and consolidated technological routes: a georeferencing and spatial analysis in Brazil, Biofuels Bioprod. Biorefin. 13 (6) (2019) 1454–1475, https://doi.org/10.1002/bbb.2041.

[45] F. Guo, J. Zhao, A. Lusi, X. Yang, Life cycle assessment of microalgae-based aviation fuel: influence of lipid content with specific productivity and nitrogen nutrient effects, Bioresour. Technol. 221 (2016) 350–357, https://doi.org/10.1016/j.biortech.2016.09.044.

[46] D.R. Shonnard, L. Williams, T.N. Kalnes, Camelina-derived jet fuel and diesel: sustainable advanced biofuels, Environ. Prog. Sustain. Energy 29 (3) (2010) 382–392, https://doi.org/10.1002/ep.10461.

[47] R.W. Stratton, H.M. Wong, J.I. Hileman, Quantifying variability in life cycle greenhouse gas inventories of alternative middle distillate transportation fuels, Environ. Sci. Technol. 45 (10) (2011) 4637–4644, https://doi.org/10.1021/es102597f.

[48] X. Li, E. Mupondwa, Life cycle assessment of camelina oil derived biodiesel and jet fuel in the Canadian Prairies, Sci. Total Environ. 481 (1) (2014) 17–26, https://doi.org/10.1016/j.scitotenv.2014.02.003.

[49] J. Han, A. Elgowainy, H. Cai, M.Q. Wang, Life-cycle analysis of bio-based aviation fuels, Bioresour. Technol. 150 (2013) 447–456, https://doi.org/10.1016/j.biortech.2013.07.153.

[50] M.O.P. Fortier, G.W. Roberts, S.M. Stagg-Williams, B.S.M. Sturm, Life cycle assessment of bio-jet fuel from hydrothermal liquefaction of microalgae, Appl. Energy 122 (2014) 73–82, https://doi.org/10.1016/j.apenergy.2014.01.077.

[51] J.A. Obnamia, H.L. MacLean, B.A. Saville, Regional variations in life cycle greenhouse gas emissions of canola-derived jet fuel produced in western Canada, GCB Bioenergy 12 (10) (2020) 818–833, https://doi.org/10.1111/gcbb.12735.

[52] D. Zemanek, P. Champagne, W. Mabee, Review of life-cycle greenhouse-gas emissions assessments of hydroprocessed renewable fuel (HEFA) from oilseeds, Biofuels Bioprod. Biorefin. 14 (5) (2020) 935–949, https://doi.org/10.1002/bbb.2125.

[53] I. Ganguly, F. Pierobon, T.C. Bowers, M. Huisenga, G. Johnston, I.L. Eastin, 'Woods-to-Wake' Life Cycle Assessment of residual woody biomass based jet-fuel using mild bisulfite pretreatment, Biomass Bioenergy 108 (2018) 207–216, https://doi.org/10.1016/j.biombioe.2017.10.041.

[54] R.E. Bailis, J.E. Baka, Greenhouse gas emissions and land use change from *Jatropha curcas*-based jet fuel in Brazil, Environ. Sci. Technol. 44 (22) (2010) 8684–8691, https://doi.org/10.1021/es1019178.

[55] A. Elgowainy, J. Han, M. Wang, N. Carter, R. Stratton, J. Hileman, A. Malwitz, S.N. Balasubramanian, Life-Cycle Analysis of Alternative Aviation Fuels in GREET, 2012.

[56] J.M. Bressanin, B.C. Klein, M.F. Chagas, et al., Techno-economic and environmental assessment of biomass gasification and Fischer-Tropsch synthesis integrated to sugarcane biorefineries, Energies 13 (17) (2020), https://doi.org/10.3390/en13174576.

[57] J. Han, L. Tao, M. Wang, Well-to-wake analysis of ethanol-to-jet and sugar-to-jet pathways, Biotechnol. Biofuels 10 (1) (2017), https://doi.org/10.1186/s13068-017-0698-z.

[58] C.R.U. Da Silva, H.C.J. Franco, T.L. Junqueira, L. Van Oers, E. Van Der Voet, J.E.A. Seabra, Long-term prospects for the environmental profile of advanced sugar cane ethanol, Environ. Sci. Technol. 48 (20) (2014) 12394–12402, https://doi.org/10.1021/es502552f.

[59] X. Xie, T. Zhang, L. Wang, Z. Huang, Regional water footprints of potential biofuel production in China, Biotechnol. Biofuels 10 (1) (2017), https://doi.org/10.1186/s13068-017-0778-0.

[60] R. Shi, S. Ukaew, D.W. Archer, et al., Life cycle water footprint analysis for rapeseed derived jet fuel in North Dakota, ACS Sustain. Chem. Eng. 5 (5) (2017) 3845–3854, https://doi.org/10.1021/acssuschemeng.6b02956.

[61] N.A. Carter, Environmental and Economic Assessment of Microalgae-Derived Jet Fuel, 2012.
[62] M.D. Staples, H. Olcay, R. Malina, et al., Water consumption footprint and land requirements of large-scale alternative diesel and jet fuel production, Environ. Sci. Technol. 47 (21) (2013) 12557–12565, https://doi.org/10.1021/es4030782.
[63] K. Blakeley, DOD alternative fuels: policy, initiatives and legislative activity, in: Alternative Fuel Use by the Department of Defense: Initiatives and Opportunities, Nova Science Publishers, Inc., 2013, pp. 1–22. https://www.novapublishers.com/catalog/product_info.php?products_id=43162.
[64] Nation Energy Administration, Development Status of Bio Aviation Fuel, 2014.
[65] International Air Transport Association, Annual Review, 2016.
[66] S.D. Jong, J.V. Stralen, M. Londo, R. Hoefnagels, A. Faaij, M. Junginger, Renewable jet fuel supply scenarios in the European Union in 2021–2030 in the context of proposed biofuel policy and competing biomass demand, GCB Bioenergy 10 (2018) 661–682.
[67] Biofuels in Aviation Major Challenges Potentials Major R&D Activities and Needs, 2010.
[68] ICAO, The Challenges for the Development and Deployment of Sustainable Alternative Fuels in Aviation, 2013.
[69] S. Nair, H. Paulose, Emergence of green business models: the case of algae biofuel for aviation, Energy Policy 65 (2014) 175–184, https://doi.org/10.1016/j.enpol.2013.10.034.
[70] A. O'Connell, M. Kousoulidou, L. Lonza, W. Weindorf, Considerations on GHG emissions and energy balances of promising aviation biofuel pathways, Renew. Sustain. Energy Rev. 101 (2019) 504–515, https://doi.org/10.1016/j.rser.2018.11.033.
[71] T. Kandaramath Hari, Z. Yaakob, N.N. Binitha, Aviation biofuel from renewable resources: routes, opportunities and challenges, Renew. Sustain. Energy Rev. 42 (2015) 1234–1244, https://doi.org/10.1016/j.rser.2014.10.095.

CHAPTER 10

Sustainability tensions and opportunities for aviation biofuel production in Brazil

Mar Palmeros Parada[a], Wim H. van der Putten[b], Luuk A.M. van der Wielen[a,c], Patricia Osseweijer[a], Mark van Loosdrecht[a], Farahnaz Pashaei Kamali[a], and John A. Posada[a]

[a]Department of Biotechnology, Delft University of Technology, Delft, The Netherlands
[b]Department of Terrestrial Ecology, Netherlands Institute of Ecology, Wageningen, The Netherlands
[c]Bernal Institute, University of Limerick, Limerick, Ireland

1. Introduction

Biobased production has been promoted as a sustainable alternative to fossil-based production in order to mitigate climate change [1]. Prominent targets for biobased production are fuels and chemicals for which there are limited alternatives, such as aviation biofuels [2]. Biomass is the only current alternative for obtaining these products, however, due to high production costs and limited availability of *sustainable* feedstock, their production remains a challenge [3]. Nevertheless, the CORSIA agreement by the United Nations' aviation agency enforces an international commitment for carbon neutral growth in the aviation sector (relative to 2020), and biobased and other sustainable aviation fuels are critical to achieve this [4].

Concerns over the sustainability of biofuels have been emerging since the production growth in the 2000s [5]. These concerns include effects on food security from the use of edible feedstock, effects of land use changes on emissions, and negative impacts on the livelihood of local communities [5–7]. While not necessarily related to all biofuels, those examples indicate that there are downsides of biobased production as well, and that tensions may emerge between different sustainability aspects, like emission reduction targets and food security impacts. As these tensions depend on local contexts [8], there is a need for comprehensive ex-ante sustainability analyses, taking into consideration the context around biofuel production.

With the growing interest in aviation biofuels, various production alternatives have been developed and assessed, indicating that aviation biofuels have the potential to reduce emissions when compared to fossil-based kerosene [9]. However, existing approaches for the design and ex-ante assessment of biofuel production tend to focus on techno-economic feasibility, climate change, and energy efficiency, and rarely address societal aspects and the

Sustainable Alternatives for Aviation Fuels
https://doi.org/10.1016/B978-0-323-85715-4.00007-0

Copyright © 2022 Elsevier Inc.
All rights reserved.

local context of the intended production chains [10]. Cavalett and Cherubini [11] investigated the impacts of aviation biofuels from forest residues in relation to the UNs Sustainable Development Goals. While their study addresses some societal implications of aviation biofuels, in their analysis not much attention is dedicated to how these goals and their measurement are relevant in the regional setting under study.

Here, we present a novel context-dependent ex-ante sustainability analysis of aviation biofuel production, which includes economic, environmental, and societal aspects. Focused on the Southeast region of Brazil, and based on inputs from local stakeholders and sustainability literature [12,13], eight aspects of sustainability were considered: climate change, commercial acceptability, efficiency, energy security, investment security, profitability, social development, and soil sustainability. For the analysis, we integrate and contrast estimates of the performance of production alternatives with regard to these aspects, which were estimated separately as part of the same research project [12–21]. Based on this contrast, sustainability tensions for the production of aviation biofuel in Southeast Brazil are discussed, and some opportunities for reconciling them in future developments are presented. In view of these findings, we provide conclusions related to the case study and the followed methodology for a more *sustainable* biobased production. Note that the followed methodology and its contribution to the field of sustainable biobased production has been recently discussed by Palmeros Parada et al. [22].

2. Methods

2.1 Production alternatives for aviation biofuel

Possible production alternatives for the case study were based on expected economic potential (the difference between sale revenues from all products and feedstock costs), production yields, and feedstock availability in Southeast Brazil, as described by Alves et al. (2017) [14]. Feedstock materials in consideration were macauba, jatropha, camelina, soybean, sugarcane, sweet sorghum, and the lignocellulosic residues of sugarcane, sweet sorghum, eucalypt, pinus, coffee, and rice. These feedstock materials were selected based on oil/sugar content, land productivity, availability in Brazil, resistance to lack of water or nutrients, production and harvesting cost, potential expansion, amongst others [14]. By-products in consideration included secondary fuel products derived from the process (such as naphtha and diesel). Higher-value biochemicals as by-product alternatives obtained from a dedicated fraction of feedstock stream were evaluated, and included intermediates for bioplastics such as ethylene, lactic acid, and succinic acid [14]. The estimated economic potential was used to narrow the range of feedstock materials to eucalypt residues, macauba, and sugarcane, and higher-value products to succinic acid only. Economic potential results are summarized in Annex 1.

Subsequently, preliminary techno-economic analyses were used to define specific combinations of feedstock and technologies for the case study, based on a production scale of 210 kton/year of aviation biofuel [17,18]. Evaluated conversion technologies for sugar feedstock materials were Direct Fermentation to alkanes (DF) and Ethanol-to-Jet (ETJ). Hydroprocessed Esters and Fatty Acids (HEFA) was considered for oily feedstock materials, and Hydrothermal Liquefaction (HTL) and Gasification Fischer-Tropsch (GFT) for lignocellulosic residues. Pretreatment alternatives were also evaluated for lignocellulosic residues, where lignin was considered for aviation biofuel production through Fast Pyrolysis (FP) and GFT, or for power co-generation. Fermentable sugars from pretreatment alternatives were considered for the production of higher-value chemicals, or for second-generation (2G) ETJ aviation biofuel in the case of bagasse. Bare equipment costs were estimated from literature data for similar technologies [23–27], and taking into account economies of scale. Variable costs were determined from mass and energy balances, using the list of prices in Annex 2. Total capital and operational expenses were estimated based on economic factors [28], which include a capital charge (i.e., annualized capital expenses) for the processing technologies considering a plant life of 15 years. Based on the results of the preliminary techno-economic analysis, the most promising production chains for the sustainability analysis described in the next section were: sugarcane processed with ETJ in combination with FP for bagasse, eucalypt residues processed with either FP or HTL, and macauba processed with HEFA in combination with HTL or FP for macauba residues. The main conversion steps for these production alternatives are summarized in Fig. 1, more process details can be found in Cornelio da Silva et al. (2016) [17] and Santos et al. (2018) [18]. As an exception, Gasification Fischer-Tropsch (GFT) is the technology considered for eucalypt conversion when evaluated for social development. GFT was considered for the social development evaluation because the availability of data and development stage of the technology were considered crucial for the analysis (see below the section on social development).

2.2 Sustainability analysis

The performance of promising production chains was evaluated considering the sustainability framework in Table 1. The sustainability aspects that conform the framework were identified from previous work in the target region [12,13], which includes interviews with stakeholders related to the potential production of aviation biofuel (such as representatives of government bodies and biomass producing organizations), a survey with experts on biofuel production, and a sustainability literature review. The sustainability aspects in this study take as benchmark the definitions from Pashaei Kamali et al. (2018) [12], which are based on G4 Sustainability Reporting Guidelines of the Global Reporting Initiative [29] and the United Nation's Food and Agriculture Organization (FAO) Sustainability Assessment of Food and Agriculture systems [30].

Some of the identified sustainability aspects for this case study were left out of the framework (i.e., accountability, cooperation and leadership, cultural diversity, equity

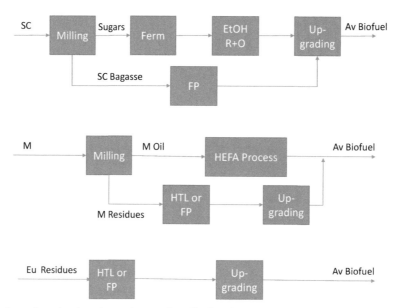

Fig. 1 Evaluated production alternatives as described in the Section 2. *Av*, aviation; *Eu*, eucalypt; *Ferm*, fermentation; *FP*, fast pyrolysis; *HEFA*, hydroprocessed esters and fatty acids; *HTL*, hydrothermal liquefaction; *M*, macauba; *R+O*, recovery and oligomerization; *SC*, sugarcane. For process details see Cornelio da Silva et al. (2016) [17] and Santos et al. (2018) [18].

and social cohesion, human health and safety, labor rights, property rights, participation, rule of law, standard of living, training and education, and working conditions) considering data availability and the design scope of this work. That is, some sustainability aspects are mostly related to the implementation of production and are beyond the scope of design choices, or for their analysis they require monitoring data that was not available (especially for macauba for which there is no commercial full-scale production). Additionally, food security, often discussed in relation to the sustainability of biofuels, was not evaluated given that stakeholders did not consider it a prominent issue in the region [13] (possibly related to reported food production surplus and land availability in Brazil [41]). Perceptions of food security impacts, particularly from international stakeholders related to the aviation sector, did emerge from the interviews and could be analyzed as an aspect of commercial acceptability [13]. However, food security perceptions as part of commercial acceptability, and which are often associated to the use of food crops, were not further investigated given that none of the considered feedstock alternatives are food crops [42]. Nevertheless, given the complexity of this topic, a dedicated study on the food security impacts derived from the use of the considered feedstocks in Southeast Brazil is suggested in future work.

Profitability, climate change, and efficiency impacts were estimated with Minimum Selling Price (MSP, the lowest price at which biofuel can be sold to cover

Table 1 Sustainability framework for the ex-ante analysis of aviation biofuel production in Southeast Brazil.

Sustainability aspects		Description	Indicator(s)	Methods	Main references
Qualitative	Commercial acceptability	Analyzed in relation to ensuring safety and a good performance of aviation biofuel	ASTM approval	Literature review and stakeholder interviews	[13,31,32]
	Energy security	Related to energy supply reliability and self-sufficiency	Potential for power generation and NREU	Literature review and stakeholder interviews	[13, 17, 18]
	Investment security	Related to the readiness level of new crops and technologies, and previous experience with potential crops	FRL and crop development status	Literature review and stakeholder interviews	[3,13,33]
	Soil sustainability	Regarding the protection and recovery of the soil in relation to biomass production.	Residue harvest	Literature review	[15,34–40]
Quantitative	Climate change	Analyzed as the GHG emissions derived from the biomass production and distribution stages, and the aviation biofuel production process	GHG emissions	Life cycle assessment	[16–18, 21]
	Efficiency	Primarily evaluated in terms of nonrenewable energy use and other mass and energy efficiency indicators related to the process	NREU	Process modeling	[17, 18]
	Profitability	Analyzed in terms of the minimum selling price of aviation biofuel required to payback production expenses, including capital and operational expenses	MSP	Techno-economic analysis	[17, 18]
	Social development	Analyzed in relation to impacts on national employment, gross domestic product and trade balance	Direct and indirect jobs, GDP contributions, and trade balance	Input-output analysis	[19]

ASTM, American Society for Testing and Materials; *FRL*, fuel readiness level; *GDP*, gross domestic product; *GHG*, greenhouse gases; *MSP*, minimum selling price; *NREU*, nonrenewable energy use.

production expenses as $/ton), GHG emissions (as g CO_2/MJ), and Nonrenewable Energy Use (NREU as kJ/MJ) as indicators. The quantitative results presented in this work are based on the detailed estimations by Cornelio da Silva (2016) for production with eucalypt and macauba using FP, HEFA, and HTL technologies [17]; and on the work by Santos et al. (2018) for sugarcane using ETJ and FP [18]. Additionally, two improvement scenarios for sugarcane are presented based on (i) the co-processing of sweet sorghum during sugarcane off-season with the same equipment and, (ii) the co-production of succinic acid from fermentable sugars [18]. The estimations of MSP, GHG emissions, and NREU in the referenced studies consider the stages of biomass production and transportation, and the conversion and upgrading to bio-kerosene. Since the carbon emitted during combustion is biogenic carbon (i.e., captured during plant growth—photosynthesis) [43], CO_2 emissions from combustion were considered as neutral in the analysis. Considering that the evaluated alternatives are multiproduct systems where most products are energy products (e.g., aviation biofuel, diesel), the allocation method for GHG emissions and NREU between products was based on energy content (economic allocation was avoided due to fluctuating market prices in the energy sector). Additionally, it has been shown that different allocation methods for sugarcane-based production, which includes nonenergy products, lead to the same conclusions in terms of GHG and NREU, differing by less than 5% [18]. Emissions from the agricultural stage are an exception and were allocated based on the economic value of by-products generated at this stage. Energy allocation would neglect differences in wood and wood residue products that have similar energy contents but very different uses and economic value. A system expansion approach was followed for bioenergy as a product, assuming it replaces the generation of power from the Brazilian grid under national mix conditions. With regard to process alternatives, the in-house production of H_2 through steam methane reforming, the heat and power generation from solid residues, and the optional cracking step were considered based on the estimations from Vyhmeister et al. (2018) [20]. However, this work does not refer to specific results obtained in that study as it was based on different indicators. Nevertheless, their conclusions regarding process options are included in the discussion of results as their analysis is based on production chain alternatives similar to the ones considered in this work.

Social development impacts are presented in terms of employment, gross domestic product (GDP), and trade balance contributions based on the macroeconomic Input-Output analysis by Wang et al. (2019) [19]. Effects with regard to these indicators are estimated for the overall economic structure of Brazil as described by the most recent national Input-Output tables [44], and include effects directly related to the production of aviation biofuel, and indirect effects that relate to intermediate inputs and activities that support production. The effects on employment, GDP, and trade balance are presented for three potential production chains as described by Wang et al. [19]: (i) sugarcane-based production with ETJ conversion for sugarcane juice and FP conversion of bagasse;

(ii) eucalypt-based production with GFT conversion; and (iii) macauba-based production with HEFA conversion for macauba oil and FP for residues. GFT is the considered technology because the Input/Output analysis was based on policy and technology development scenarios, for which other technologies were discarded based on data availability and development stage. It is expected that the difference between GFT considered in the social development analysis, and FP and HTL for the rest of the indicators, does not strongly affect the overall comparison considering the large effect of the feedstock production stage on social development impacts, such as employment creation [45]. In the work by Wang et al. (2019), two different estimations are available for the three production chains, differing only on the projected aviation biofuel demand (i.e., 360 kton and 540 kton) [19]. In this work the average of these two estimations of employment, GDP, and trade balance impacts is presented per kton of aviation biofuel (the difference between estimations is less than 3%).

Commercial acceptability, energy security, investment risks, and soil sustainability were qualitatively investigated based on recent literature reports for the considered feedstock and technology alternatives, as seen in Table 1. **Commercial acceptability** was explored as an aspect of the sustainability of aviation biofuel production, and considering the concerns of stakeholders in the aviation sector regarding regulations and safety perceptions [13]. This aspect was explored in terms of the approval status by the ASTM, in alignment with the Brazilian National Agency of Petroleum, Natural Gas and Biofuels [46]. ASTM sets quality standards for "drop-in" aviation biofuels, and certification is granted to a specific aviation biofuel depending on the production processes to obtain it. Certification thus assures that the fuel has the same safety and performance, and can use the same infrastructure as conventional kerosene [47]. To put the results from the exploration of commercial acceptability in a visual form, alternatives that imply ASTM approved technologies were considered as having a positive score. A neutral qualification was given to alternatives with technologies in queue for approval, while a negative score on this aspect was considered for technologies that are not yet in consideration for ASTM approval.

Energy security was explored in terms of contribution to energy reliability and self-sufficiency considering the concerns of government and biofuel stakeholders about these aspects, and who referred to energy supply problems in the past [13]. Therefore, to analyze energy security, energy use derived from process simulations was used as a relative indication of the performance of conversion technologies on this aspect (i.e., a negative score for the alternative with highest NREU and a positive score for the alternative with lowest NREU) [17,18]. The potential of the different alternatives for power generation (expected to contribute to energy reliability [13]) was taken as an indicator of energy security performance related to each feedstock. A positive qualification was given when a feedstock alternative implied the availability of residues for co-generation regardless of the process configuration. A neutral score was considered when availability depended on the process configuration (there was no alternative with a negative effect on this aspect).

Investment security was explored depending on the readiness level of a conversion technology and feedstock. This aspect was considered according to the responses of stakeholders from the government, technology companies, and research institutes. Some of these stakeholders referred to farmers who perceived risk in unproven technologies (including feedstock materials), especially those for which they had no relatable experience [13]. For technology alternatives, the fuel readiness level scale (FRL, 1–9) was used as a reference. FRL is a risk management framework to specifically track the research and development stage of alternative fuels, considering the technology to produce it, manufacturing capacity, and compatibility with existing infrastructure [33]. The analysis takes as reference the conclusions from Mawhood et al. (2016) [3], and it is complemented with more recent information about the considered technologies [31,48,49]. For feedstock biomass, the FRL scale from the Commercial Aviation Alternative Fuels Initiative was used as a benchmark [50], taking recent literature into account [51–54]. Then, a positive score was considered for feedstock biomass that already reached a full-scale commercial deployment, a neutral effect for feedstock biomass in precommercial testing, and a negative one for feedstock biomass at the preliminary evaluation stage.

Soil sustainability was investigated following stakeholders' concerns regarding the protection and recovery of natural resources, especially with regard to deforestation and the degradation of land [13]. Most interviewed stakeholders showed concern about this aspect, including respondents from the government, aviation and technology companies, and research institutes [13]. Soil sustainability was studied through a review of the literature. For sugarcane, a recent and extensive review on the agronomic and environmental implications of residue removal in Brazil [34] was used as main reference for our analysis. For eucalypt, different studies in the context of Brazil [35–40,55] were consulted, as well as other studies regarding forests in other contexts [56–59]. Extensive budgets were made for biomass and nutrients present in the various components of the trees (wood, bark, branches, leaves) depending on stand age, geographic region, and tree species and cultivars [15]. All these factors were of influence on the conclusions on harvest residues. However, as for sugarcane, there were no studies that provide an integral assessment of all components of soil sustainability.

3. Discussion on sustainability performance of production alternatives

In this section, the evaluation of the considered production alternatives is discussed according to the sustainability framework presented in Table 1.

3.1 Quantitative aspects

Climate change. Aviation biofuel produced from macauba oil and residues is estimated to be the least emitting alternative, with about 90% lower GHG emissions when compared to conventional kerosene; eucalypt alternatives are second best with emission savings of 75%–90% (Fig. 2A). For eucalypt, higher GHG emissions were estimated with

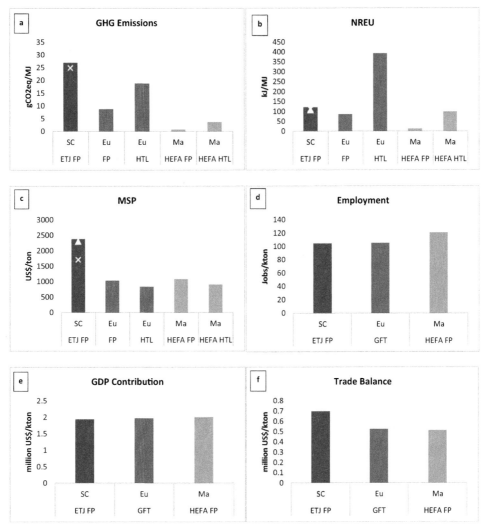

Fig. 2 Performance of potential production chains with regard to GHG emissions (A) as indicator of climate change, with GHG emissions from fossil kerosene at 87.5 g CO_2/MJ [60]; NREU (B) as indicator of efficiency, with 1200 kJ/MJ of NREU required for fossil kerosene production; MSP (C) as indicator of profitability, with conventional kerosene price in the range of 311–722 $/ton the past 3 years [61]; and employment (D), GDP contribution (E), and trade balance (F) as indicators of social development. In A, B, and C a triangle marker (▲) indicates the improvement scenarios with sweet sorghum during sugarcane off-season; a cross (×) indicates the scenario with a fraction of the sugar for succinic acid production. *ETJ*, ethanol to jet; *Eu*, eucalypt; *FP*, fast pyrolysis; *GFT*, gasification Fischer-Tropsch (see Section 2); *HEFA*, hydroprocessed esters and fatty acids; *HTL*, hydrothermal liquefaction; *Ma*, macauba; *SC*, sugarcane.

HTL than FP due to greater natural gas requirements for producing H_2 [17]. Sugarcane-based production results in about 60%–70% lower GHG emissions than fossil-based production depending on the process configuration [18]. In-house power generation and hydrogen production improve the performance on environmental indicators, while a cracking step that increases the production yield has a small impact [20]. A consequential life cycle analysis (LCA), which also takes into account indirect effects such as land use changes and product replacement, indicates that ETJ aviation biofuel from sugarcane juice has a potential for negative emissions of about -10 g CO_2/MJ when assuming the replacement of natural gas power from the grid [21]. While this number does not mean that CO_2 is captured, it indicates a potential for GHG mitigation, or fewer emissions in a context beyond aviation biofuel (i.e., considering power generation for the grid) [16,21]. However, the effects of using by-products beyond the presented production chains, such as the actual provisioning of bioenergy to the regional power system, need to be investigated in more detail.

Energy efficiency. All production chains require lower nonrenewable energy use per unit of aviation biofuel than conventional kerosene. The processing of macauba and eucalypt residues with HEFA and FP is more energy efficient than alternatives with HTL and sugarcane (Fig. 2B). The lower efficiency of HTL compared to FP is due to higher energy requirements for H_2 production [17]. The lower efficiency of sugarcane alternatives is derived from the biomass growth stage, considering that all the energy use from this stage is accounted for the sugarcane feedstock, while for eucalypt it is allocated between by-products (e.g., wood and residues) [17,18]. Regarding process options, in-house power and hydrogen production in thermochemical routes improve the process efficiency, but are economically unfavorable [20].

Profitability. Production based on eucalypt residues and macauba shows a lower minimum selling price, indicating a higher profitability potential than with sugarcane [17,18]. As expected, all alternatives perform worse than conventional kerosene (Fig. 2C). Aviation biofuel MSP from the processing of eucalypt and macauba is in the range of 850–1100 $/ton. For processing lignocellulosic residues, HTL shows a lower MSP than FP, although the difference is small when compared to sugarcane ETJ conversion (1720–2390 $/ton). The low profitability potential of sugarcane ETJ is a result of lower conversion yields and the high capital expenses related to the seasonality of sugarcane. In the improvement scenarios, sugarcane ETJ MSP can be reduced by 3%–28% by processing sweet sorghum during sugarcane off-season and by producing higher-value chemicals [18]. However, the estimated MSP for these alternatives remains higher than the MSP for eucalypt- and macauba-based production (Fig. 1C).

Social development. Macauba-based production shows 17% more employment generation than the other crops, while the difference between alternatives is less than 5% in terms of GDP contributions (Fig. 2D and E). For both employment and GDP, direct effects are largely due to feedstock production as expected, and indirect effects are primarily related to the trade sector [19]. When considering that aviation biofuel may displace part of the production of conventional kerosene, an input-out analysis reveals that overall net jobs and added value (i.e., GDP) can be generated by the transition to aviation biofuel [19]. Regarding trade balance impacts, eucalypt- and macauba-based production resulted in about 34% less imports than with sugarcane (Fig. 2F). The difference lies in the larger inputs from the chemical sector associated to the production chain based on sugarcane. Furthermore, based on the existing economic structure in Brazil, it is estimated that more imported goods, such as industrial chemicals, would be required for the production of aviation biofuel than for conventional kerosene [19]. A possibility to avoid this import increase would be to stimulate the national production of (bio-) chemicals together with the development of aviation biofuel. Lastly, these comparisons are made with available data, with macauba production still under development [54]. It can be expected that as macauba production matures, production costs will drop as has already happened with other mature crops, e.g., sugarcane [62]. This possibility needs to be further investigated as macauba-based production could result in lower direct effects on employment and GDP, and trigger different indirect effects than those presented here.

3.2 Qualitative aspects

Commercial acceptability. From the considered alternatives, only HEFA, ETJ, and GFT aviation biofuels have been approved for commercial use by the American Society for Testing and Materials (ASTM, in alignment with the Brazilian National Agency of Petroleum, Natural Gas and Biofuels) [31,46], indicating that these technologies are more commercially acceptable than the other alternatives. FP has been reported to be in queue for certification but no advances have been reported recently [32,63,64], while HTL is the farthest behind [32]. Because certification assures that a fuel has the same safety and performance, and can be distributed and used with the same infrastructure as conventional kerosene [47], the commercial acceptability of HTL biofuel is considered the lowest when compared to the other technology alternatives (Fig. 3). To get ASTM approval, HTL developers have to directly invest in certification. Certification can take 3–5 years and costs 10–15 million dollars on average [32], and for it sufficient volumes for testing are needed. Therefore, certification implies investing time and resources to scale-up the technology [32], which will constrain start-up ventures.

Energy security. Brazilian aviation biofuel production can reduce the need for kerosene imports, with about 20% of kerosene being imported in Brazil (1.3 million m^3 were imported in 2016 [65]). Hence, more significant contributions can be expected from conversion alternatives with higher efficiency, i.e., FP and HEFA (Fig. 3). Significant to the case is the potential to benefit regional power reliability through co-generation from biomass or process residues, considering stakeholders' concerns regarding energy security (i.e., related to past drought-driven power shortages) [13].

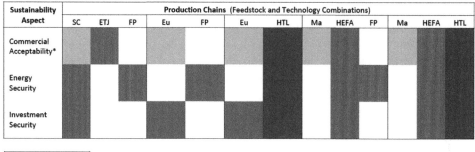

Fig. 3 Qualitative comparison of the performance of the aviation biofuel production alternatives presented per production chain. Production chains (five in total) are evaluated in terms of **commercial acceptability, energy security**, and **investment risk**, considering the combination of a feedstock and one or two technologies (3×2 or 3×3 cells, respectively). The sustainability aspects were analyzed in relative terms as described in the Section 2. *Soil sustainability is not presented as there is not enough data available for a comparison. *ETJ*, ethanol to jet; *Eu*, eucalypt; *FP*, fast pyrolysis; *HEFA*, hydroprocessed esters and fatty acids; *HTL*, hydrothermal liquefaction; *Ma*, macauba; *N/A*, not available; *SC*, sugarcane.

Energy balances suggest that process energy self-sufficiency and power surplus for the grid can be achieved through co-generation from sugarcane residues [18], which is already the case in many sugarcane mills in Brazil [66]. In the case of eucalypt and macauba, a dedicated fraction of the biomass for co-generation would be required to reach energy self-sufficiency, implying a lower aviation biofuel production per amount of processed feedstock [17]. Therefore, sugarcane alternatives are considered as having a relative positive impact when compared to the other feedstock materials (Fig. 3).

Investment security. Investment security was explored in terms of technologies and feedstock biomass. With regards to technologies, HEFA aviation biofuel is the alternative that implies less investment risk with a fuel readiness level (FRL) of up to 8, indicating that HEFA biofuels are certified and commercially available [3]. ETJ fuels recently received ASTM approval, bringing them to an FRL of 7 [31], and slightly behind some HEFA fuels. For FP, there are some ventures in the process of ASTM certification [3,48], indicating a FRL of 6. However, HTL for aviation biofuel production has only been tested at lab scale [49], and it is therefore considered to imply more investment risk at an FRL of 4. For feedstock biomass, investment security was explored in terms of supply certainty and the familiarity of farmers with the crops [13]. Sugarcane and eucalypt, despite not being originally from Brazil, are well established crops in the region, covering developed markets such as sugar, ethanol, charcoal, and wood [51,52], and implying a relatively high investment security. Macauba, although native to Brazil, has not been studied or developed at the same level. Currently, there are a few macauba demonstration plantations being started in Minas Gerais; however, there is still a need for research to develop a production chain (e.g., develop new varieties and plantation management practices) [53,54]. Therefore, investing in macauba in the short term would imply a relatively higher risk for biorefinery operators related to supply uncertainty, as well as for farmers who have neither experience nor access to management practices for production.

Soil sustainability. Soil sustainability was reviewed regarding the effect of residue harvest for biobased production, although there is limited information about macauba (most is known about sugarcane followed by eucalypt). In a recent review where effects of yield, nutrient recycling, soil carbon stocks, GHG emissions, and soil erosion were considered, it was concluded that leaving 7 ton/ha of sugarcane straw is recommendable in order to sustain soil properties [34]. Usually, sugarcane straw yields can vary, as much as 8–30 ton/ha, so it is not simply a matter of leaving half the straw in the field [34]. Remaining straw also comes at a cost, as it may increase certain pests and weeds, and nutrient addition is required as only some 31% of N and 23% of P in the straw will be released for use by plants [34]. Eucalypt, as macauba and other trees, consist of stems, bark, branches, and leaves. The eucalypt wood is 77% of the total tree biomass, but it contains 39% of the nutrients when considering wood and harvest residues together [56]. When stands age, the proportion of wood to total biomass increases, and more nutrients get removed when harvesting, although there are differences among species [35,36] and selection lines within species [37]. Also, the type of residue management in a replanted eucalypt plantation has effects on productivity. For example, 8 years after planting, biomass production was 88% when harvest residues were removed compared to when harvest residues were retained [38], and even decreased to 63% when also the litter was removed [38,39]. Therefore, residue management in tree plantations, such as eucalypt and macauba, appears to be crucial for sustainability. Keeping harvest residues on the fields will be an effective way to maintain soil organic matter levels for all crops. However, in contrast to sugarcane, little information is available on amounts of residues that need to be left behind for eucalypt, and effects depend on the age of the stands when harvested. Recommendations for forests with long rotation cycles range from 20% to 50% of residues and are merely based on expert judgment [57–59]. Therefore, in all cases of biomass production, soil sustainability will depend on leaving behind harvest residues, and further integral studies need to establish rotation lengths and other management practices in order to enhance sustainability impacts.

4. Sustainability tensions and opportunities

Tensions emerge with regard to different sustainability aspects. Prominently, all options yield much lower emissions than fossil-based kerosene but all are more costly (over $300/

ton more than the average kerosene price of the past 3 years [61]). Analyzing the other sustainability aspects reveals other tensions as well. In this section, these tensions and some opportunities for further developments on aviation biofuels in the region are discussed. We discuss tensions related to the technical alternatives for production, to the implementation of production itself, and to the ex-ante analysis of sustainability (Fig. 4).

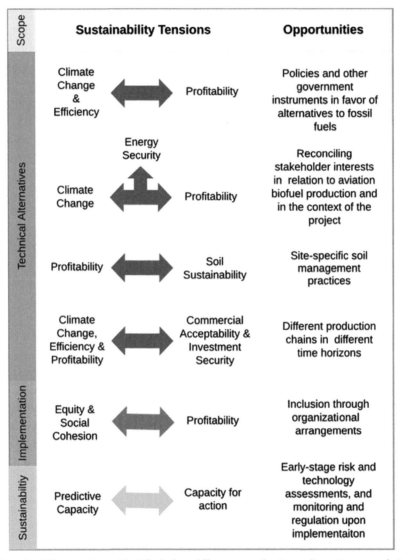

Fig. 4 Sustainability tensions identified for different production alternatives, and identified opportunities in the context of Brazilian aviation biofuel production. Sustainability aspects on opposite sides of arrows are in tension in the context of aviation biofuel production. The *colored column* in the left indicates the scope in which the tension emerges: technical aspects in *blue*, production implementation in *magenta*, and sustainability analysis in *yellow*.

Technology alternatives: Looking at policy contexts. All studied options lead to lower emissions and less NREU than conventional kerosene, however at higher expenses. When looking at technology alternatives to process lignocellulosic residues and produce in-house power and hydrogen, the most favorable alternative in economic terms (HTL) is the least favorable with respect to climate change and energy efficiency. An opportunity for resolving this tension is to explore alternative approaches for the generation of hydrogen. Steam methane reforming considered in the present study is the most common and economic option but it is one of the main contributors of NREU and emissions in the case of HTL [17]. Interesting alternatives that can be further explored are, for example, the thermochemical conversion of a fraction of the biomass for producing H_2, or even the electrolysis and photolysis of water run on renewable energy [67]. Alternatives for hydrogen production have received substantial attention at the policy level worldwide and in Brazil specifically, with an upcoming National Hydrogen Plan and pilot projects for renewable hydrogen being developed [68].

Furthermore, the presented profitability estimations did not account for GHG emission costs, which have become more relevant since the 2015 Paris Agreement [69]. Prominently, Brazil recently passed the National Biofuel Policy (RenovaBio) to promote the reduction of GHG emissions by the country's fuel sector [70]. As part of this policy, a market for certificates representing GHG emissions savings (relative to fossil fuel emissions) is being launched. Certificates are to be issued by biofuel producers and bought by distributors who have to meet decarbonization targets [71]. As result, GHG emission savings will yield a profit for biofuel producers. Mechanisms like this can therefore open opportunities for aviation biofuels by making them financially more competitive [14,18], especially those biofuels that yield lower emissions (i.e., from macauba with HEFA and FP, and eucalypt with FP).

Technology alternatives: Reconciling stakeholders' interests. A tension emerges between different product alternatives as each favor the interests of different stakeholders: Higher-value products like succinic acid can be produced from a dedicated part of the feedstock stream, resulting in more profitability for investors. However, this option comes at the cost of aviation biofuel production capacity per amount of processed feedstock, requiring more feedstock to meet the emission reduction targets of the aviation sector. Alternatively, power generation can be favored over higher-value products or aviation biofuel by dedicating a fraction or all of lignocellulosic residues for co-generation. Bioenergy can thus be part of distributed power generation in the region for the sake of energy security, as it is in the interest of the regional government. These interests represent sustainability aspects favored by different stakeholders depending on the values and beliefs of the group they represent [72]. Therefore, a sustainability analysis on its own cannot indicate which alternative is the best or the worst. Instead, a sustainability analysis that explicitly identifies sustainability tensions, as presented in this work, can contribute to a negotiation process with all stakeholders to define acceptable conditions (e.g., a minimum contribution to the regional power supply per production plant), or even a common objective for developing a production chain. Such openness and inclusion of stakeholders, with, e.g., social learning and responsible innovation tools, could reduce the ambiguity associated to diverging values of stakeholders [72], and strengthen the stakeholder network for the development of more sustainable and responsible biobased production [73,74].

Technology alternatives: Site-specific soil management practices. A clear tension exists between soil sustainability and harvesting as much biomass as possible for increasing productivity, and thus profitability [15,34]. Defining an optimal amount of residues to leave on the field, as well as other improved practices regarding rotation length, can contribute to solve this tension while also accounting negative consequences of leaving harvest residues in the field (i.e., pest and weed management) [34]. Also, fertilization is needed as in all cases nutrients are removed when harvesting, and not all the nutrients from leftover residues become available to the next crop [34]. However, nitrogen fertilizer is costly in terms of GHG emissions and energy efficiency [75]. Therefore, planning of biomass crop plantations for biofuels requires site-specific recommendations accounting for, e.g., soil type, land surface steepness, climate, length of the rotation, and how these factors influence residue retention and its effect on soil quality and soil functioning, as well as on pest and weed management.

Technology alternatives: Explicit time horizons. Aviation biofuel production based on macauba and eucalypt residues results in more potential benefits in terms of climate change, profitability, and social development. However, they imply a lower investment security than other alternatives. Macauba implies a high investment risk in the short term as production is still under development, and harvest only starts after more than 6 years from planting [76]. Eucalypt, although widely available in the region, implies processing technologies (i.e., FP and HTL) that are still under development, resulting in a lower commercial acceptability and higher investment risks than sugarcane processing technologies.

An opportunity to deal with the tension between climate change, profitability and social development on one hand, and commercial acceptability and investment risk on the other, is to consider the time horizon of projects. Also, it has to be bared in mind that a single crop-and-technology combination does not need to supply all aviation biofuel demand in the region at once. In this way, production based on macauba, with HEFA for processing oil and FP or HTL for residues, could be considered as an alternative in the long term. Sugarcane ETJ and eucalypt FP aviation biofuels could be considered for meeting emission reduction targets in a shorter term. In the case of aviation biofuel from sugarcane juice, the total capital investments could be lower if ethanol mills are already in place, requiring extra capital expenses for ETJ only. This would make sugarcane an easier option. Additionally, the improvement scenarios presented for sugarcane (i.e., production of higher-value products and second crop during off-season) and optimized plantation management options (related to, e.g., nutrient recycling and carbon storage in the soil) could be explored in more detail to improve the system performance on climate change, profitability, and soil sustainability. Nevertheless, stimulating the development of aviation biofuel production implies encouraging producers to switch from their usual crop or product. For example, in the case of sugarcane aviation biofuel, introducing feed-in tariffs in combination with a gasoline tax can encourage its large-scale production and use [77].

Implementation: Organizational arrangements. Although impacts on equality and social cohesion were not evaluated with regard to the different alternatives (see Section 2), a tension between these aspects and profitability was identified during our analysis. In the emergence of production chains for commodity products, like aviation biofuel, economies of scale tend to favor land concentration and vertical integration models (i.e., where the production plant owner also (co-)owns other stages of the production chain, like biomass production) [78]. These production models are in tension with equity and social cohesion aspects since they could lead to the exclusion of smallholders (e.g., family farmers, small-scale local companies) from the production chain [79–81]. An opportunity however, are the business models of nontraditional mill owners, or new entrants, who base their production on arrangements with feedstock producers, as reported for sugarcane expansion areas like Goiás [82]. While new entrants favor these partnership models due to the lower capital requirements for production (i.e., no need to acquire land) [81], these models also open the possibility for the inclusion of smallholder farmers, reconciling aspects of equality and social cohesion with entrepreneurship concerns. To encourage partnership models, there is a need to support organizational arrangements among producers (e.g., cooperatives and farmers associations), and the development of contracts that give revenue certainty to farmers and feedstock security to biorefinery operators [82,83]. While more research is needed in this end, partnership models with such organizational arrangements could result in benefits for rural smallholders with respect to income and stability opportunities, and support the preservation of local knowledge and culture. These outcomes would be an important advantage of aviation biofuels when compared with fossil kerosene.

Sustainability analysis: Knowledge and capacity for action. There is an intrinsic tension when analyzing the sustainability impacts of a technology: In early stages of development, there is more space for improving an innovation (e.g., a technology or a crop) but little information is available; at later stages of development, there is more information about its impacts but it is more difficult to change it. Therefore, ex-ante analyses as presented here imply inherent uncertainties related to limited data and knowledge about the performance and consequences of production. For example, in this study there are uncertainties related to conversion yields and GHG emissions at large scale, indirect land use changes, and long-term consequences for the sustainability of soils. This quandary is an instance of the famous Collingridge dilemma, which states that at early development stages of a technology there is limited knowledge about its impacts, but later when it is implemented there is limited capacity to change it [84].

A straight forward solution to this dilemma is increasing the predictive capacity of ex-ante analyses, for example by incorporating risk analyses to support decision-making, as done for nanomaterials [85,86]. In the case of aviation biofuels, there are already a few studies looking at the uncertainties associated to aviation biofuels production, mostly focused on economic and technological uncertainties [14,87]. This type of analyses could be further extended to cover other relevant aspects of a specific biofuel production chain. A way to deal with knowledge gaps and unexpected events is to develop monitoring schemes along the development and implementation of the technology, leaving the possibility to change its direction [72,88]. Overall, combining strategies for increasing knowledge and capacity for action is a way to deal with the limitations of ex-ante sustainability analyses.

5. Conclusions

We presented a novel ex-ante analysis of the sustainability of aviation biofuel that includes a discussion of sustainability tensions and opportunities for its production in Southeast Brazil. Our analysis shows that macauba-based production with HEFA, followed by thermochemical conversion of lignocellulosic residues, performs better than sugarcane alternatives in terms of climate change, efficiency, profitability, and social development. However, choosing the macauba-based alternative over others implies facing a relatively low commercial acceptability and high investment risks. Therefore, we conclude that sugarcane ATJ aviation biofuel is the most opportune feedstock for the production of aviation biofuel in the short term, while eucalypt processing with FP and macauba processing with HEFA and HTL seem as better alternatives in the longer term. To improve the profitability of sugarcane, the production of higher-value products and the processing of a second crop in order to complement off-season production dips will be beneficial. These improvements could be combined with plantation management practices (e.g., optimized nutrient recycling) to ameliorate sugarcane production effects on soil sustainability and GHG emissions, which is applicable to all feedstock biomass. Additionally, to improve the efficiency and climate change performance of thermochemical alternatives, hydrogen generation options based on renewable energy should be explored. As different by-product alternatives can be in the interests of different stakeholders (e.g., improving the economic performance of the production chain or contributing to the energy security of the region), the decision over by-products should be open to participation of relevant stakeholders. With regard to the implementation of production, it was found that producer-operator partnerships can open opportunities for the inclusion of smallholders in the region. Promoting these partnerships and strengthening the role of smallholders through, e.g., organizational arrangements, can serve to bring equality and social cohesion into the development of the production chain. Lastly, we conclude that emerging fuel and carbon policies may provide opportunities for the development of biofuel production.

The presented approach allowed integrating considerations of the local context and stakeholders for an ex-ante sustainability analysis. Engagements with stakeholders allowed to identify relevant sustainability aspects for the case study, and to specify them with regard to the local context. While it was not possible to evaluate all identified sustainability aspects, the recognition of these issues allowed to understand sustainability tensions related to the considered production alternatives, and to identify opportunities for further developments. This understanding will provide a first step toward reducing the ambiguity associated to diverging values of stakeholders, and support the strengthening of a stakeholder network for the development of more sustainable biobased production. For achieving this, social learning and responsible innovation tools can be useful. Overall, the presented approach may be also applicable to other regions and other production chains in support of a more sustainable transition away from fossil resources.

Annex 1 Economic potential (US$ kg^{-1} feedstock) of various production chain alternatives, depending on feedstock type and by-product based on the results in Ref [89].

HVC	Macauba Min	Macauba Max	Jatropha Min	Jatropha Max	Camelina Min	Camelina Max	Soybean Min	Soybean Max	Sugarcane Min	Sugarcane Max	Sweet sorghum Min	Sweet sorghum Max
SA	0.19	0.29	−0.04	0.03	0.15	0.32	0.11	0.28	0.23	0.39	0.21	0.38
ET	0.06	0.15	−0.12	−0.06	0.03	0.20	0.00	0.16	0.06	0.24	0.06	0.23
EtOH	0.00	0.09	−0.16	−0.10	−0.03	0.15	−0.05	0.11	0.00	0.17	0.00	0.16
LA	0.13	0.22	−0.08	−0.02	0.09	0.26	0.06	0.22	0.14	0.32	0.14	0.30
1-BUT	0.02	0.12	−0.14	−0.08	0.00	0.17	−0.03	0.14	0.02	0.20	0.02	0.19
IsoPRO	0.00	0.09	−0.16	−0.10	0.00	0.17	−0.06	0.11	0.00	0.17	0.00	0.16
3-HPA	0.05	0.14	−0.13	−0.07	0.02	0.19	−0.01	0.15	0.05	0.23	0.05	0.22
2,5-FDCA	0.04	0.13	−0.14	−0.07	0.01	0.18	−0.02	0.15	0.04	0.21	0.04	0.20
1,3-PDO	0.06	0.15	−0.12	−0.06	0.03	0.20	0.00	0.17	0.07	0.24	0.06	0.23
1,4-BDO	0.05	0.14	−0.13	−0.06	0.02	0.19	−0.01	0.16	0.05	0.23	0.05	0.22

HVC	Sugarcane residues Min	Sugarcane residues Max	Sweet sorghum residues Min	Sweet sorghum residues Max	Eucalyptus residues Min	Eucalyptus residues Max	Pine residues Min	Pine residues Max	Coffee residues Min	Coffee residues Max	Rice residues Min	Rice residues Max
SA	0.21	0.37	0.18	0.32	0.21	0.34	0.18	0.31	0.19	0.33	0.19	0.33
ET	0.08	0.23	0.04	0.19	0.08	0.22	0.06	0.19	0.06	0.20	0.06	0.20
EtOH	0.02	0.17	−0.01	0.13	0.03	0.16	0.01	0.14	0.00	0.14	0.01	0.15
LA	0.15	0.30	0.11	0.25	0.15	0.28	0.12	0.25	0.13	0.27	0.13	0.27
I-BUT	0.04	0.19	0.01	0.15	0.05	0.19	0.03	0.16	0.02	0.16	0.03	0.17
IsoPRO	0.01	0.17	−0.01	0.13	0.03	0.16	0.00	0.13	0.00	0.14	0.01	0.14
3-HPA	0.06	0.22	0.03	0.18	0.07	0.21	0.05	0.18	0.05	0.19	0.05	0.19
2,5-FDCA	0.05	0.21	0.02	0.17	0.07	0.20	0.04	0.17	0.04	0.18	0.04	0.18
1,3-PDO	0.08	0.23	0.05	0.19	0.09	0.22	0.06	0.19	0.06	0.20	0.07	0.20
1,4-BDO	0.07	0.30	0.04	0.25	0.08	0.28	0.05	0.25	0.05	0.26	0.06	0.26

Ranges express the minimum and maximum economic potential obtained considering the conversion yields with different technology alternatives.

1-BUT, 1-butanol; *1,3-PDO*, 1,3 propanediol; *1,4 BDO*, 1,4-butanediol; *2,5-FDCA*, ET: ethylene; *3-HPA*, 3-hydroxypropionic acid; *HVC*, High-value chemical; *EtOH*, ethanol; *IsoPRO*, Isopropanol; *LA*, lactic acid; *SA*, succinic acid.

Annex 2 List of prices used for the techno-economic estimations, adapted from Ref. [90], with prices updated to 2015, in US$ ton^{-1}, and based on the Brazil market, considering crude oil barrel price 64 US$ bbl^{-1}.

Compound	Price (US$ ton^{-1})	Specifications[a]	References
Sugarcane	22.3		[91]
Transportation of sugarcane	6.2	10 km with 40 ton truck, bundles density—400 kg m^{-3}	[92]
Sugarcane trash	16.9		
Transportation of sugarcane trash	9.8	10 km with 40 ton truck, bundles density—175 kg m^{-3}	[92]
Sweet sorghum	27.0		[89]
Transportation of sweet sorghum	10.4	22 km 40 ton truck, bundles density—350 kg m^{-3}	[92]
Sweet sorghum grains	78.4		[93]
LPG	234.8	Prices of May 2015	[94]
Naphtha	598.1		
Jet fuel	605.2		
Transportation of jet fuel—Sao Paulo	14.8	150 km with train	[92]
Transportation of jet fuel—Rio de Janeiro	26.6	570 km with train	[92]
Diesel		Price of May 2015	[94]
Acetic acid	672.6		[95]
Furfural	957.5		[96]
S sulfur	151.1		[97]
Lignin[b]	400		Estimated [90]
Sugarcane juice[c]	631.8		Estimated [90]
Transportation of juice (65°Brix)	6.5	20 km with 35 ton tank-truck	[92]
Enzyme for biomass hydrolysis	156.6	Price per ton of ethanol	[98]
Cooling water	0.1		[99]
Chilled water	0.5		[100]
Natural gas	104.7	LHV of CH$_4$ considered 40.7 MJ kg^{-1}	[101]
Process water	0.25		Estimated [90]
Solids disposal in landfill	0.84		[102]
Operators salary	10.9	US$ h^{-1}	[103]

Catalysts	Price (US $ ton^{-1})	WSHV (h^{-1}) w/w	Lifetime (years)	Type	References
Ethanol dehydration	411,905	5 [104]	3	[105]	[106]
Ethylene condensation and oligomerization	252,934	2	5	[105]	[106]

Annex 2 List of prices used for the techno-economic estimations, adapted from Ref. , with prices updated to 2015, in US$ ton^{-1}, and based on the Brazil market, considering crude oil barrel price 64 US$ bbl^{-1}—cont'd

Catalysts	Price (US $ ton^{-1})	WSHV (h^{-1}) w/w	Lifetime (years)	Type	References
Olefins hydrogenation	245,723	3	5	[105]	[106]
Farnesene hydrocracking	39,354	2	5 [107]	[108]	[107]
PSA packing	5079	0.685	3	[102]	[102]
Hydrotreating catalyst	39,354	1st—1.5; 2nd—0.5	2	[109]	[110]
H$_2$ SMR	38,084	1.4	3	[102]	[102]
Water gas shift	20,315	0.07	3	[111]	[102]
Fischer–Tropsch	15,760	2.22	3	[111]	[111]

[a]Distances are estimated once location of plant is established in Campinas, Sao Paulo. Feedstock transportation distance is estimated with land productivity and average feedstock annual capacity of all scenarios. Transportation method and cost methodology follows from Ref. [92].
[b]Maximum selling price of lignin, considering that it will be sold for a polyurethane manufacturer with a project payback time of 10 years, IRR at 12% and polyurethanes sold at market price.
[c]Maximum selling price of juice, considering that it will be sold to a succinic acid (SA) manufacturer with an annual capacity of 42.3 kton SA yr^{-1}, with a project payback time of 10 years, IRR at 12% and succinic acid sold at 2356 US$ ton^{-1}. Process yields, OPEX and CAPEX methodology follow from *Efe et al.* [112].

Acknowledgments

This study was carried out within the BE-Basic R&D Program, which was granted a FES subsidy from the Dutch Ministry of Economic Affairs. We would like to thanks all the partners of the Horizontal International Project of the BE Basic Foundation, in particular Pella Brinkman and Zhizhen Wang for the useful discussions during the preparation of this article.

Competing interests statement

The authors declare no competing interests.

References

[1] S. Pfau, J. Hagens, B. Dankbaar, A. Smits, Visions of sustainability in bioeconomy research, Sustainability 6 (2014) 1222–1249.
[2] I. Tsiropoulos, et al., Emerging bioeconomy sectors in energy systems modeling—integrated systems analysis of electricity, heat, road transport, aviation, and chemicals: a case study for the Netherlands, Biofuels Bioprod. Biorefin. 12 (2018) 665–693.
[3] R. Mawhood, E. Gazis, S. de Jong, R. Hoefnagels, R. Slade, Production pathways for renewable jet fuel: a review of commercialization status and future prospects, Biofuels Bioprod. Biorefin. 10 (2016) 462–484.
[4] ICAO United Nations, Carbon Offsetting and Reduction Scheme for International Aviation (CORSIA), 2019. https://www.icao.int/environmental-protection/CORSIA/Pages/default.aspx.

[5] M.W. Rosegrant, S. Msangi, Consensus and contention in the food-versus-fuel debate, Annu. Rev. Environ. Resour. 39 (2014) 271–294.

[6] S. Bouzarovski, M.J. Pasqualetti, V.C. Broto, The Routledge Research Companion to Energy Geographies, Taylor & Francis, 2017.

[7] B. Aha, J.Z. Ayitey, Biofuels and the hazards of land grabbing: tenure (in)security and indigenous farmers' investment decisions in Ghana, Land Use Policy 60 (2017) 48–59.

[8] R.A. Efroymson, et al., Environmental indicators of biofuel sustainability: what about context? Environ. Manag. 51 (2013) 291–306.

[9] R.S. Capaz, J.E.A. Seabra, Life cycle assessment of biojet fuels, in: C.J. Chuck (Ed.), Biofuels for Aviation, Academic Press, 2016, pp. 279–294, https://doi.org/10.1016/B978-0-12-804568-8.00012-3 (Chapter 12).

[10] M. Palmeros Parada, P. Osseweijer, J.A. Posada Duque, Sustainable biorefineries, an analysis of practices for incorporating sustainability in biorefinery design, Ind. Crop. Prod. 106 (2017) 105–123.

[11] O. Cavalett, F. Cherubini, Contribution of jet fuel from forest residues to multiple Sustainable Development Goals, Nat. Sustain. 1 (2018) 799–807.

[12] F. Pashaei Kamali, J.A.R. Borges, P. Osseweijer, J.A. Posada, Towards social sustainability: screening potential social and governance issues for biojet fuel supply chains in Brazil, Renew. Sustain. Energy Rev. 92 (2018) 50–61.

[13] M. Palmeros Parada, L. Asveld, P. Osseweijer, J.A. Posada, Setting the design space of biorefineries through sustainability values, a practical approach, Biofuels Bioprod. Biorefin. 12 (2018) 29–44.

[14] C.M. Alves, et al., Techno-economic assessment of biorefinery technologies for aviation biofuels supply chains in Brazil, Biofuels Bioprod. Biorefin. 11 (2017) 67–91.

[15] P. Brinkman, R. Postma, W. van der Putten, A. Termorshuizen, Influence of Growing Eucalyptus Trees for Biomass on Soil Quality, 2017. https://pure.knaw.nl/portal/files/5893927/HIP_Eucalyptus_final.pdf.

[16] R.S. Capaz, J.E.A. Seabra, P. Osseweijer, J.A. Posada, Life cycle assessment of renewable jet fuel from ethanol: an analysis from consequential and attributional approaches, in: Papers of the 26th European Biomass Conference: Setting the course for a biobased economy, ETA-Florence Renewable Energies, 2018.

[17] C. Cornelio da Silva, L.A.M. van der Wielen, J.A. Posada, S.I. Mussatto, Techno-Economic and Environmental Analysis of Oil Crop and Forestry Residues based Biorefineries for Biojet Fuel production in Brazil, Delft University of Technology, 2016. http://resolver.tudelft.nl/uuid:1dd8082f-f4a5-4df6-88bb-e297ed483b54. (Accessed March 2022).

[18] C.I. Santos, et al., Integrated 1st and 2nd generation sugarcane bio-refinery for jet fuel production in Brazil: techno-economic and greenhouse gas emissions assessment, Renew. Energy 129 (2018) 733–747.

[19] Z. Wang, F. Pashaei Kamali, P. Osseweijer, J.A. Posada, Socioeconomic effects of aviation biofuel production in Brazil: a scenarios-based Input-Output analysis, J. Clean. Prod. 230 (2019) 1036–1050.

[20] E. Vyhmeister, G.J. Ruiz-Mercado, A.I. Torres, J.A. Posada, Optimization of multi-pathway production chains and multi-criteria decision-making through sustainability evaluation: a biojet fuel production case study, Clean Technol. Environ. Policy 20 (2018) 1697–1719.

[21] R.S. Capaz, J.A. Posada, P. Osseweijer, J.E.A. Seabra, The carbon footprint of alternative jet fuels produced in Brazil: exploring different approaches, Resour. Conserv. Recycl. 166 (2021) 105260.

[22] M. Palmeros Parada, W.H. van der Putten, L.A.M. van der Wielen, P. Osseweijer, M. van Loosdrecht, F. Pashaei Kamali, J.A. Posada, OSiD: opening the conceptual design of biobased processes to a context-sensitive sustainability analysis, Biofuels Bioprod. Biorefin. 14 (2021) 961–972, https://doi.org/10.1002/bbb.2216.

[23] D. Kumar, G.S. Murthy, Impact of pretreatment and downstream processing technologies on economics and energy in cellulosic ethanol production, Biotechnol. Biofuels 4 (2011) 27.

[24] D. Humbird, et al., Process Design and Economics for Biochemical Conversion of Lignocellulosic Biomass to Ethanol: Dilute-Acid Pretreatment and Enzymatic Hydrolysis of Corn Stover, 2011, https://doi.org/10.2172/1013269. https://www.osti.gov/biblio/1013269.

[25] M.O.S. Dias, et al., Simulation of integrated first and second generation bioethanol production from sugarcane: comparison between different biomass pretreatment methods, J. Ind. Microbiol. Biotechnol. 38 (2011) 955–966.

[26] J. Kautto, M.J. Realff, A.J. Ragauskas, Design and simulation of an organosolv process for bioethanol production, Biomass Conv. Bioref. 3 (2013) 199–212.

[27] C.N. Hamelinck, G. van Hooijdonk, A.P. Faaij, Ethanol from lignocellulosic biomass: techno-economic performance in short-, middle- and long-term, Biomass Bioenergy 28 (2005) 384–410.

[28] W.D. Seider, J.D. Seader, D.R. Lewin, S. Widagdo, Product and Process Design Principles: Synthesis, Analysis and Design, John Wiley & Sons, 2008.

[29] Global Reporting Initiative, G4 Sustainability Reporting Guidelines, Global Reporting Initiative, 2015.

[30] Food and Agriculture Organization of the United Nations, SAFA Guidelines: Sustainability Assessment of Food and Agriculture Systems, FAO, 2014.

[31] ASTM International, ASTM D7566-19 Standard Specification for Aviation Turbine Fuel Containing Synthesized Hydrocarbons, ASTM International, 2019, https://doi.org/10.1520/D7566-19.

[32] US DOE, Alternative Aviation Fuels: Overview of Challenges, Opportunities, and Next Steps, 2017, https://doi.org/10.2172/1358063. http://www.osti.gov/servlets/purl/1358063/.

[33] R. Altman, Sustainable aviation alternative fuels: from afterthought to cutting edge, in: O. Inderwildi, S.D. King (Eds.), Energy, Transport, & the Environment, Springer, London, 2012, pp. 401–434.

[34] J.L.N. Carvalho, et al., Agronomic and environmental implications of sugarcane straw removal: a major review, GCB Bioenergy 9 (2017) 1181–1195.

[35] R.B. Harrison, G.G. Reis, M.D.G.F. Reis, A.L. Bernardo, D.J. Firme, Effect of spacing and age on nitrogen and phosphorus distribution in biomass of *Eucalyptus camaldulensis*, *Eucalyptus pellita* and *Eucalyptus urophylla* plantations in southeastern Brazil, For. Ecol. Manag. 133 (2000) 167–177.

[36] F.C. Zaia, A.C. Gama-Rodrigues, Nutrient cycling and balance in eucalypt plantation systems in north of Rio de Janeiro state, Brazil, Rev. Bras. Ciênc. Solo 28 (2004) 843–852.

[37] C.C. Rosim, et al., Nutrient use efficiency in interspecific hybrids of eucalypt, Rev. Ciênc. Agron. 47 (2016) 540–547.

[38] J.H.T. Rocha, et al., Forest residue maintenance increased the wood productivity of a Eucalyptus plantation over two short rotations, For. Ecol. Manag. 379 (2016) 1–10.

[39] J.L.M. Gonçalves, et al., Soil fertility and growth of *Eucalyptus grandis* in Brazil under different residue management practices, South. Hemisphere For. J. 69 (2007) 95–102.

[40] R.C. Fialho, Y.L. Zinn, Changes in soil organic carbon under eucalyptus plantations in Brazil: a comparative analysis, Land Degrad. Dev. 25 (2014) 428–437.

[41] J. Woods, et al., Land and bioenergy, in: G.M. Souza, L.V. Reynaldo, C.A. Joly, L.M. Verdade (Eds.), Bioenergy & Sustainability: Bridging the Gaps, Scientific Committee on Problems of the Environment, 2015, pp. 258–300.

[42] K.L. Kline, et al., Reconciling food security and bioenergy: priorities for action, GCB Bioenergy (2016), https://doi.org/10.1111/gcbb.12366.

[43] H. Jeswani, Carbon footprint of biofuels, in: S. Massari, G. Sonnemann, F. Balkau (Eds.), Life Cycle Approaches to Sustainable Regional Development, Routledge, 2017.

[44] Brazilian Institute of Geography and Statistics, Portal Do IBGE, 2017. https://ww2.ibge.gov.br/home/default.php.

[45] R. Diaz-Chavez, et al., Social considerations, in: L.M. Sibanda, M. Mapako (Eds.), Bioenergy & Sustainability: Bridging the Gaps, Scientific Committee on Problems of the Environment, 2015, pp. 528–553.

[46] Agencia Nacional de Petroleo, Gas Natural and Biofuels, Biocombustíveis de Aviação, 2016. http://www.anp.gov.br/biocombustiveis/biocombustiveis-de-aviacao.

[47] L. Cortez, et al., Perspectives for sustainable aviation biofuels in Brazil, Int. J. Aerosp. Eng. 2015 (2015).

[48] K. Borislava, Current status of alternative aviation fuels, in: DLA Energy Worldwide Energy Conference, US Defense Logistics Agency, 2017.

[49] P. Biller, A. Roth, Hydrothermal liquefaction: a promising pathway towards renewable jet fuel, in: M. Kaltschmitt, U. Neuling (Eds.), Biokerosene: Status and Prospects, Springer, Berlin Heidelberg, 2018, pp. 607–635, https://doi.org/10.1007/978-3-662-53065-8_23.

[50] J.I. Hileman, et al., Near-Term Feasibility of Alternative Jet Fuels, 2009. https://stuff.mit.edu/afs/athena/dept/aeroastro/partner/reports/proj17/altfuelfeasrpt.pdf.

[51] A.C. Sant'Anna, A. Shanoyan, J.S. Bergtold, M.M. Caldas, G. Granco, Ethanol and sugarcane expansion in Brazil: what is fueling the ethanol industry? Int. Food Agribus. Manag. Rev. 19 (2016) 163–182.

[52] D.E. McMahon, R.B. Jackson, Management intensification maintains wood production over multiple harvests in tropical Eucalyptus plantations, Ecol. Appl. 29 (2019).

[53] C.A. Colombo, L.H.C. Berton, B.G. Diaz, R.A. Ferrari, Macauba: a promising tropical palm for the production of vegetable oil, OCL 25 (2018) D108.

[54] A. Cardoso, et al., Opportunities and challenges for sustainable production of *A. aculeata* through agroforestry systems, Ind. Crop. Prod. 107 (2017) 573–580.

[55] R.L. Cook, D. Binkley, J.L. Stape, Eucalyptus plantation effects on soil carbon after 20 years and three rotations in Brazil, For. Ecol. Manag. 359 (2016) 92–98.

[56] J. Hernández, A. del Pino, M. Hitta, M. Lorenzo, Management of forest harvest residues affects soil nutrient availability during reforestation of *Eucalyptus grandis*, Nutr. Cycl. Agroecosyst. 105 (2016) 141–155.

[57] B.D. Titus, D.G. Maynard, C.C. Dymond, G. Stinson, W.A. Kurz, Wood energy: protect local ecosystems, Science 324 (2009) 1389–1390.

[58] P. Lamers, E. Thiffault, D. Paré, M. Junginger, Feedstock specific environmental risk levels related to biomass extraction for energy from boreal and temperate forests, Biomass Bioenergy 55 (2013) 212–226.

[59] J. de Jong, C. Akselsson, G. Egnell, S. Löfgren, B.A. Olsson, Realizing the energy potential of forest biomass in Sweden—how much is environmentally sustainable? For. Ecol. Manag. 383 (2017) 3–16.

[60] S. de Jong, et al., Life-cycle analysis of greenhouse gas emissions from renewable jet fuel production, Biotechnol. Biofuels 10 (2017) 64.

[61] IndexMundi, Jet Fuel—Daily Price—Commodity Prices, 2019. https://www.indexmundi.com/commodities/?commodity=jet-fuel&months=60.

[62] J.D. van den Wall Bake, M. Junginger, A. Faaij, T. Poot, A. Walter, Explaining the experience curve: cost reductions of Brazilian ethanol from sugarcane, Biomass Bioenergy 33 (2009) 644–658.

[63] CAFI, Fuel Qualification, http://www.caafi.org/focus_areas/fuel_qualification.html#.

[64] ICAO United Nations, Conversion Processes, https://www.icao.int/environmental-protection/GFAAF/Pages/Conversion-processes.aspx.

[65] Agencia Nacional de Petroleo, Gas Natural and Biofuels, Fuel Production and Supply Opportunities in Brazil, 2017. http://www.anp.gov.br/images/publicacoes/Fuel_Production_and_Supply_Opportunities_in_Brazil.pdf.

[66] L.A.H. Nogueira, R.S. Capaz, Ethanol from sugarcane in Brazil: economic perspectives A2, in: A. Pandey, R. Höfer, M. Taherzadeh, K.M. Nampoothiri, C. Larroche (Eds.), Industrial Biorefineries & White Biotechnology, Elsevier, 2015, pp. 237–246 (Chapter 4B).

[67] P. Nikolaidis, A. Poullikkas, A comparative overview of hydrogen production processes, Renew. Sustain. Energy Rev. 67 (2017) 597–611.

[68] Deutsche Gesellschaft für Internationale Zusammenarbeit, Mapeamento Do Setor de Hidrogênio Brasileiro, Deutsche Gesellschaft für Internationale Zusammenarbeit GmbH, Bonn and Eschborn, 2021. https://www.energypartnership.com.br/fileadmin/user_upload/brazil/media_elements/Mapeamento_H2_-_Diagramado_-_V2h.pdf.

[69] European Commission. Paris Agreement, Climate Action—European Commission, 2016. https://ec.europa.eu/clima/policies/international/negotiations/paris_en.

[70] Agencia Nacional de Petroleo, Gas Natural and Biofuels, RenovaBio, 2018. http://www.anp.gov.br/producao-de-biocombustiveis/renovabio.

[71] Ministerio de Minas e Energia, RenovaBio, 2018. http://www.mme.gov.br/documents/10584/55980549/RenovaBio.pdf/e89e1dc0-69a3-4f8c-907e-382b1235dd67; jsessionid=F14B86C1B9F4B9030F3F1B2001956F92.srv155.

[72] L. Asveld, D. Stemerding, Social learning in the bioeconomy, the ecover case, in: I. Van de Poel, L. Asveld, D.C. Mehos (Eds.), New Perspectives on Technology in Society: Experimentation Beyond the Laboratory, Routledge, 2018.

[73] J. Mossberg, P. Söderholm, H. Hellsmark, S. Nordqvist, Crossing the biorefinery valley of death? Actor roles and networks in overcoming barriers to a sustainability transition, Environ. Innov. Soc. Trans. 27 (2018) 83–101.

[74] H. Hellsmark, J. Mossberg, P. Söderholm, J. Frishammar, Innovation system strengths and weaknesses in progressing sustainable technology: the case of Swedish biorefinery development, J. Clean. Prod. 131 (2016) 702–715.

[75] J. Han, A. Elgowainy, H. Cai, M.Q. Wang, Life-cycle analysis of bio-based aviation fuels, Bioresour. Technol. 150 (2013) 447–456.

[76] A.S. da César, F.A. de Almeida, R.P. de Souza, G.C. Silva, A.E. Atabani, The prospects of using *Acrocomia aculeata* (macaúba) a non-edible biodiesel feedstock in Brazil, Renew. Sustain. Energy Rev. 49 (2015) 1213–1220.

[77] J.A. Moncada, et al., Exploring the emergence of a biojet fuel supply chain in Brazil: an agent-based modeling approach, GCB Bioenergy 11 (2019) 773–790.

[78] F. Chaddad, Agriculture in Southeastern Brazil: vertically integrated agribusiness, in: The Economics and Organization of Brazilian Agriculture, Academic Press, 2015, pp. 73–110 (Chapter 4).

[79] S. Latorre, K.N. Farrell, J. Martínez-Alier, The commodification of nature and socio-environmental resistance in Ecuador: an inventory of accumulation by dispossession cases, 1980–2013, Ecol. Econ. 116 (2015) 58–69.

[80] L. Levidow, Les bioraffineries éco-efficientes. Un techno-fix pour surmonter la limitation des ressources? Écon. rural. Agric. aliment. territ. (2015) 31–55, https://doi.org/10.4000/economierurale.4718.

[81] F. Kaup, Empirical research—setor sucroenergético in Brazil—from the experts' mouths, in: F. Kaup (Ed.), The Sugarcane Complex in Brazil: The Role of Innovation in a Dynamic Sector on Its Path Towards Sustainability, Springer International Publishing, 2015, pp. 63–260, https://doi.org/10.1007/978-3-319-16583-7_4.

[82] A. Marques Postal, Acceso a cana-de-acucar na expansao sucroenergetica brasileira do pos 2000: O caso de Goias, 2014. http://repositorio.unicamp.br/handle/REPOSIP/286417.

[83] K. Watanabe, D. Zylbersztajn, Building supply systems from scratch: the case of the castor bean for biodiesel chain in Minas Gerais, Brazil, Int. J. Food Syst. Dyn. 3 (2013) 185–198.

[84] D. Collingridge, The Social Control of Technology, Frances Pinter, 1980.

[85] B. Fadeel, et al., Advanced tools for the safety assessment of nanomaterials, Nat. Nanotechnol. 13 (2018) 537.

[86] A.P. van Wezel, et al., Risk analysis and technology assessment in support of technology development: putting responsible innovation in practice in a case study for nanotechnology: putting Responsible Innovation in Practice, Integr. Environ. Assess. Manag. 14 (2018) 9–16.

[87] E.B. Connelly, L.M. Colosi, A.F. Clarens, J.H. Lambert, Risk analysis of biofuels industry for aviation with scenario-based expert elicitation, Syst. Eng. 18 (2015) 178–191.

[88] W. Liebert, J.C. Schmidt, Collingridge's dilemma and technoscience: an attempt to provide a clarification from the perspective of the philosophy of science, Poiesis Prax. 7 (2010) 55–71.

[89] C. Alves, et al., Techno-economic assessment of refining technologies for aviation biofuels supply chains in Brazil, Biofuels Bioprod, Biorefin. 11 (2017), https://doi.org/10.1002/bbb.1711.

[90] C.I. Santos, et al., Integrated 1st and 2nd generation sugarcane bio-refinery for jet fuel production in Brazil: techno-economic and greenhouse gas emissions assessment, Renew. Energy 129 (2018) 733–747, https://doi.org/10.1016/j.renene.2017.05.011.

[91] M. Toledo, Crisis in Brazilian Sugarcane Industry with Closures and Redundancies, 2015.

[92] A.M. Pantaleo, N. Shah, The Logistics of Bioenergy Routes for Heat and Power, INTECH Open Access Publisher, 2013.

[93] MFRural, Sweet Sorghum Grains for Animal Feed, <http://www.mfrural.com.br/busca.aspx?palavras=sorgo+sacarino>, 2016.

[94] IndexMundi, Crude Oil (Petroleum); Dated Brent Daily Price, <http://www.indexmundi.com/commodities/?commodity=crude-oil-brent&months=60>, 2016.

[95] Alibaba, Acetic Acid, <http://www.alibaba.com/product-detail/acetic-acid_60121789009.html?spm=a2700.7724857.29.28.bqea9l&s=p>, 2016.

[96] S. de Jong, et al., The feasibility of short-term production strategies for renewable jet fuels—a comprehensive techno-economic comparison, Biofuels Bioprod. Biorefin. 9 (2015) 778–800.

[97] M.O.S. Dias, et al., Simulation of integrated first and second generation bioethanol production from sugarcane: comparison between different biomass pretreatment methods, J. Ind. Microbiol. Biotechnol. 38 (2010) 955–966, https://doi.org/10.1007/s10295-010-0867-6.

[98] M.O. Dias, et al., Integrated versus stand-alone second generation ethanol production from sugarcane bagasse and trash, Bioresour. Technol. 103 (2012) 152–161.

[99] L. Mesa, et al., Techno-economic evaluation of strategies based on two steps organosolv pretreatment and enzymatic hydrolysis of sugarcane bagasse for ethanol production, Renew. Energy 86 (2016) 270–279.

[100] R. Basto, Integrated Reactor and Separator for Microbial Production of Diesel and Jet Biofuels—Equipment Design for Pilot and Production Scale (PDENG Bioprocess Engineering thesis), TU Delft—Biotechnology, 2015.

[101] Investing.com, Gás Natural Futuros, <http://br.investing.com/commodities/natural-gas-streaming-chart>, 2016.

[102] R.M. Swanson, J.A. Satrio, R.C. Brown, Techno-Economic Analysis of Biofuels Production Based on Gasification, Technical Report NREL, 2010.

[103] indeed, Machine Operator Salary in Brazil, IN, <http://www.indeed.com/salary/q-Machine-Operator-l-Brazil,-IN.html>, 2016.

[104] M.W. Peters, J.D. Taylor, M. Jenni, L.E. Manzer, Henton, D.E., Google Patents, 2010.

[105] W.-C. Wang, L. Tao, Bio-jet fuel conversion technologies, Renew. Sustain. Energy Rev. 53 (2016) 801–822, https://doi.org/10.1016/j.rser.2015.09.016.

[106] K. Atsonios, M.-A. Kougioumtzis, K.D. Panopoulos, E. Kakaras, Alternative thermochemical routes for aviation biofuels via alcohols synthesis: process modeling, techno-economic assessment and comparison, Appl. Energy 138 (2015) 346–366.

[107] S.B. Jones, et al., Process Design and Economics for the Conversion of Algal Biomass to Hydrocarbons: Whole Algae Hydrothermal Liquefaction and Upgrading, Pacific Northwest National Laboratory, 2014.

[108] Neftehim, Heavy Residues Hydrocracking, <http://nefthim.com/manual/heavy-residues-hydrocracking/>.

[109] S.B. Jones, J.L. Male, Production of Gasoline and Diesel from Biomass via Fast Pyrolysis, Hydrotreating and Hydrocracking: 2012 State of Technology and Projections to 2017, PNNL-22133, 2012.

[110] S. Jones, et al., Process Design and Economics for the Conversion of Lignocellulosic Biomass to Hydrocarbon Fuels: Fast Pyrolysis and Hydrotreating Bio-oil Pathway, National Renewable Energy Laboratory (NREL), Golden, CO, 2013.

[111] Y. Zhu, et al., Techno-economic Analysis for the Thermochemical Conversion of Biomass to Liquid Fuels, Pacific Northwest National Laboratory (PNNL), Richland, WA, 2011.

[112] Ç. Efe, L.A. van der Wielen, A.J. Straathof, Techno-economic analysis of succinic acid production using adsorption from fermentation medium, Biomass Bioenergy 56 (2013) 479–492.

Index

Note: Page numbers followed by *f* indicate figures and *t* indicate tables.

9780323857154